4D Printing of Foods

4D Printing of Foods

Fundamentals, Applications and Prospects

C. Anandharamakrishnan
CSIR - National Institute for Interdisciplinary Science and Technology (NIIST)
Thiruvananthapuram, Kerela, India

Jeyan A. Moses
National Institute of Food Technology, Entrepreneurship
and Management - Thanjavur (NIFTEM - T), Thanjavur, Tamil Nadu, India

P. Santhoshkumar
University of Foggia, Foggia, Italy

WILEY

This edition first published 2026

Registered Offices
John Wiley & Sons, Inc., 111 River Street, Hoboken, NJ 07030, USA
John Wiley & Sons Ltd., New Era House, 8 Oldlands Way, Bognor Regis, West Sussex, PO22 9NQ, UK

For details of our global editorial offices, customer services, and more information about Wiley products visit us at www.wiley.com.

The manufacturer's authorized representative according to the EU General Product Safety Regulation is Wiley-VCH GmbH, Boschstr. 12, 69469 Weinheim, Germany, e-mail: Product_Safety@wiley.com.

Wiley also publishes its books in a variety of electronic formats and by print-on-demand. Some content that appears in standard print versions of this book may not be available in other formats.

Library of Congress Cataloging-in-Publication Data Applied for:

Hardback ISBN: 9781394305308
ePDF: 9781394305315
ePub: 9781394305254
oBook: 9781394305285

Cover Design: Wiley
Cover Image: © Chanwit/stock.adobe.com

Set in 9.5/12.5pt STIXTwoText by Straive, Pondicherry, India
Printed and bound by CPI Group (UK) Ltd, Croydon, CR0 4YY

C9781394305308_131125

Contents

Preface

Over the years, food printing has gained prominent research and commercial interests. Different printing methodologies have evolved and the application range has enormously diversified. To support these endeavors, significant advancements are happening in hardware, software, and overall process control. Post-printing and post-processing approaches are also being increasingly explored and several newer approaches to quality assessment have gained popularity. Considering the fourth dimension, in the past decade, the science of 4D printing has emerged as a buzzword! Taking its roots from nonfood applications, researchers are developing innovative concepts in stimulus and material design, modeling, and simulation. The stimulus responsiveness of different food systems is being investigated in detail, and printer configurations and process variations to accommodate 4D printing have gained attention.

As an exclusive resource in this field, this book aims to present such fundamentals and dive into the fascinating field of 4D printing of foods. Sector-specific applications are detailed, highlighting the wide range of interdisciplinary applications and the many concepts that can be applied to food systems. An emphasis is given on the role of smart/intelligent materials in food 4D printing and the major application range currently explored: changes in shape, color, nutrient content, texture, taste, flavor/aroma, and live cell cultures. An exclusive chapter details the range of stimuli in the 4D printing of foods, spanning from physical to chemical to biological stimuli, also presenting futuristic perspectives. Following this, the fundamentals and current trends in 5D, 6D, and 7D printing technologies are presented. The need and approaches for this evolution are detailed, emphasizing potential food applications.

Consumer perception toward food printing technologies is deliberated as an exclusive chapter, highlighting case studies and further detailing consumer behavior and food choices toward food printing technologies. Importantly, changing consumer dynamics, indicating widespread preferencing for such novel technologies is highlighted. Toward the end, chapters of the book focus on the current global status of food printing technologies, presenting bibliometric analyses of scientific publications and patents, research funding, and current challenges, majorly focusing on the techno-economic feasibility aspect. The concluding chapter details the scope of digitalization and provides insights into the future of food printing technologies, covering the most recent and futuristic applications, including digitalized kitchens, robotic integration, and other avenues linked to information and communication technologies.

We are confident that this authored book will be an excellent resource for beginners and practitioners in food printing technology, who are keen on expanding applications to the fourth dimension.

We wish our readers a happy reading!

C. Anandharamakrishnan
Jeyan A. Moses
P. Santhoshkumar

1

Introduction to Food Printing Technology

CHAPTER MENU

1.1 Introduction

Integrating food personalization with additive manufacturing (AM) has been a notable breakthrough in the culinary industry lately. Implementing AM technologies, such as three-dimensional (3D) printing in the food industry, has greatly expanded the potential for culinary creativity beyond traditional boundaries. AM facilitates the precise fabrication of 3D foods by utilizing digital models, enabling consumers to create attractive shapes and structures of food products that were yet unachievable through conventional food manufacturing techniques. Integrating novel technologies and gastronomy science enables intricate customization of food products, accommodating individual preferences spanning from flavor and consistency to nutritional composition. In contemporary times, food scientists modify specific components at a molecular scale, enabling them to augment

4D Printing of Foods: Fundamentals, Applications and Prospects, First Edition.
C. Anandharamakrishnan, Jeyan A. Moses, and P. Santhoshkumar.

food's taste and nutritional composition to cater to individuals' dietary needs and preferences.

Moreover, controlling food texture through 3D printing parameters presents new possibilities for sensory exploration, offering experiences beyond traditional cooking methods' capabilities (D'Souza et al., 2024). Customizing food products to meet specific nutritional requirements can revolutionize dietary practices by offering personalized solutions for health concerns such as allergies, intolerances, and dietary restrictions. Additionally, 3D printing can enhance sustainability in food production by reducing waste and optimizing resources (Chen et al., 2023; Hooi Chuan Wong et al., 2022). Integrating 3D printing with culinary arts represents a substantial shift in our perception and interaction with food beyond basic technological advancements. This chapter examines the evolving environment where innovation and gourmet experiences converge, providing a valuable understanding of the future trajectory of personalizing food through novel manufacturing techniques.

1.2 Advances in Food Customization

Food customization refers to modifying food to conform to a consumer's specific requirements or preferences; it can accommodate a wide range of tastes and nutrition based on dietary requirements. It guarantees contentment and inclusiveness by enabling individuals to customize their dining experience based on their preferences and dietary needs. Customization accommodates diverse dietary preferences as well as limitations (Anandharamakrishnan et al., 2022). Whether an individual follows a vegetarian, vegan, gluten free, or allergy diet, the ability to customize allows restaurants and food services to provide appropriate choices that provide to these requirements. In addition to improving customer satisfaction, this practice fosters inclusivity by enabling all individuals to engage in dining experiences easily. This phenomenon also exemplifies the need for customized experiences in the consumer market. The demand for customized solutions that correspond to consumers' unique preferences and lifestyles is growing sharply.

For instance, customization enables restaurants to distinguish themselves and appeal to a wider range of customers (Hwang et al., 2021). Culinary professionals and food innovators are urged to engage in the exploration of various ingredients, flavors, and cooking methods in order to supply to the varied demands of their customers. This dynamic strategy ensures that menus remain continually updated and captivating, encouraging customers to return and generate positive word-of-mouth recommendations. The essence of food customization lies in its adaptability and ability to address the demands of customers promptly. It not only improves the eating experience by guaranteeing that meals are pleasurable and appropriate but also fosters a more inclusive and customer-focused methodology within the food sector. Consequently, personalization has emerged as a fundamental element of contemporary food service, enhancing customer pleasure and economic performance (Kolb et al., 2014). In recent decades, it has been predominantly utilized to formulate specialized diets for dysphagia patients, astronauts, athletes, parturients, youngsters, military people (Subhasri et al., 2023) and for any other specific job or age requirements.

1.2.1 Customization Approaches

There are diverse approaches encountered to customize foods; conventional approaches, such as the addition of additives, can enhance flavor, color and textures, packaging, and plating modification to attract consumers (Derossi et al., 2020). Alternatively, food structuring techniques have gained prominent focus in food customization. Structuring techniques are gelation, spherification, multidimensional food printing, micronization, encapsulation, texturization, and sous-vide cooking (Figure 1.1). A concise summary of these processes is presented below:

- Gelation enables the creation of textures in various foods, such as yogurt, cheese, confections, and tofu, as well as in the cooking processes of eggs and meat. It aims to convert food systems from a liquid form to a solid gel. The viscoelastic characteristics of gels significantly influence texture, appearance, and mouthfeel. The gelation process is predominantly affected by the type of gelling agents employed, their characteristics, and the synergistic interactions among different components (Banerjee and Bhattacharya 2012). Gelling agents derived from proteins and polysaccharides are among the most often utilized. Moreover, hydrocolloids serve as crucial structuring agents, esteemed for their functional attributes, including solubility and viscosity, and are fundamental to numerous conventional gelation-based foods (Bangar et al., 2022).

Figure 1.1 Popular food structuring and customization techniques. *Source*: Subhasri et al. (2022) / with permission of Elsevier.

- Spherification, a pioneering technique in molecular gastronomy and one of the most ancient gelation-based structuring techniques, has achieved extensive utilization in food science. This method entails encapsulating a liquid within an ionic biopolymer solution (Tireki, 2025). The capacity to encapsulate a liquid core within a slender, practically translucent outer membrane has resulted in significant consumer approval. Also, it serves as the foundation for producing diverse forms of customized foods, including caviar, eggs, gnocchi, and ravioli. The approach aims to create robust gels through crosslinking interactions between polysaccharides and alkaline earth metals, predominantly calcium (Mohan et al., 2024; Subhasri et al., 2022).
- Form and composition are essential elements in the design and formulation of food products. Texturization explains the process of modifying food structure through the application of significant mechanical stress and shear, resulting in the formation of textured products through the realignment of components and the establishment of new interactions (Devkota et al., 2025). Extrusion is a popular texturization technique that is used for a variety of products, including pasta, breakfast cereals, savory snacks, and textured proteins. The increasing customer desire for ready-to-cook (RTC) foods has made texturized products a convenient and healthier alternative, consistent with contemporary demands for convenience and nutrition (Schmid et al., 2022).
- Sous-vide cooking depends on regulated cooking and pasteurization methods. The procedure entails cooking food, either independently or with brine, in vacuum-sealed pouches at either low or high preset temperatures for brief or prolonged durations, followed by rapid cooling and storage (Onyeaka et al., 2022). This method has acquired popularity for cooking meat, vegetables, and shellfish at exact temperatures tailored to each product while facilitating preservation. Scientifically, the decreased oxygen levels in vacuum-sealed pouches preserve the food's nutritional integrity and sensory attributes by safeguarding bioactive substances (Tura et al., 2025). Microbial proliferation in food is affected by water content and product type; thus, it is crucial to choose suitable cooking and preservation parameters for high-water activity meals to guarantee safety and quality (Latoch et al., 2023).
- Micronization denotes the process of reducing particle size, wherein liquid particles are commonly reduced through techniques such as microfluidization, high-pressure homogenization, and ultrasonic homogenization, while solid particles are size reduced using methods including jet milling, ball milling, colloidal milling, hammer milling, and tribomechanical micronization. Decreasing particle size enhances surface area, hence improving absorption, solubility, and release behaviors of food materials or molecules (Dhiman & Prabhakar, 2021).
- Encapsulation is the process of enclosing an active substance, termed the "core" or internal phase, within a material known as the "wall" or coating, also identified as the membrane, shell, carrier material, or matrix. In this context, the food sector presents numerous possible applications, such as modifying the condition of the active agent; improving stability, mobility, safety, textural or visual attributes and facilitating controlled release (Santhoshkumar & Moses, 2025b). Popular encapsulation techniques include spray drying, emulsification, freeze drying, spray chilling, centrifugal extrusion, hot-melt extrusion, coacervation, solvent evaporation, liposome entrapment,

fluid bed coating, and co-extrusion (Anandharamakrishnan & Parthasarathi, 2019). A main benefit of such structural alteration is that micro and nanoemulsions generated during encapsulation can be transformed into dried powder via spray drying or freeze-drying or utilized in their liquid state. Contemporary methods such as microfluidics are also being investigated to create sophisticated nutrient delivery systems (Logesh et al., 2021).

- Multidimensional food printing holds significant potential for providing customized shapes, textures, and personalized nutrition. It involves a deep understanding of the printability of the material, which requires knowledge of food chemistry and the processes and techniques used to create tailored healthy food. This technique allows the creation of visually appealing food products for adults and children by adding layers of food materials. It enhances the texture, sensory qualities, and overall appeal of the food, making it more attractive and enjoyable to eat (Anandharamakrishnan et al., 2022). This approach is better detailed in the upcoming chapters of this book.

1.3 Additive Manufacturing and 3D Printing

AM has evolved in the last couple of decades, and its usage has accelerated in the past decade. This is closely associated with the fourth industrial revolution (Industry 4.0) (Raja Santhi and Muthuswamy 2023). Material printing is a manufacturing process that can make a complex product/component, typically through layer-by-layer addition of materials. The swift and substantial progress of AM materials and techniques presents significant opportunities for structural applications across diverse fields, including aerospace, biomedical, electronics, nuclear industries, flexible and wearable electrical devices, soft sensors, actuators and robotics, jewelry, artistic adornments, and building constructions (Fidan et al., 2023). Recent years have witnessed the widespread popularity and application range of AM in different industry sectors. AM, often referred to as 3D printing, creates a 3D object from a computer-aided design (CAD) model. To advance the field of AM research, new dimensions of printing have evolved, and these include the concepts of 4D, 5D, 6D, 7D, 8D, 9D, and 10D. These advancements signify a significant expansion in capabilities and possibilities within the realm of AM. It is important to note that some of these dimensions are in their nascent stage in terms of technology adoption and commercialization in different industry sectors.

The food industry is rapidly adopting 3D printing to manufacture food for different customization applications. Overall, the applications of 3D printing have evolved from customization to personalization (nutrients), food package designs and intelligent sensing systems, space or special foods, and nutraceutical delivery, to name a few. Emerging food research is increasingly exploring dimensions beyond the conventional three (length, width, height), delving into 4D, 5D, 6D, and even 7D realms. These dimensions promise to revolutionize how customers perceive, produce, and consume food. Research in these dimensions is poised to innovate food technology, enhance culinary experiences, and address critical sustainability challenges in the global food supply chain. However, food being biologically intricate, exploration of such frontier dimensions and areas of research often comes with several challenges.

1.4 Fundamental Concepts of 3D Food Printing

Among the food printing technologies, 3D food printing is an increasingly popular way of customizing food, where objects can be created in 3D designs. This approach involves a series of sequential steps, as illustrated in Figure 1.2. The four stages majorly involved in 3D food printing are (i) material selection and preprocessing, (ii) model creation, (iii) printing, and (iv) postprocessing. The choice of materials in 3D printing relies on the specific printer and the desired characteristics of the final product. The printing material supply/food ink may range from solids to liquids to gels to semisolid or semiliquids,

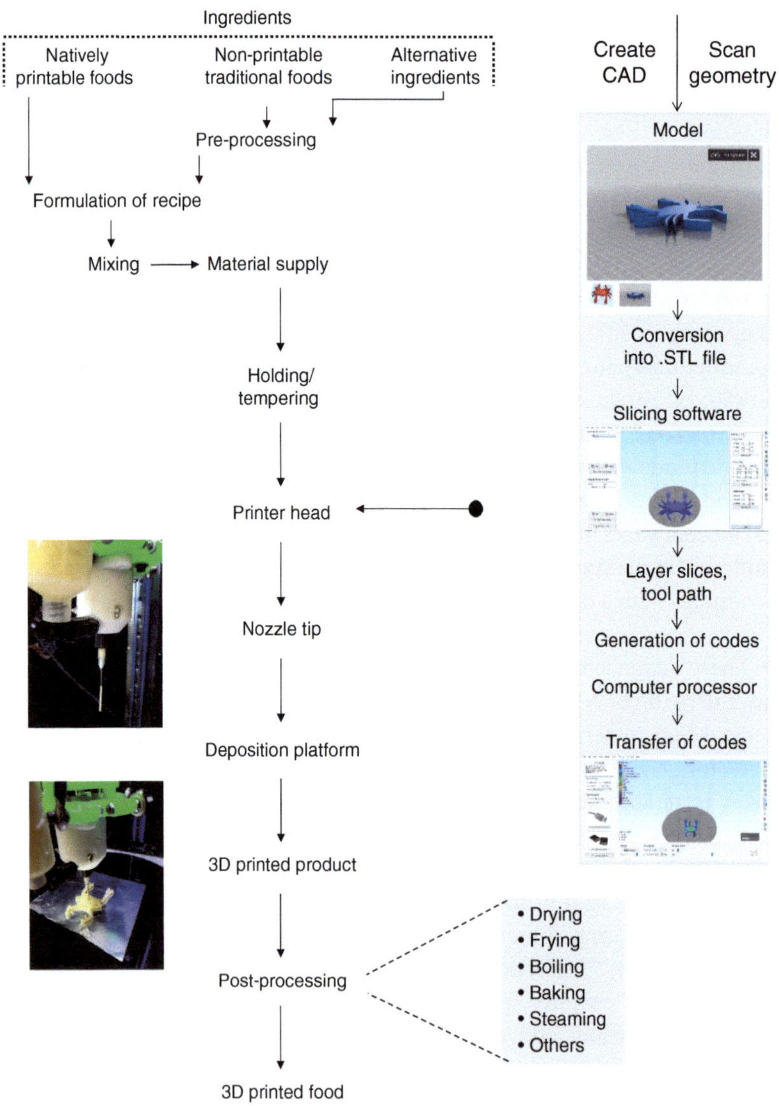

Figure 1.2 Typical 3D food printing process. *Source*: Nachal et al. (2019) / with permission of Springer Nature.

depending on material or product requirements (Domżalska & Jakubczyk, 2024). Some food materials are not natively printable but may be printed with the use of suitable preprocessing techniques, such as the addition of hydrocolloids/binders to improve their printability (Hu et al., 2023) or subjected to thermal or nonthermal treatments. During the preprocessing stage, the material undergoes a transformation to achieve the desired physical properties before it is sent for printing. Then, typically, a Standard Triangle Language (STL) file is generated from CAD software or online cloud platform to create the desired 3D shape. Optionally, a 3D scanner can be employed to directly scan a physical object and obtain its 3D geometry. Subsequently, the file gets sliced into layers using a slicer software (such as Simplify3D or PrusaSlicer, etc), which facilitates communication with the 3D printer. The 3D object is formed by adding material layer by layer using a 3D printer configuration. This process is optimized by adjusting the nozzle size, printing speed, motor speed, and other such variables, specific to the printer (Schuller et al., 2024).

Typically, if the material undergoes hot extrusion printing, the resulting printed object does not need any additional postprocessing techniques to be transformed into its final ready-to-eat (RTE) form (Santhoshkumar et al., 2025). However, in the case of room temperature extrusion, RTC products are obtained, and additional postprocessing processes are required to convert the printed object into its cooked form. Common postprocessing techniques are popular food processing unit operations such as steaming, baking, frying, drying, boiling, and freezing, and the selection depends on the product and its applicability (Theagarajan et al., 2021). Recently, Tong et al. (2024) reported a laser-integrated approach for postprocessing. In addition to these conventional techniques that happen as the next exclusive unit operation after the printing process gets completed, of late, techniques such as laser heating and blowing hot air are being considered to transform materials into their cooked forms as and when deposited onto the printing platform as individual layers. Eventually, many integrated post-processing approaches were recently reported in 3D bioprinting applications (Jeong et al., 2024).

1.5 Nomenclature/Terminologies Related to Food Printing

3D printing was first documented in 1986, and it has since evolved into a diverse set of printing technologies that are employed across several industries. Considering publication trends, 3D printing is also referred to as "additive manufacturing" and "rapid prototyping" (Alexander et al., 2021), although, in their actual context, there are specific definitions that differentiate them. For instance, the concept of rapid prototyping better suits a process in which "a model" of the "actual product" is printed for speedy release, demonstration, or other applications. Therefore, it is understood that the expansion of 3D printing technologies has led to an increase in the number of names used to describe them. Some of these phrases are registered trademarks, nonstandard terms, or inadequately defined in different fields, particularly in the automobile and medical sectors.

Given this, it is important to note that there are established norms for the terminologies used in the field of 3D printing. In the field of AM, international standards are given by ASTM International (formerly known as the American Society for Testing and Materials). The recently updated standard document in March 2022, titled "ASTM ISO/ASTM52900:2021

Standard Terminology for Additive Manufacturing – General Principles – Fundamentals and Vocabulary," details the classification of AM and their important definitions (ASTM, 2022). Table 1.1 presents the seven established types of 3D printing technology as defined by ISO/ASTM 52900:2021 and accompanied by various supplementary categories. Printing materials and processes are presented in ASTM standard "ASTM D7188-05(2019) Standard Terminology for Printing Inks, Materials, and Processes," last updated in January 2019 (ASTM, 2019). The different terminologies related to AM in different contexts are presented in Table 1.2.

Also, as of early 2025, there are no such specific standards or paperwork in the field of 3D food printing, and ASTM International terminologies are considered in food printing as

Table 1.1 Generalized terms for 3D printing as per ASTM international.

Standard classification	Commercial and other terms	Description
Binder jetting	• ProJet Color Jet Printing (CJP)	• Liquid agents are selectively dropped onto powder media • Subsequent infiltration or heating may be required
Direct energy deposition	• Laser-Engineered Net Shape (LENS) • Electron Beam Additive Manufacture (EBAM)	• Focused application of energy and material electively melted and fused on a build platform or part
Material extrusion	• Fused Deposition Modeling (FDM) • Fused Filament Fabrication (FFF)	• Material is dispensed, often through a heated nozzle, onto a build platform
Material jetting	• Nanoparticle Jetting (NPJ) • Drop-on-Demand (DOD) • PolyJet • ProJet Multijet Printing (MJP)	• A print head dispenses droplets of media, usually a photopolymer, onto a build platform where each layer is solidified or cured
Powder bed fusion	• Selective Laser Sintering (SLS) • Selective Laser Melting (SLM) • Direct Metal Printing (DMP) • Direct Metal Laser Sintering (DMLS) • Electron Beam Melting (EBM) • Multijet Fusion (MJF)	• Powder media is deposited on a build platform and subsequently bonded together through a heating process
Sheet lamination	• Laminated Object Manufacturing (LOM)	• Discrete layers of material are fused or glued together to form an object
Vat photopolymerization	• Stereolithography (SLA) • Direct Light Processing (DLP) • Continuous Liquid Interface Production (CLIP)	• Liquid photopolymer is selectively exposed to a light source, facilitating layer-by-layer curing

Source: Alexander et al. (2021) / Springer Nature / CC BY 4.0.

Table 1.2 Alternative terminologies for additive manufacturing.

Alternative terminologies	Context
3D Printing (3DP)	Global
Rapid prototyping/manufacturing	Global
Free-Form Fabrication (FFF)	United States (USA)
Solid Freeform Fabrication (SFF)	United States (USA)
Direct Digital Manufacturing (DDM)	United States (USA)
Generative manufacturing	Germany
e-Manufacturing	Germany
Constructive manufacturing	Germany
3D Additive Layer Manufacturing (ALM)	European Aeronautic Defense and Space (EADS)

Source: Choong (2022) / with permission of Elsevier.

well. In modern times, research has expanded its scope; for instance, multidimensional or 4D/5D/6D printing to enhance its capabilities and functionalities, also considering multimaterial and multinozzle printing. As advancements in 3D printing technologies take shape, the significance of standardized language must be emphasized for global consistency.

1.6 Food Printability

Printability is an initial assessment that determines whether a material is appropriate for 3D printing or if it needs extra additives or treatments. When considering printability, materials can be categorized into three types: natively printable, non-natively printable, and alternative ingredients (Nachal et al., 2019). The following sections present aspects of the printability of materials, fabrication approaches, and their printing types.

1.6.1 Materials Classification

In 3D food printing, the nutritional composition influences the printability of food materials. For example, typically, fiber-rich materials are unsuitable for printing, making them printable requires the addition of suitable additives/preprocesses (Wilson et al., 2023). On the other hand, materials packed with starch are often natively printable (Santhoshkumar et al., 2024a). Figure 1.3 categorizes the naturally printed substances according to their nutritional makeup.

- *Natively printable*: The term "natively printable" refers to the ability of a material composition to be extruded or printed into the shape of a 3D object without the need for any additional substances/processes. Materials must be tailored to meet printing requirements; for instance, the addition of sufficient levels of water to the printing material supply.

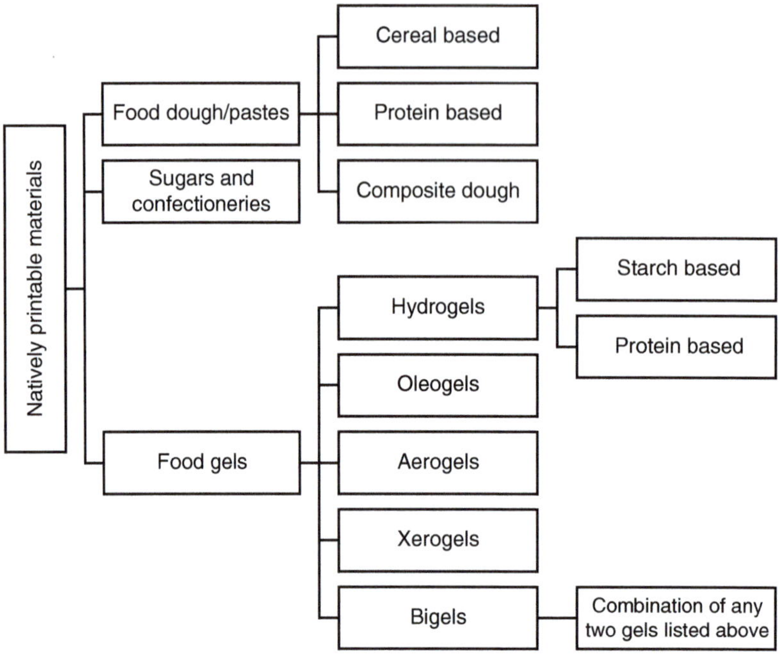

Figure 1.3 Types of natively printable materials. *Source*: Anandharamakrishnan et al. (2022) / with permission of John Wiley & Sons.

- *Natively nonprintable*: In several other cases, the material necessitates the addition of external additives, such as hydrocolloids, in order to transform it from a nonprintable state to a printable one, presenting the next category of materials. Hydrocolloids, such as carrageenan, gum arabic, xanthan gum, and gelatin, can be added to enhance the rheological behavior of the material supply for improving printability of the food materials (Santhoshkumar et al., 2025). This approach is of particular interest in the case of extrusion-based 3D printing, which is by far the most popular and widely researched food printing approach. Nonprintable materials can also undergo various pretreatment processes, including homogenizing, size reduction, thermal, and nonthermal treatments to transform them into printable substances.

  Furthermore, an additional approach involves combining non-natively printable material with natively printable material in equal or greater proportions. This combination can also alter the printability of materials. For instance, idli (an indigenous food of South India) batter has a liquid consistency that cannot be easily printed and would not have proper postprinting stability. The addition of pearl millet flour was found to change the material's consistency. Under optimized ratios, an indigenous food could be conveniently 3D printed (Raja et al., 2023). Similarly, by adding banana pulp, which is natively nonprintable, with pearl millet flour, it was possible to produce a 3D print of a snack (Santhoshkumar et al., 2024a). Wheat flour is natively printable in its dough form and was added with button mushroom (Keerthana et al., 2020) and in another study with ground chicken meat (Wilson et al., 2020) to develop printing inks with the "desired" printable consistency. In addition, using different types of flour in specific proportions resulted in synergistic effects on the ability to print the material. For instance, composite flour, which includes millet, gram, and seed flour mixed with water, creates a printable form of material that is high in fiber (Keerthana et al., 2020). Each of these composites has unique nutritional characteristics in terms of food nutrition. When combined, they yield a nutrient-dense material supply for 3D food printing. However, one key aspect that must be considered is the impact of such combinations on the sensorial acceptability of the resulting products.

- *Alternative ingredients*: This includes algae, fungi, seaweeds, lupines, and insects, which have garnered significant interest for their potential to create healthier goods. Such ingredients are abundant in protein, dietary fiber, and/or several beneficial bioactives. While this also underscores the sustainability dimension of 3D food printing as a technology with the versatility to include newer food sources, it is important to note that the actual definitions for such "alternative" or "novel" foods must essentially vary with time (Reuben et al., 2025).

Interestingly, recent studies focus on newer raw materials, and thus emergent is the concept of "food inks," which are materials in liquid-like consistencies. Other associated formulations include bigels and bio-inks, both of which are associated with extrusion-based printing, and in cases where a powder-like material is used for the printing process, a suitable approach, such as binder jetting, is used for the fabrication process. With a better understanding of newer food sources, composite feed types, and novel/hybrid printing technologies, as a rapidly evolving field, food printing and the understanding of "material printability" are expected to widen in the future.

1.7 3D Food Printing Types

Food printing is a process that involves both large-scale and small-scale elements of material resources. In the aspect of 3D food printing, it is important to consider the nutritional, engineering, and physiological properties of food materials and their desired application. For instance, in polymer printing, temperature addition is a critical parameter to melt and form the shape of the 3D structure. This approach, if applied to food printing, would denature or destroy the nature of the food. That is, although it may be printable, the approach would fail from the nutritional perspective. Therefore, it is imperative to consider the nature and final application of food matrices as key criteria for 3D printing fabrication approaches. Therefore, among the seven popular approaches to 3D printing, only four of them are applied in food systems. These include extrusion, inkjet, binder jetting, and selective sintering-based 3D printing methods (Le-Bail et al., 2020).

1.7.1 Extrusion-Based 3D Printing

Extrusion-based 3D printing is the most popular used method in food printing. The approach is also fundamental to bioprinting approaches, such as the printing of living cells in 3D structures. These have applications in many industries, including food, biomedical, pharmaceutical, and biotechnology fields (Tibrewal et al., 2023). Ideally, the substance used should be a non-Newtonian fluid. That is, a shear-thinning material becomes "thinner" when subjected to shear. This method utilizes a robotic arm equipped with a cylindrical device, commonly known as a syringe, which traverses a surface and dispenses a substance through a nozzle. The sequential deposition of layers is accomplished by positioning the cylinder at predetermined positions based on a 3D model. This type of extrusion printer commonly known as delta type 3D printer, employs food pastes with a high level of viscosity (Le-Bail et al., 2020). The food ink is expelled/extruded through a die or printing head and is required to possess a viscosity/mechanical characteristic that enables it to be vertically assembled as self-supporting layers (Santhoshkumar & Moses, 2025a,b).

Based on extrusion-based 3D food printing in the context of material delivery, it was majorly classified into three: syringe, screw, and air pressure-based mechanical-driven mechanisms (Guo et al., 2019). Further, syringes are categorized into piston and air-pressure-driven mechanisms through the nozzle. To find a better mechanism in 3D food printing, a simulation study was conducted; in their study, syringe-based mechanisms showed better results than screw-based 3D printing (Guo et al., 2019), considering material delivery. Screw-based mechanisms are considered to be only used for high mechanical strength and viscous materials, whereas syringes can also possibly be used in ink and gel printing. More recently, food bio-inks use piston-syringe-based mechanisms for printing (Tibrewal et al., 2023).

The extrusion printing technique is also classified based on its processing condition (majorly, the temperature involved), that is, as room-temperature, hot, and gel-forming extrusion. Room temperature extrusion, also known as cold extrusion, refers to the printing of food materials without the addition of temperatures or external heating sources (Figure 1.4a); in contrast with the addition of temperature, the printing process is called hot extrusion printing (Figure 1.4b) (Kavimughil et al., 2022b). Interestingly, in the third category, the materials changed their form of nature during the printing process. Classic examples are hydrogel, oleogel, xerogel, and cryogel printing applications (Figure 1.4c). Ideally, during the heating process, hydrophobic contacts, disulfide bonds, and gelation are facilitated through the unfolding of molecules and the exposing of hydrophobic patches and thiol groups, which are capable of forming gels (Sharma et al., 2023), and the exact underlying chemistry–physics is specific to the material and its composition.

Room-temperature extrusion involves forcing food materials through a narrow die at ambient temperatures, typically requiring further postprocessing steps to create the final product. These steps might include drying, baking, cooking, or additional seasoning, cooling, freezing, steaming and coating to achieve the desired texture, flavor, and microbial safety. This method, in the printing stage, is energy efficient and suitable for heat-sensitive ingredients. However, the necessity for a subsequent postprocessing stage to make the product ready for consumption may pose additional challenges. In contrast, hot extrusion integrates heating into the extrusion process, allowing the production of RTE items immediately as they exit the extruder, benefiting from both the cooking and shaping in a single and continuous process (Waseem et al., 2024).

Recently, hybrid processes involving extruding food material at ambient temperature, creating layers as it emerges from the nozzle, and simultaneous laser or alternative heating components impart heat to the material on the platform are gaining interest. This method guarantees that the food is thoroughly cooked, boosting the final product's texture, flavor and improving the adhesion between layers (Blutinger et al., 2021). This approach facilitates

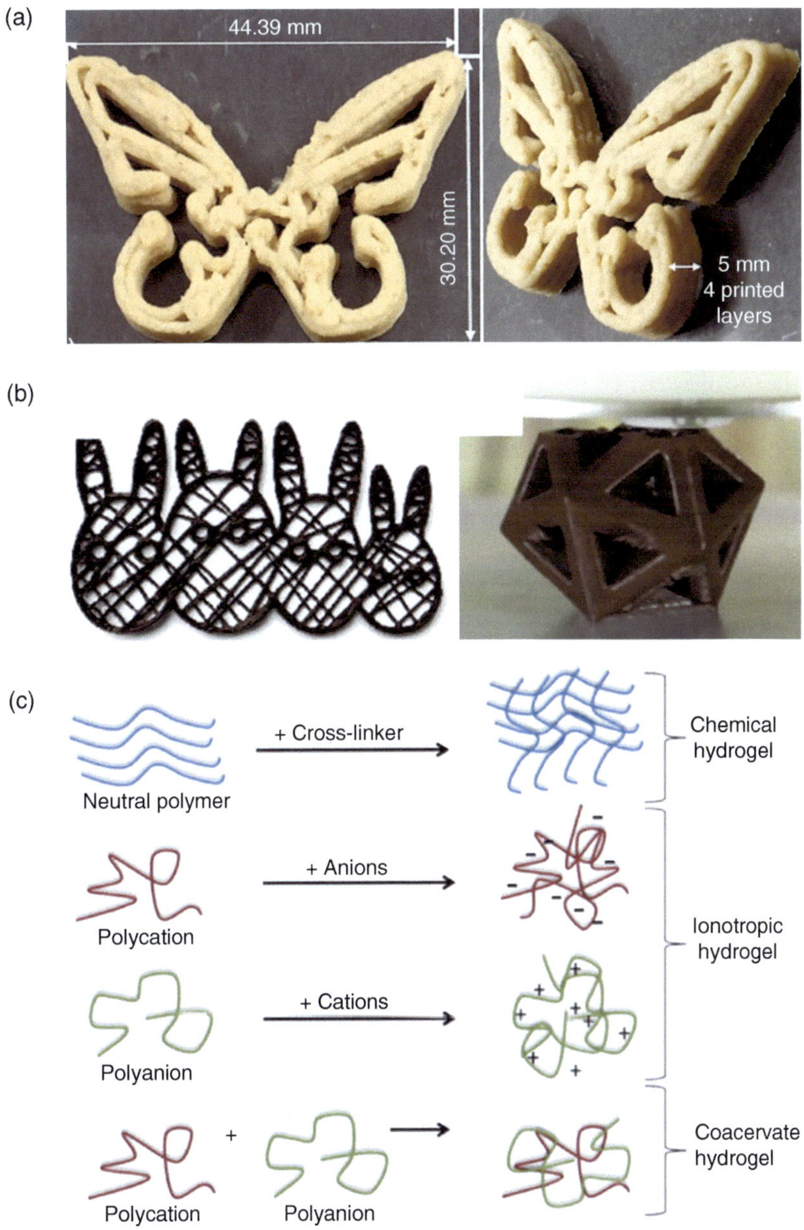

Figure 1.4 Examples of extrusion-based 3D food-printed samples. (a) Cold extrusion: fiber-rich snack. *Source:* Keerthana et al. (2020) / with permission of Elsevier; (b) hot melt extrusion: chocolate. *Source:* Hao et al. (2019); Lanaro et al. (2019) / with permission of Elsevier; (c) gel-forming extrusion: hydrogel. *Source:* Godoi et al. (2016) / with permission of Elsevier.

the production of RTE food from the extruder, merging the energy economy of room-temperature extrusion with the instantaneous cooking advantages of hot extrusion. Although these are specific to the material being used, such processes can provide major benefits in time, resource, and space savings in commercial operation lines.

Extrusion-based techniques have significantly contributed to the advancement of 3D printing technology, particularly in food applications. The effectiveness and adaptability of these technologies have been thoroughly documented and examined up to pilot-scale-level production. Currently, the progression of 3D food printing has now stepped to industrial-scale level production. These techniques are adaptable and can accommodate diverse food components, facilitating the creation of an extensive array of complex food products (Portanguen et al., 2019; Taneja et al., 2022). Moreover, an important benefit of extrusion-based 3D printing is its ability to maintain the nutritional integrity of the printed food by preventing nutrient denaturation. This attribute is especially crucial as it guarantees that the printed food retains its nutritional advantages, rendering it appropriate for various dietary requirements and tastes. In addition, the capability to preserve nutritional quality has expanded the use of extrusion-based 3D printing in bioculture printing. Advancements in bio-cell printing have shown the ability to replicate the texture and structure of flesh and plant-based meat substitutes. These advances are noteworthy as they expand the potential for developing sustainable and healthy alternatives to conventional meat products. The versatility, cost-effectiveness, and capacity to retain nutrients of extrusion-based processes have established them as the preferred option for food applications, leading to advancements in bioprinting to develop cultured meat (Guo et al., 2024), meat analogs (Wen et al., 2023), functional food development (Tomašević et al., 2021), nutraceutical delivery (Kavimughil et al., 2022a), and space food (Santhoshkumar et al., 2023). Overall, there has been a growing trend in research to integrate automation and robots into 3D food printing (Jazmin et al., 2025). This focus is mostly on achieving mass customization and producing intricate designs and also encompasses incorporating artificial intelligence (AI), surpassing the limitations of conventional extrusion-based 3D printing methods. These technological developments are fueling substantial expansion in the realm of 3D food printing research.

1.7.2 Inkjet 3D Printing

Inkjet printing employs a heat head to dispense droplets into specific locations, effectively coating surfaces or embellishing food items such as cookies, cakes, and pizzas. This method commonly employs either thermal or piezoelectric heads. Thermal inkjet printers utilize an electrically heated print head to generate pressure pulses that propel droplets from the nozzle (Uddin et al., 2022). There exist two distinct inkjet printing techniques: continuous jet printing and drop-on-demand printing. Continuous jet printing involves the continuous ejection of ink through a piezoelectric crystal that vibrates at a consistent frequency. Conductive agents are frequently used to achieve the necessary flow of ink. A valve is responsible for regulating the release of ink from the printing heads in drop-on-demand printing, ensuring it is expelled at the required pressure. While drop-on-demand systems often have slower printing speeds compared to continuous jet systems, they provide superior resolution and precision. Inkjet printers are generally unsuitable for creating intricate food structures since they are designed to work with substances that have low viscosity (>0.03 Pa/s). As a result of this impact, this method has been referred to as 2D food printing (Vadodaria & Mills, 2020). Foods such as chocolate, liquid dough, sugar icing, meat paste, cheese, jams, gels, and other similar items (Pitayachaval et al., 2018) have been inkjet 3D printed. Some examples are presented in Figure 1.5 (I).

Figure 1.5 (I) Inkjet 3D food printing applications: (a) decoration, (b) surface filling, (c) cavity deposition. *Source:* Godoi et al. (2016) / Elsevier); (II) binder-jetted 3D-printed foods: (a) sugar cube, (b) chocolate, (c) coffee and cocoa-printed structures. *Source:* Holland et al. (2019) / with permission of Elsevier; (III) examples of sintering-based 3D-printed foods. Selective laser sintering: (a) sugar, (b) nesquik, (c) curry cube, paprika pyramid, cinnamon cylinder, and pepernoot pentagon. *Source:* Godoi et al. (2016) / with permission of Elsevier; selective hot air sintering and melting: (d) process, (e) toroidal coil sculpture – granulated sugar, (f) complex sugar geometry, (g) sugar candies. *Source:* Mantihal et al. (2020) / with permission of Elsevier.

1.7.3 Binder Jetting

Binder jet 3D printing utilizes powder bed printing technology, where a liquid binder is accurately applied to clump together adjacent powder particles, forming a 3D-printed structure. During the process of binder jetting, particles can come together and form larger clusters through several mechanisms, including sintering, chemical reactions, and the creation of liquid bridges. This 3D printing approach offers several notable advantages when compared to other 3D printing technologies. These include rapid production, the capability to produce goods with a high concentration of solid material, and the capacity to structure objects using flexible material compositions (Stevens et al., 2020). Utilizing this approach, the most popular applications are in the nonfood segment, especially in biomedical research for scaffold preparation and in pharmaceuticals to fabricate tablets. For instance, the first 3D-printed tablet (Spritam©) was fabricated using this printing approach (Sen et al., 2021). The approach also extends to other nonedible materials, such as stainless steel and building materials (Mostafaei et al., 2021).

So far, only a limited number of research reports have documented the potential of utilizing food substances such as cellulose, xanthan gum, starch, sugar, calcium caseinate, and medium-chain triglycerides powder for binder jet 3D printing applications (Holland et al., 2018, 2019; Vadodaria & Mills, 2020; Zhu et al., 2022). Additionally, a recent patent describes a recipe for a protein bar manufactured with whey protein and egg white albumin using binder jet 3D printing technology (Zhu et al., 2022). Representative images of binder jetting 3D-printed chocolate and cocoa powder are shown in Figure 1.5 (II). Overall, as compared to inkjet printing, with a wider applicability of material types, the binder jetting approach currently has a wider application range.

1.7.4 Selective Sintering

Complex shapes can be created by selectively sintering sugars and sugar-rich powders. Typically, after a layer of freshly deposited powder is uniformly spread, a sintering source, such as hot air or a laser, will move along the *X* and *Y* axes to heat and bind powder particles, resulting in the formation of a solid, cohesive layer. The procedure entails the continuous deposition of new material particles onto the melted surface until a 3D object is completely constructed. The sintered material forms the component, while the unsintered powder remains to support the structure and may be recycled (Sun et al., 2015). The powder bed is heated to a temperature slightly below the material's melting point to minimize thermal distortion and enhance the adhesion of the new layer to the previous one.

Selective sintering printing is classified into two types according to the sintering sources employed: (i) selective laser sintering (SLS) and (ii) selective hot air sintering and melting (SHASAM). This printing method benefits enhanced printing freedom and speed, eliminating the need for additional postprocessing. In addition, the printing method allows for modifying each sintered powder layer to produce food with distinct ingredients in each layer, creating more intricate food products. It enables the rapid construction of intricate food items within a brief period without subsequent curing. Although the approach has proven effective capabilities in sintering metals, ceramics, polymers, and other materials, its application in the food industry is limited due to material property limits. Currently, the primary application of this approach is to print food ingredients composed of sugar or lipids, which possess a low melting point and can be easily fused (Figure 1.5 (III)). In addition, the materials employed in selective sintering printing are restricted to powders, making them inappropriate for some food products like moisture material. This is because their properties do not align with the needs of the printing process (Jiang et al., 2022).

1.7.5 Bioprinting

Bioprinting refers to the application of printing techniques to construct and organize molecules, cells, tissues, and biodegradable biomaterials into designated arrangements to execute one or more biological tasks (Yilmaz et al., 2021). The principal objective of bioprinting is to fabricate viable, functional tissues, and organs for transplantation by combining living cells, scaffolds with 3D-linked architectures, and bioactive compounds. Thus, 3D bioprinting necessitates the concurrent manipulation of living cells and diverse substances (Betancourt & Chen, 2022). Conventional 3D printing methods were originally created for quick prototyping, industrial manufacturing tools, and the production of consumer goods. Nonetheless, the extreme conditions linked to these technologies render them inappropriate for printing living cells and various biological materials, thereby limiting their application in biomedicine to "passive" structures such as bare scaffolds and porous substrates that facilitate cell adhesion and proliferation (Yilmaz et al., 2021).

Food bioprinting is an innovative methodology that merges tissue engineering with food technology to revolutionize conventional food production and consumption ways. This novel method could transform food production by mimicking the structural intricacy and diversity of real meat through advanced manufacturing methods created in tissue engineering. As the most popular application in the realm of conventional meat alternatives, lab-grown meat or cell-cultured meat is increasingly being explored for sustainable and ethical food production and consumption. Herein, food bioprinting utilizes 3D printing technology in conjunction with biological and synthetic components derived from such alternative sources to produce edible products. Researchers are leveraging the benefits of 3D bioprinting to investigate the extensive potential of lab-grown muscle-based foods as sustainable substitutes for conventional animal agriculture and aquaculture (Curel et al., 2024). Although the product's texture and other sensorial attributes are similar to conventional meat, on date, the approach is regarded to be very expensive and has its own resource intensiveness. Further, these products come with newer regulatory and commercialization challenges.

1.7.6 Others

The aforementioned printing approaches are popularly applied in food printing applications. Apart from these printing procedures, there are various approaches used for other material types; however, due to the technical constraints in terms of the nature of the food, printers, material availability, etc., other types of printers/printing approaches have not yet showed successful outcomes in food printing. Further research is required on other types of printers in food applications, including software and food materials development for the developed printers. Subsequently, the diversity of printers was widely investigated in nonfood applications; those printers need to be altered for this sector; however, substantial research is required.

1.8 Hardware Requirements and 3D Printer Configurations

3D printing, especially regarding food, has become a significant instrument for swift prototyping and inventive culinary uses. This technique provides multiple benefits, such as minimizing material consumption, preventing waste, conserving energy and time, and enhancing cost-effectiveness. Fundamentally, 3D printing functions by sequentially depositing material in layers based on a specified design. The procedure commences with the transformation of a 3D model into a STL file, which acts as the basis for beginning and overseeing the printer's functions along its coordinate axes (Nachal et al., 2019). Multiple configurations are utilized in 3D printing, such as Cartesian, Delta, Polar, and Selective Compliance Assembly Robot Arm (SCARA) systems, each possessing unique functionalities. Cartesian and Delta configurations are the predominant choices for 3D food printing, owing to their accuracy in methodically piling materials along the *X*, *Y*, and *Z* axes. These technologies provide meticulous regulation of the deposition of food-grade materials such as chocolate, dough, or purees, guaranteeing superior results (Anandharamakrishnan et al., 2022).

The SCARA configuration is increasingly favored for its capacity to enhance printing dimensions and practical functionalities, rendering it appropriate for larger-scale or more intricate food-printing applications. Recent improvements in SCARA-based systems have facilitated the creation of large 3D printers capable of handling complex designs and increased production (Dinakaran et al., 2023; Pan et al., 2024). The hardware specifications for 3D food printers often encompass food-grade extruders, temperature regulation systems for managing materials with particular thermal characteristics, and nontoxic, food-safe components to guarantee hygiene. The printer must include software that can manage intricate STL files and convert them into accurate material deposition patterns. Collectively, these hardware and configuration components facilitate advanced food production methods, fostering individualized nutrition, complex food designs, and sustainable culinary practices.

1.9 Software Requirements

One of the most suitable software is essential in 3D food printing to communicate with the printer and support overall process control. The procedure begins with creating a 3D object (in STL file format) utilizing software such as AutoCAD or SolidWorks or cloud platforms. Slicing software used to communicate with printers (Table 1.3) processes the file, converts it into Geometric code (G code), and sends it to the printer. For instance, PrusaSlicer is a software recently applied in multimaterial food printing, which requires advanced technical knowledge to operate. In addition, G-code editors (e.g., FullControl, NC Viewer) provide options to directly manipulate toolpath and motion in multimaterial printing. This slicing software uses G and M codes (machine code) to communicate printer (G codes – direct the motors and printing arms to their specified positions; M codes – machine functions). For example, the most commonly used general commands are G_0 (rapid linear motion) and G_1 (linear motion), specifically used to regulate the movement of the arms. AI is recently used to convert 2D images into 3D files in STL format. For instance, Tinkercad is an open-sourced cloud platform that can create STL files easily without the high-end software like AutoCAD or Solidworks. Furthermore, through the slicing software, users can set essential printing parameters such as printing speed, retraction speed, layer height, temperature, nozzle size,

Table 1.3 Examples of slicing software used in 3D printing.

Software/cloud	Subscription ranges	Level of users: Beginners	Intermediate	Advanced
UltiMaker Cura	Free	■	■	■
CraftWare	Free	■	■	■
KISSlicer	Free and paid versions	■	■	■
Repetier host	Free		■	■
Simplify3D	Paid	■	■	■
ReplicatorG	Free	■	■	■
Slic3r	Free			■
OctoPrint	Free		■	■
3Dprinter OS	Free and paid versions			■
MatterControl	Free	■	■	■
Formide	Free	■		
AstroPrint	Free			■
SelfCAD	Free and paid versions	■	■	■
MakerBot Print	Free	■		
Tinkerine Suite	Free	■		

Source: Anandharamakrishnan et al. (2022) / with permission of John Wiley & Sons.

outer shells, and infill density. This meticulous regulation guarantees superior, precise prints (Anandharamakrishnan et al., 2022). Additionally, 3D scanners and open-sourced websites are available to get the STL file of the shapes. For example, through the scanner, one can get the 3D STL file of the respective scanned shapes. Additionally, Thingiverse, Pinshape, and Cults offer openly accessible 3D object files in STL format. For postprinting, stability analysis can be examined by using image analysis approaches (Raja et al., 2023; Santhoshkumar et al., 2024a) and other nondestructive quality evaluation techniques.

In the future, building software for 3D food printing will need to integrate regular 3D printing software features with particular adaptations for food safety, material qualities, and culinary creativity. The user interface should be intuitive, catering to both professionals and hobbyists. The key features should include cloud-based recipe management that can store, edit, and share; customization options for various food shapes and textures; and a real-time visual preview. Ultimately, the software should offer a seamless and creative food 3D printing experience, blending technological precision with culinary artistry.

1.10 Conclusion

3D printing has made great progress in several industrial sectors and is now making inroads into the food sector. It facilitates the creation of intricate and customized edible structures, yet it faces challenges due to the complex properties of food materials. Researchers are currently exploring and expanding the potential applications of 3D food printing, aiming to position it as a viable alternative for large-scale industrial production. This greater efficiency can lower manufacturing expenses and boost the overall quality and uniformity of food. The progress of technology continues to propel the development of food printing, improving its functionalities and broadening its range of possible uses. Integrating automation, robotics, and AI enhances the capabilities of 3D food printing, facilitating significant advancements in large-scale customization and intricate food design. In the future, food printing has the potential to revolutionize dietary habits by creating a more technologically integrated and customized experience. Due to advancements in research and heightened corporate investment, the notion of a "digitalized diet" and "printer @ home" is more viable. This opens up new opportunities for sustainable and healthy food outlets in the Industry 4.0 era and tailored food solutions in the future.

References

Alexander, A. E., Wake, N., Chepelev, L., Brantner, P., Ryan, J., & Wang, K. C. (2021). A guideline for 3D printing terminology in biomedical research utilizing ISO/ASTM standards. *3D Printing in Medicine*, *7*(1), 8. doi:10.1186/s41205-021-00098-5

Anandharamakrishnan, C., Moses, J. A., & Anukiruthika, T. (2022). *3D printing of foods*. John Wiley & Sons Ltd.

Anandharamakrishnan, C., & Parthasarathi, S. (2019). In C. Anandharamakrishnan & S. Parthasarathi (Eds.), *Food nanotechnology: Principles and applications* (1st ed.). CRC Press.

ASTM. (2019). *Standard terminology for printing inks, materials, and processes (D7188-05(2019))*. ASTM https://doi.org/10.1520/D7188-05R19

ASTM. (2022). *ISO/ASTM52900-21. Additive manufacturing—general principles—fundamentals and vocabulary*. ASTM. https://doi.org/10.1520/F3177-21

Banerjee, S., & Bhattacharya, S. (2012). Food gels: gelling process and new applications. *Critical Reviews in Food Science and Nutrition*, *52*(4), 334–346. doi:10.1080/10408398.2010.500234

Bangar, S. P., Esua, O. J., Sharma, N., & Thirumdas, R. (2022). Ultrasound-assisted modification of gelation properties of proteins: A review. *Journal of Texture Studies*, *53*(6), 763–774. doi:10.1111/jtxs.12674

Betancourt, N., & Chen, X. (2022). Review of extrusion-based multi-material bioprinting processes. *Bioprinting*, *25*, e00189. doi:10.1016/j.bprint.2021.e00189

Blutinger, J. D., Tsai, A., Storvick, E., Seymour, G., Liu, E., Samarelli, N., Karthik, S., Meijers, Y., & Lipson, H. (2021). Precision cooking for printed foods via multiwavelength lasers. *npj Science of Food*, *5*(1), 24. doi:10.1038/s41538-021-00107-1

Chen, L., Debono, D., & Hemsley, B. (2023). A bite closer: Using 3D food printing to achieve Sustainable Development Goals 2, 3, 9 and 17. *International Journal of Speech-Language Pathology*, *25*(1), 58–61. doi:10.1080/17549507.2022.2132290

Choong, Y. Y. C. (2022). Additive manufacturing for digital transformation. In *Digital manufacturing* (pp. 145–182). Elsevier. https://doi.org/10.1016/B978-0-323-95062-6.00002-4

D'Souza, C., Adkari, A., & Alahakoon, D. (2024). Coupling AI with empirical research – A case of 3D printed food technology. *Food Quality and Preference*, *120*, 105229. doi:10.1016/j.foodqual.2024.105229

Derossi, A., Husain, A., Caporizzi, R., & Severini, C. (2020). Manufacturing personalized food for people uniqueness. An overview from traditional to emerging technologies. *Critical Reviews in Food Science and Nutrition*, *60*(7), 1141–1159. doi:10.1080/10408398.2018.1559796

Devkota, L., Kumar, G., Zhang, P., Raghuwanshi, V. S., & Dhital, S. (2025). Legume protein composition influences texturization during high temperature protein-starch complexation. *Food Hydrocolloids*, *164*, 111213. doi:10.1016/j.foodhyd.2025.111213

Dhiman, A., & Prabhakar, P. K. (2021). Micronization in food processing: A comprehensive review of mechanistic approach, physicochemical, functional properties and self-stability of micronized food materials. *Journal of Food Engineering*, *292*, 110248. doi:10.1016/j.jfoodeng.2020.110248

Dinakaran, V. P., Balasubramaniyan, M. P., Muthusamy, S., & Panchal, H. (2023). Performa of SCARA based intelligent 3 axis robotic soft gripper for enhanced material handling. *Advances in Engineering Software*, *176*, 103366. doi:10.1016/j.advengsoft.2022.103366

Domżalska, Z., & Jakubczyk, E. (2024). Characteristics of Food Printing Inks and Their Impact on Selected Product Properties. *Foods*, *14*(3), 393. https://doi.org/10.3390/foods14030393

Fidan, I., Huseynov, O., Ali, M. A., Alkunte, S., Rajeshirke, M., Gupta, A., Hasanov, S., Tantawi, K., Yasa, E., Yilmaz, O., Loy, J., Popov, V., & Sharma, A. (2023). Recent Inventions in Additive Manufacturing: Holistic Review. *Inventions*, *8*(4), 103. doi:10.3390/inventions8040103

Godoi, F. C., Prakash, S., & Bhandari, B. R. (2016). 3d printing technologies applied for food design: Status and prospects. *Journal of Food Engineering*, *179*, 44–54. doi:10.1016/j.jfoodeng.2016.01.025

Guo, C.-F., Zhang, M., & Bhandari, B. (2019). A comparative study between syringe-based and screw-based 3D food printers by computational simulation. *Computers and Electronics in Agriculture*, *162*, 397–404. doi:10.1016/j.compag.2019.04.032

Guo, X., Wang, D., He, B., Hu, L., & Jiang, G. (2024). 3D bioprinting of cultured meat: A promising avenue of meat production. *Food and Bioprocess Technology*, *17*(7), 1659–1680. doi:10.1007/s11947-023-03195-x

Gurel, M., Rathod, N., Cabrera, L. Y., Voyton, S., Yeo, M., Ozogul, F., & Ozbolat, I. T. (2024). A narrative review: 3D bioprinting of cultured muscle meat and seafood products and its potential for the food industry. *Trends in Food Science & Technology*, *152*, 104670. doi:10.1016/j.tifs.2024.104670

Hao, L., Li, Y., Gong, P., & Xiong, W. (2019). Material, process and business development for 3D chocolate printing. In *Fundamentals of 3D food printing and applications* (pp. 207–255). Elsevier. https://doi.org/10.1016/B978-0-12-814564-7.00008-0

Holland, S., Foster, T., & Tuck, C. (2019). Creation of food structures through binder jetting. In *Fundamentals of 3D food printing and applications* (pp. 257–288). Elsevier. https://doi.org/10.1016/B978-0-12-814564-7.00009-2

Holland, S., Tuck, C., & Foster, T. (2018). Selective recrystallization of cellulose composite powders and microstructure creation through 3D binder jetting. *Carbohydrate Polymers*, *200*, 229–238. doi:10.1016/j.carbpol.2018.07.064

Hooi Chuan Wong, G., Pant, A., Zhang, Y., Kai Chua, C., Hashimoto, M., Huei Leo, C., & Tan, U.-X. (2022). 3D food printing–sustainability through food waste upcycling. *Materials Today Proceedings*, *70*, 627–630. doi:10.1016/j.matpr.2022.08.565

Hu, H., Guo, J., Huang, Z., Huang, Q., & Lu, X. (2023). Effect of different hydrocolloids on the physicochemical, printing, and digestion properties of 3D printed purple sweet potato. *Food Hydrocolloids*, *144*, 109058. doi:10.1016/j.foodhyd.2023.109058

Hwang, J., Kim, S., & Lee, Y.-K. (2021). Mass customization in food services. *International Journal of Hospitality Management*, *93*, 102750. doi:10.1016/j.ijhm.2020.102750

Jazmin, K., & S., M., Daniela, L., Alvarez, M. M., Molina, A., & Chairez, I. (2025). Collaborative heterogeneous mini-robotic 3d printer for manufacturing complex food structures with multiple inks and curved deposition surfaces. *Micromachines*, *16*(3), 264. doi:10.3390/mi16030264

Jeong, Y. G., Yoo, J. J., Lee, S. J., & Kim, M. S. (2024). 3D digital light process bioprinting: Cutting-edge platforms for resolution of organ fabrication. *Materials Today Bio*, *29*, 101284. doi:10.1016/j.mtbio.2024.101284

Jiang, Q., Zhang, M., & Mujumdar, A. S. (2022). Novel evaluation technology for the demand characteristics of 3D food printing materials: A review. *Critical Reviews in Food Science and Nutrition*, *62*(17), 4669–4683. doi:10.1080/10408398.2021.1878099

Kavimughil, M., Leena, M. M., Moses, J. A., & Anandharamakrishnan, C. (2022a). 3D printed MCT oleogel as a co-delivery carrier for curcumin and resveratrol. *Biomaterials*, *287*, 121616. doi:10.1016/j.biomaterials.2022.121616

Kavimughil, M., Leena, M. M., Moses, J. A., & Anandharamakrishnan, C. (2022b). Effect of material composition and 3D printing temperature on hot-melt extrusion of ethyl cellulose based medium chain triglyceride oleogel. *Journal of Food Engineering*, *329*, 111055. doi:10.1016/j.jfoodeng.2022.111055

Keerthana, K., Anukiruthika, T., Moses, J. A., & Anandharamakrishnan, C. (2020). Development of fiber-enriched 3D printed snacks from alternative foods: A study on button mushroom. *Journal of Food Engineering*, *287*, 110116. doi:10.1016/j.jfoodeng.2020.110116

Kolb, M., Blazek, P., & Streichsbier, C. (2014). Food customization: an analysis of product configurators in the food industry (pp. 229–239). 10.1007/978-3-319-04271-8_20

Lanaro, M., Desselle, M. R., & Woodruff, M. A. (2019). 3D printing chocolate. In *Fundamentals of 3D food printing and applications* (pp. 151–173). Elsevier. https://doi.org/10.1016/B978-0-12-814564-7.00006-7

Latoch, A., Głuchowski, A., & Czarniecka-Skubina, E. (2023). Sous-vide as an alternative method of cooking to improve the quality of meat: A review. *Food*, *12*(16), 3110. doi:10.3390/foods12163110

Le-Bail, A., Maniglia, B. C., & Le-Bail, P. (2020). Recent advances and future perspective in additive manufacturing of foods based on 3D printing. *Current Opinion in Food Science*, *35*, 54–64. doi:10.1016/j.cofs.2020.01.009

Logesh, D., Vallikkadan, M. S., Leena, M. M., Moses, J. A., & Anandharamakrishnan, C. (2021). Advances in microfluidic systems for the delivery of nutraceutical ingredients. *Trends in Food Science & Technology*, *116*, 501–524. doi:10.1016/j.tifs.2021.07.011

Mantihal, S., Kobun, R., & Lee, B. (2020). International journal of gastronomy and food science 3D food printing of as the new way of preparing food: A review. *International Journal of Gastronomy and Food Science*, *22*, 100260. doi:10.1016/j.ijgfs.2020.100260

Mohan, M., Jayashree, E., Alfiya, P. V., & Anees, K. (2024). Encapsulation technologies for the delivery of spice extractives: An overview. *International Journal of Advanced Biochemistry Research*, *8*(3), 39–51. doi:10.33545/26174693.2024.v8.i3a.673

Mostafaei, A., Elliott, A. M., Barnes, J. E., Li, F., Tan, W., Cramer, C. L., Nandwana, P., & Chmielus, M. (2021). Binder jet 3D printing–Process parameters, materials, properties, modeling, and challenges. *Progress in Materials Science*, *119*, 100707. doi:10.1016/j.pmatsci.2020.100707

Nachal, N., Moses, J. A., Karthik, P., & Anandharamakrishnan, C. (2019). Applications of 3D printing in food processing. *Food Engineering Reviews*, *11*(3), 123–141. doi:10.1007/s12393-019-09199-8

Onyeaka, H., Nwabor, O., Jang, S., Obileke, K., Hart, A., Anumudu, C., & Miri, T. (2022). Sous vide processing: A viable approach for the assurance of microbial food safety. *Journal of the Science of Food and Agriculture*, *102*(9), 3503–3512. doi:10.1002/jsfa.11836

Pan, J., Chen, X., Zhu, Y., Xu, B., Li, C., Khin, M. N., Cui, H., & Lin, L. (2024). Design and development of dual-extruder food 3D printer based on selective compliance assembly robot arm and printing of various inks. *Journal of Food Engineering*, *370*, 111973. doi:10.1016/j.jfoodeng.2024.111973

Pitayachaval, P., Sanklong, N., & Thongrak, A. (2018). A review of 3D food printing technology. *MATEC Web of Conferences*, *213*, 01012. doi:10.1051/matecconf/201821301012

Portanguen, S., Tournayre, P., Sicard, J., Astruc, T., & Mirade, P.-S. (2019). Toward the design of functional foods and biobased products by 3D printing: A review. *Trends in Food Science & Technology*, *86*, 188–198. doi:10.1016/j.tifs.2019.02.023

Raja Santhi, A., & Muthuswamy, P. (2023). Industry 5.0 or industry 4.0S? Introduction to industry 4.0 and a peek into the prospective industry 5.0 technologies. *International Journal on Interactive Design and Manufacturing*, *17*, 947–979. doi:10.1007/s12008-023-01217-8

Raja, V., Moses, J. A., & Anandharamakrishnan, C. (2023). Effect of 3D printing conditions and post-printing fermentation on pearl millet fortified idli. *Journal of the Science of Food and Agriculture*, *103*(5), 2401–2412. doi:10.1002/jsfa.12410

Reuben, S. J., Chandra, S., Praveena, B., Santhoshkumar, P., & Moses, J. (2025). 3D printing of alternative proteins: Approaches, challenges and advances. *Sustainable Materials and Technologies*, *43*, e01285. doi:10.1016/j.susmat.2025.e01285

Santhoshkumar, P., & Moses, J. A. (2025a). Impact of konjac glucomannan on hot extrusion 3D printability of cake gel. *International Journal of Biological Macromolecules*, *305*, 141162. doi:10.1016/j.ijbiomac.2025.141162

Santhoshkumar, P., & Moses, J. A. (2025b). Nanotechnology in the Food Industry. In C. Anandharamakrishnan, J. A. Moses, & M. Leena (Eds.), *Nanotechnology for Sustainable Food Packaging* (1st ed., pp. 1–13). Wiley. doi:10.1002/9781119875154.ch1

Santhoshkumar, P., Negi, A., & Moses, J. A. (2023). 3D printing for space food applications: Advancements, challenges, and prospects. *Life Sciences in Space Research*. doi:10.1016/j.lssr.2023.08.002

Santhoshkumar, P., Raja, V., Priyadarshini, S. R., & Moses, J. A. (2024a). Evaluating the 3D printability of pearl millet flour with banana pulp blends. *Journal of the Science of Food and Agriculture*. doi:10.1002/jsfa.13389

Santhoshkumar, P., Raja, V., Priyadarshini, S. R., & Moses, J. A. (2024b). Impact of baking and steaming of 3D printed pearl millet - Banana pulp blends on physicochemical and sensory attributes. *International Journal of Gastronomy and Food Science*, *37*, 100962. doi:10.1016/j.ijgfs.2024.100962

Santhoshkumar, P., Swathika, B. S., Negi, A., & Moses, J. A. (2025). Improving gel performance and hot extrusion three-dimensional printability of blueberry gel with guar gum. *Food Biophysics*, *20*(2), 1–15. doi:10.1007/s11483-025-09964-9

Schmid, E., Farahnaky, A., Adhikari, B., & Torley, P. J. (2022). High moisture extrusion cooking of meat analogs: A review of mechanisms of protein texturization. *Comprehensive Reviews in Food Science and Food Safety*, *21*(6), 4573–4609. doi:10.1111/1541-4337.13030

Schuller, T., Jalaal, M., Fanzio, P., & Galindo-Rosales, F. J. (2024). Optimal shape design of printing nozzles for extrusion-based additive manufacturing. *Additive Manufacturing*, *84*, 104130. doi:10.1016/j.addma.2024.104130

Sen, K., Mehta, T., Sansare, S., Sharifi, L., Ma, A. W. K., & Chaudhuri, B. (2021). Pharmaceutical applications of powder-based binder jet 3D printing process – A review. *Advanced Drug Delivery Reviews*, *177*, 113943. doi:10.1016/j.addr.2021.113943

Sharma, N., Bansal, V., & Sahu, J. K. (2023). Plant-based gels. In *Engineering plant-based food systems* (pp. 131–150). Elsevier. https://doi.org/10.1016/B978-0-323-89842-3.00003-8

Stevens, E., Kimes, K., Martin, J., & Chmielus, M. (2020). Multi-material binder jet printing of functional Ni-Mn-Ga alloys. *Microscopy and Microanalysis*, *26*(S2), 2942–2944. doi:10.1017/S1431927620023296

Subhasri, D., Dutta, S., Leena, M. M., Moses, J. A., & Anandharamakrishnan, C. (2022). Gastronomy: An extended platform for customized nutrition. *Future Foods*, *5*, 100147. doi:10.1016/j.fufo.2022.100147

Subhasri, D., Moses, J. A., & Anandharamakrishnan, C. (2023). New food structures and their influence on nutrition, health and well-being. In *Food structure engineering and design for improved nutrition, health and well-being* (pp. 17–39). Elsevier. https://doi.org/10.1016/B978-0-323-85513-6.00014-1

Sun, J., Peng, Z., Zhou, W., Fuh, J. Y. H., Hong, G. S., & Chiu, A. (2015). A review on 3D printing for customized food fabrication. *Procedia Manufacturing*, *1*, 308–319. doi:10.1016/j.promfg.2015.09.057

Taneja, A., Sharma, R., Ayush, K., Sharma, A., Mousavi Khaneghah, A., Regenstein, J. M., Barba, F. J., Phimolsiripol, Y., & Sharma, S. (2022). Innovations and applications of 3-D printing in food sector. *International Journal of Food Science & Technology*, *57*(6), 3326–3332. doi:10.1111/ijfs.15691

Theagarajan, R., Nimbkar, S., Moses, J. A., & Anandharamakrishnan, C. (2021). Effect of post-processing treatments on the quality of three-dimensional printed rice starch constructs. *Journal of Food Process Engineering*, *44*(9), e13772. doi:10.1111/jfpe.13772

Tibrewal, K., Dandekar, P., & Jain, R. (2023). Extrusion-based sustainable 3D bioprinting of meat & its analogues: A review. *Bioprinting*, *29*, e00256. doi:10.1016/j.bprint.2022.e00256

Tireki, S. (2025). Basic spherification as a gelation application in the kitchen: Diffusion coefficient estimation and appearance quality of beverage pearls. *Physics of Fluids*, *37*(3). doi:10.1063/5.0246163

Tomašević, I., Putnik, P., Valjak, F., Pavlić, B., Šojić, B., Bebek Markovinović, A., & Bursać Kovačević, D. (2021). 3D printing as novel tool for fruit-based functional food production. *Current Opinion in Food Science*, *41*, 138–145. doi:10.1016/j.cofs.2021.03.015

Tong, Q., Jiang, Y., Xiao, S., Meng, Y., & Dong, X. (2024). Research on improving the structural stability of surimi 3D printing through laser cooking techniques. *Journal of Food Engineering*, *375*, 112075. doi:10.1016/j.jfoodeng.2024.112075

Tura, M., Tappi, S., Valli, E., Salas, P. G., Suleiman, R., Thabet, C., Essaidi, I., Muyonga, J. H., Akumu, G., Bahri, W. R., Ghanem, R., Ben Souissi, J., Okoth, M. W., Thuku, L., Maitha, I. M., Mounir, M., Girmay, G., Weldegerima, L., Gebreegziabher, Z., ... Toschi, T. G. (2025). A selection of African traditional and emerging foods: Processing, composition, nutritional, sensory and nutraceutical aspects. *Current Research in Food Science*, *101112*. doi:10.1016/j.crfs.2025.101112

Uddin, M. J., Hassan, J., & Douroumis, D. (2022). Thermal inkjet printing: prospects and applications in the development of medicine. *Technologies, 10*(5), 108. doi:10.3390/technologies10050108

Vadodaria, S., & Mills, T. (2020). Jetting-based 3D printing of edible materials. *Food Hydrocolloids, 106,* 105857. doi:10.1016/j.foodhyd.2020.105857

Waseem, M., Tahir, A. U., & Majeed, Y. (2024). Printing the future of food: The physics perspective on 3D food printing. *Food Physics, 1,* 100003. doi:10.1016/j.foodp.2023.100003

Wen, Y., Chao, C., Che, Q. T., Kim, H. W., & Park, H. J. (2023). Development of plant-based meat analogs using 3D printing: Status and opportunities. *Trends in Food Science & Technology, 132,* 76–92. doi:10.1016/j.tifs.2022.12.010

Wilson, A., Anukiruthika, T., Moses, J. A., & Anandharamakrishnan, C. (2020). Customized shapes for chicken meat–based products: Feasibility study on 3D-printed nuggets. *Food and Bioprocess Technology, 13.*

Wilson, A., Anukiruthika, T., Moses, J. A., & Anandharamakrishnan, C. (2023). Preparation of fiber-enriched chicken meat constructs using 3D printing. *Journal of Culinary Science & Technology, 21*(1), 127–138. doi:10.1080/15428052.2021.1901817

Yilmaz, B., Al Rashid, A., Mou, Y. A., Evis, Z., & Koç, M. (2021). Bioprinting: A review of processes, materials and applications. *Bioprinting, 23,* e00148. doi:10.1016/j.bprint.2021.e00148

Zhu, S., Vazquez Ramos, P., Heckert, O. R., Stieger, M., van derx Goot, A. J., & Schutyser, M. (2022). Creating protein-rich snack foods using binder jet 3D printing. *Journal of Food Engineering, 332,* 111124. doi:10.1016/j.jfoodeng.2022.111124

2

3D Food Printing and Emerging Applications

CHAPTER MENU

2.1 Introduction

3D printing has expanded to extensive applications, transforming domains such as healthcare, manufacturing, and construction. In healthcare, it permits the fabrication of bespoke prosthetics, implants, and surgical instruments tailored to individual patients and their requirements. Architects and engineers employ 3D printing to produce intricate building models and components, facilitating innovative designs and sustainable construction methodologies. This technology is also reshaping the automotive and aerospace sectors, where it enables the manufacture of lightweight, high-performance components. Furthermore, artists and designers utilize 3D printing to create elaborate sculptures, jewelry, and fashion accessories, expanding the limits of creativity (Gibson et al., 2021). This versatile and multifaceted technology continues to broaden its influence across various industries, promoting innovation and efficiency.

3D food printing is moving toward industrial use, propelled by the demand for superior quality products and customized manufacturing methods. From one perspective, the

4D Printing of Foods: Fundamentals, Applications and Prospects, First Edition.
C. Anandharamakrishnan, Jeyan A. Moses, and P. Santhoshkumar.

primary focus of 3D food printing was on using extrusion-based techniques to produce foods that are rich in nutrients. Another goal of implementing this technology in the food industry is to create visually attractive products that improve consumer perception, taking into account elements such as taste, appearance, texture, and overall desirability. For instance, the baking industry finds this method useful for making esthetically pleasing baked items such as cakes, cookies, and confections. 3D food printing has made it possible to manufacture edible goods such as candies, gummies, and chocolates using hot melt extrusion processes (Keerthy et al., 2025). Moreover, this approach has been expanding to include gels for customizing 3D-printed foods. Digital gastronomy, edible robots and electronics, sustainable packaging that can decompose naturally, attractive package casings, personalized food creations, and the precise delivery of bioactive substances are just a few growing areas that utilize 3D food printing (Kavimughil et al., 2023; Nida et al., 2022; Santhoshkumar et al., 2024; Subhasri et al., 2022; Ukkunda et al., 2025).

Comprehensively, 3D food printing has prospects for wide utility in advanced areas of application, including medicine, sports, military, and space. It can also be used in the preparation of food for patients suffering from dysphagia to come up with meals that have a close resemblance of the typical food consistency, thus making the meal more enjoyable for the person; in space food preparation, it enables astronauts to cook meals right on the spacecraft, which is quite convenient, especially for long duration space missions (Santhoshkumar et al., 2023). In the military and sports industries, 3D food printing can be created in forms containing dense levels of nutrients and energy. These foods can be modified to provide the specific dietary needs of soldiers and athletes. This chapter focuses on the state of the art of 3D food printing and its various applications, highlighting its potential for transforming food manufacturing and customization applications.

2.2 Emerging Concepts in 3D Food Printing

In recent years, there have been attempts to invent industry-scale 3D food printers, and researchers are actively exploring ways to expand the capabilities of these printers beyond the three-dimensional realm. The expansion of dimensions regarding the food's shapes, colors, and nutritional values is a growing research and commercial interest in this field. Whilst the challenges associated with these innovations also increase, there is ample scope for development. With 3D food printing, it is possible to completely transform the culinary industry by providing unmatched levels of customization and personalization. Importantly, current advancements in 3D food printing focus on sustainability, for instance, by considering alternative protein sources, such as plant-based materials, insects, legumes, cereals and lab-grown meat, to produce ecofriendly food products (Reuben et al., 2025). This strategy also entails waste valorization and a circular economy to reduce and efficiently utilize food waste. "On-demand" food manufacturing offers the advantage of reducing inventory requirements and minimizing spoiling in restaurants and food services. Similarly, "on-site" food manufacturing can eliminate the need for longer and more complex food supply chains. Multisensory creations allow the formulation of new tastes and textures in foods that cannot be created otherwise, and chefs and artists can make edible creations for people with healthy ambitions, through food safety and quality as a priority. New applications of 3D printing have further extended from culinary uses to targeted drug delivery and manufacturing of edible electronics for implant or invasive applications. Besides these general uses, materials have been incorporated during the printing of 3D structures with methods referred to as single material and multimaterial and in terms of movement of the nozzle as coaxial, dual-axial, adaptive nozzle, and quad-screw nozzle types (Fujiwara et al., 2025; Tang et al., 2024). Such advancements enable the design of multiple sensuous features in one food product, encouraging consumers to better accept 3D-printed foods.

2.2.1 Multimaterial Printing

Food can be manufactured utilizing multiple materials through the 3D printing approach. It is a novel approach in 3D food printing when different layers are stacked to provide multisensorial attributes in a single bite. This logic can consider varieties of material additions in a single meal and enhance the nutritional and sensory profiles of printed products (Tang et al., 2024). Recently, multimaterial printing has been explored for developing plant-based meat (Caron et al., 2024) and personalized nutritious products with high consumer acceptability (Lee et al., 2023). Although the applications of this concept span a wide range, including new forms and textures in the culinary arts, there are obstacles that need to be overcome, including compatibility of materials, challenges in selecting suitable printers and software, regulatory concerns, food safety assurance, and addressing the consumer acceptance in order to achieve successful implementation of 3D food printing in food industry (Tian et al., 2024). Notwithstanding these difficulties, using multiple materials in 3D food printing is noteworthy progress in food science, gastronomy, technology, engineering, and culinary arts. To date, the available approaches in multimaterial 3D printing in food and bioprinting applications are illustrated in Figure 2.1.

2.2.1.1 Dual/Multinozzle Axial Printing

Dual or multinozzle axial printing involves using multiple extruders to deposit different food materials simultaneously or in a certain order (Ali et al., 2016). This technique enables the production of intricate and personalized food items, facilitating diverse textures, flavors, and nutritional compositions in a single printed food product. Through the utilization of many nozzles, it becomes feasible to accurately position various ingredients or printing inks, supporting the creation of intricate designs and complex structures that would be challenging or even unattainable using conventional food preparation methods (Fujiwara et al., 2025; Lee et al., 2024). This technique has the potential to improve the attractiveness and gastronomic experience of printed foods, providing fresh opportunities for creativity in the food sector. Direct-ink writing-based 3D approaches can facilitate the development of multimaterial food printing (Tian et al., 2024). Such approaches for preparing multimaterials can be extended for 3D food printing applications (Figure 2.2).

Recently, in multimaterial printing, the concept of single nozzle usage has been introduced. Herein, the pressure maintenance on both sides can support layer deposition on the printing platform. The advanced version of this nozzle, called the adaptive nozzle, has the

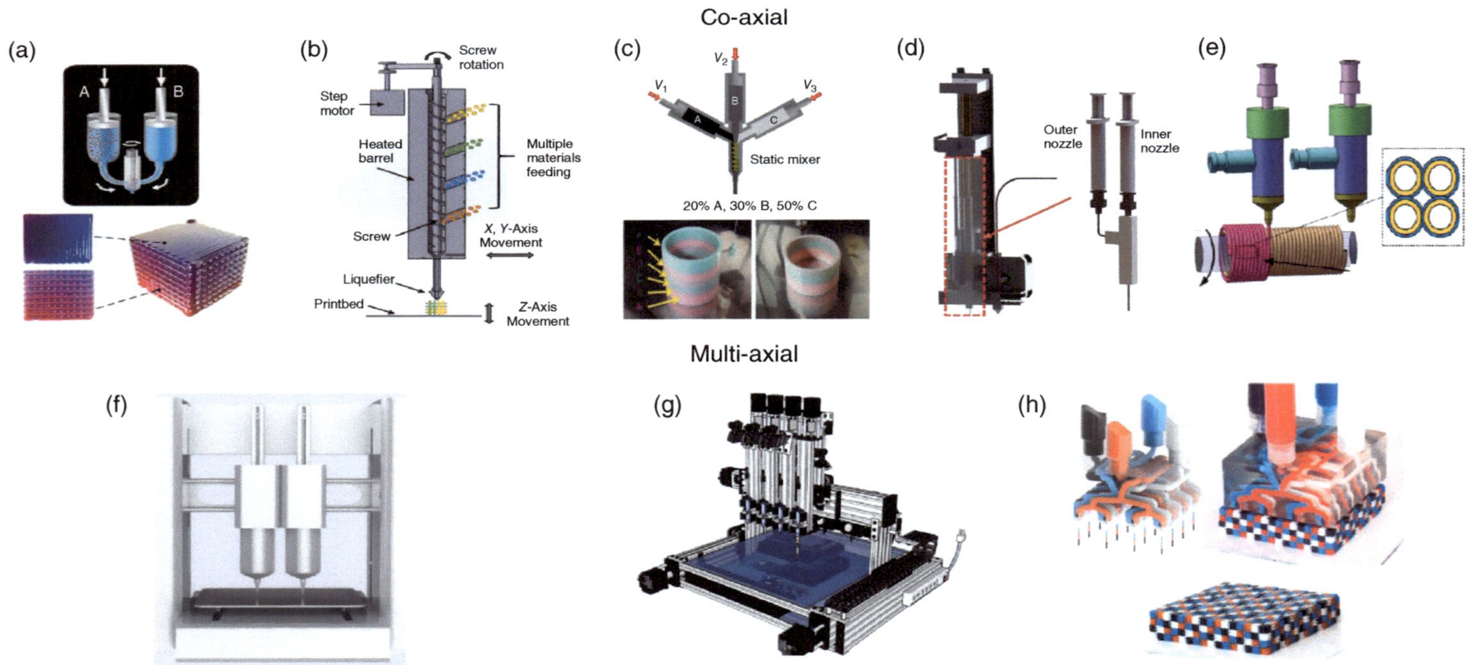

Figure 2.1 Multimaterial extrusion-based 3D printing approaches. (a) Gradient printing; (b) single screw extrusion with multiple feed inlet; (c) triple materials through single nozzle; (d) bioprinter coaxial nozzle; (e) coaxial nozzle-assisted printing of vascular; (f) dual-print head; (g) multinozzle; and (h) four different material printing. *Source*: Modified from Tang et al. (2024) / with permission of Elsevier.

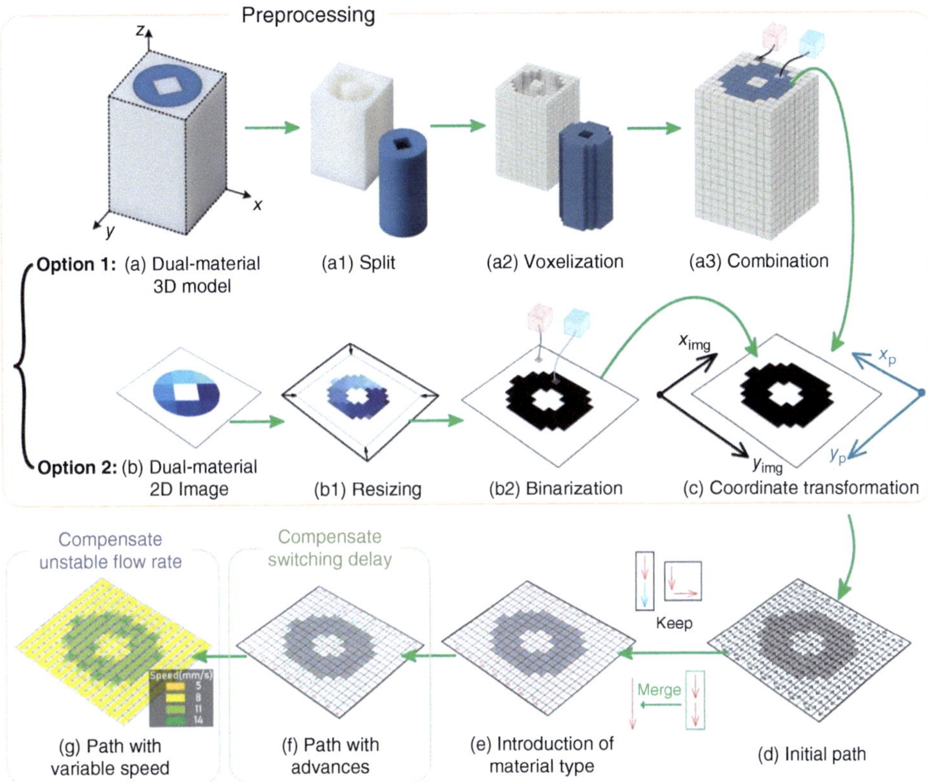

Figure 2.2 Flowchart explains the path generation strategy for multimaterial printing: (a) process of importing dual-material 3D model; (b) process of importing dual-material 2D image; (c) the transformation from the coordinate system $(x,y)_{img}$ to $(x,y)_{printer}$; (d) schematic of the initial path; (e) introduction of material type, different colors correspond to different materials; (f) path with advanced switching points, green path indicates switching in advance; (g) speed heatmap with adaptive movement speed. *Source*: Tian et al. (2024) / Taylor & Francis Group / CC BY 4.0.

capability to automatically change its size as directed by the software and codes. Nevertheless, this concept must be supplemented with extensive material characterization and indeed, handling two different material consistencies simultaneously is more challenging. It is also to be noted that the connection between filaments in the single nozzle type is better than that in the multinozzle type (Figure 2.3). However, as desired, the appeal and taste of customized products developed of the latter score better in sensory acceptability than that of the single material printing (Kokane et al. 2024).

2.2.1.2 Coaxial Nozzle Printing

Coaxial nozzle printing is a precise form of 3D food printing technology in which two or more materials are extruded concurrently using a nozzle setup arranged concentrically (i.e., one inside the other). Within the realm of food printing, this method enables the fabrication of intricate food formations that can incorporate two or more elements in a single printing process. In other words, a coaxial nozzle has the capability to deposit a filling material while being surrounded by a distinct outer material, all in a single uninterrupted

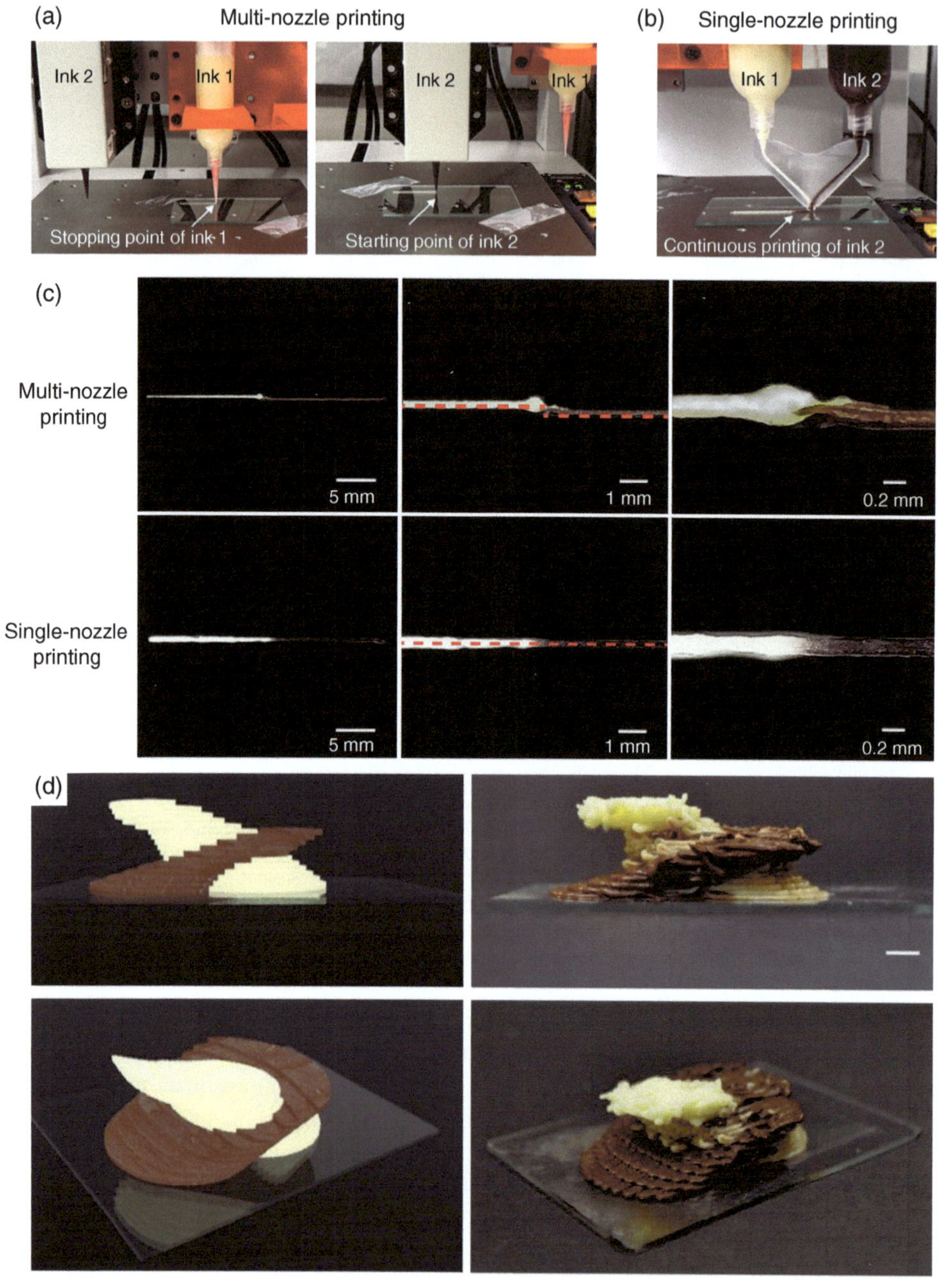

Figure 2.3 Multinozzle and single-nozzle printing of different materials. (a) Photographic image of multinozzle printing; (b) photographic image of single-nozzle printing; (c) filaments joining deposition views; and (d) photographic image of multimaterial 3D-printed chocolate. *Source*: Lee et al. (2024) / with permission of Elsevier.

operation. This technique improves the capacity to create culinary items with several layers or fillings with diverse textures and flavors while ensuring exact control over the arrangement and amount of each ingredient especially in dysphagia foods (Marques et al., 2025). The coaxial nozzle printing in food makes it possible to produce distinctive and refined food products that integrate esthetics, taste, and nutritional value, fostering new avenues for creativity in food printing. It can also be extended to adapt to encapsulate food functional compounds in micro and macro ranges (Kong et al., 2025). For instance, coaxial printing for nutraceutical delivery systems has used nanoemulsion gels carrying nutraceuticals in the inside filament and an outside coated paste as a 3D-printed product (Jeon et al., 2021); in another study, 3D food printing combined with nanotechnology author studied for codelivery of curcumin and resveratrol (Leena et al., 2022). Similarly, this approach makes cultured meat product development more convenient in cellular agriculture applications (Lanz et al., 2024; Schüler et al., 2024). Also, the approach can significantly contribute to developing nutrient-rich or fortified products and healthy, medical, special foods. However, there is ample opportunity for research, and the needs are many; studies must consider bioprinting and biomaterial printing applications as well (Jang et al., 2018; Wu et al., 2024).

2.3 Application Segments – 3D Food Printing

The application of 3D food printing began with the customization of food structures, progressing to personalized nutrient compositions, visually appealing packaging, smart and intelligent packaging, healthier food options, specialized diets, and, ultimately, the delivery of nutraceuticals. Today, the perspective on applicability is rapidly evolving with newer concepts and focuses. This section presents an overview of the popular application range of 3D food printing. For the purpose of providing organized content, the authors have categorized these applications considering the uniqueness of concepts and application-focus involved.

2.3.1 Personalizing Nutrients and Nutraceuticals

Personalized nutrition is an important application segment in 3D food printing, and this aims at the creation of customized food products. 3D printing has introduced a novel avenue for food personalization applications. Personalization is achieved not only through the goal of being health conscious but also by providing "enjoyment" and "involvement" to consumers through printed foods. This includes promoting stronger social connections through effective communication of food messaging and sensory experiences (Dankar et al., 2018). The material sources for personalized products can be plants, animals, or any possible food source. In order to fulfill specific dietary requirements, it is necessary to select sources rich in targeted nutritional and/or nutraceutical compositions. Therefore, this includes edible insects, oils, lipids, enzymes, hydrocolloids, and plants (Abedini et al., 2024). The presence of macronutrients and micronutrients in the chosen substance also affects the ability to print food material. For example, materials that are high in fiber content pose challenges when it comes to printing, but those that are rich in starch can be relatively conveniently 3D printed.

Personalized nutrition has advanced through several means; today, microbes are viewed as a promising food source for the future. Their cultivation process, in many ways, including fermentation, boosts the nutritional content of the printed items by developing microorganisms and their byproducts (Chen et al., 2023; Raja et al., 2023b). Subsequently, most studies have focused on the application of 3D printing in creating food for those with dysphagia. This technology can produce readily swallowed food that resembles regular food in terms of flavor and nutritional composition (Qiu et al., 2023; Raheem et al., 2021; Zhang et al., 2024). Utilizing 3D food printing to modify food texture is a prominent concept in food manufacturing (Chao et al., 2024). 3D food printing in the field of fortifying macro and micronutrients, incorporating drugs into medicinal food, and incorporating probiotics and nutraceuticals into functional food preparations is highly advantageous (Eswaran et al., 2023; Karavasili et al., 2020; Kavimughil et al., 2022). Figure 2.4 presents previously described investigations for different food printing applications.

2.3.2 Customizing Appearance

As the 3D printing technology expands to several industries, customization has emerged as a primary and significant benefit. Its usage in the food sector allows for the customization of foods in several aspects, including shape, color, size, nutrients, texture, and taste. This includes novel methods for customized delivering nutrients in a manner that ensures safety and enhances consumer appeal whilst offering benefits in terms of improved bioaccessibility/bioavailability. 3D food printing offers a more convenient means of specifying the shape and dimensions of the food product compared to conventional modification techniques such as molding. Nevertheless, the application of 3D food printing in the food industry is currently important in the research stage and necessitates significant innovations for large-scale production. However, on the one side, the technology is transitioning from the laboratory to pilot-scale applications, meanwhile, the development of large-scale food 3D printers and processes is being widely explored. For example, recently, Revo Foods has introduced a 3D food printer with full-scale industrial capacity (Derossi et al., 2024).

Customization in the food business presents substantially greater difficulties compared to other industries. Thus far, out of the previously documented methods, 3D printing has been shown to be a more feasible option for adapting to the customizing process compared to the other methods (Ottolenghi et al., 2024). The process of customization is initiated by the customer and serves to increase the degree of consumer satisfaction in both the product and the market. Therefore, the market transformed into a market that operates based on customer demand. In addition to the tactile appeal of the product, 3D printing is also employed for the creation of esthetically pleasing packaging casings (Ottolenghi et al., 2024). Furthermore, several surveys and research investigations have indicated that 3D-printed forms exhibit greater degrees of sensory acceptability with regard to their visual appeal (Caulier et al., 2020; Dankar et al., 2018; Silva et al., 2024). Silva et al. (2024) reported that the future food supply chain will be closely linked with 3D food printing because customers have a strong preference for 3D-printed products than traditional ones. Recent advancements in 3D food printing have allowed its expansion to a range of printing dimensions, such as 4D, 5D, and 6D. Increasing the dimension also amplifies the intricacy of the process; conversely, the degree of desirableness among the consumer also rises. Importantly,

Figure 2.4 Examples of foods developed for personalization nutrition and nutraceutical delivery through 3D food printing. (a) Edible growth; (b) texture-modified foods; (c) pasta; (d) vitamin D-fortified orange concentrated snacks; (e) vitamins-fortified candies; (f) minerals-fortified candies. *Source*: Eswaran et al. (2023) / Elsevier / CC BY-NC-ND 4.0; (g) probiotic-incorporated snacks. *Source*: Yoha et al. (2021) / with permission of Elsevier; (h) fibers-enriched snacks. *Source*: Keerthana et al. (2020) / with permission of Elsevier; (i) protein-enriched meats. *Source*: Yang et al. (2022) / with permission of Elsevier; (j) microgel loaded with astaxanthin. *Source*: Zhang et al. (2022a) / with permission of Elsevier; (k) oleogel loaded with curcumin and resveratrol. *Source*: Kavimughil et al. (2022) / with permission of Elsevier; and (l) chocolates incorporated with drug. *Source*: Karavasili et al. (2020) / with permission of Elsevier.

intelligent and innovative designs and functionalities of food products can be achieved using 4D, 5D, and 6D printing technologies.

2.3.3 Customizing Release of Target Molecules

Customization of shape/design, size, and infill (physical characteristics) can also link with texture, flavor, and aroma. Alongside color and other attributes, the sensory perception of food products by consumers is significantly influenced by their flavor profile. Through the process of flavor customization, each bite of the 3D-printed product can be customized to yield a distinct flavor. Flavor customization is an intriguing field of research in food printing. The existing methodologies in 3D food printing, particularly multimaterial 3D food printing, have the capability to yield multiflavored culinary arts within a single product. Furthermore, apart from multiple flavors, incorporating the number of dimensions may potentially result in a greater degree of consumer acceptance. Within the realm of 3D food printing for flavor customization, a recent study employed printing gels/inks as personalized flavor constituents (Sharma et al., 2024). Recently, in the context of alternative meat, this technology was explored to mimic the texture and taste profiles of conventional meat (Wen et al., 2023). Such applications can be extended to different other product types and to meet specific requirements.

3D food printing supports customized nutrition by regulating the levels of nutrients released into the body. This notion is analogous to encapsulation, which refers to deliberately targeting medication release to a specific location, such as the gut or colon. If these nutrients exhibit any light sensitivity, temperature sensitivity, or pH sensitivity criteria, they may not reach the intended location in their stable and bioactive form. To address this issue, encapsulation methods have been employed. In 3D food printing, the bioactive components have been enclosed within gels such as hydrogel, oleogel, or similar materials. The selection of this material is determined by the specific demands while following the recommended nutritious loading level stipulated by the regulatory agency. For instance, in a nutraceutical delivery application, resveratrol and curcumin were incorporated in an oleogel prepared from multichain triglycerides oil and printed in a 3D shape using holt-melt extrusion printing (Kavimughil et al., 2023). The same authors in a different study (Kavimughil et al., 2022) observed that these oleogels loaded with nutraceuticals exhibited increased bioaccessibility and an altered release of free fatty acids. Similarly, many studies have been conducted on the use of oleogels (Gao et al., 2022; Shang et al., 2023), hydrogels (Leena et al., 2022), and bigels (Okonkwo et al., 2022; Swathika et al., 2025) in nutraceutical delivery applications. Moreover, 3D printing methods have also been documented in drug delivery applications in the pharmaceutical sector (Beg et al., 2020; Goole & Amighi, 2016; Prasad & Smyth, 2016).

2.3.4 Healthier Food Options

3D printing can be used to efficiently create healthier food items, such as low-sugar, vitamin-rich, low glycemic index (GI), fiber rich, and others, in addition to foods with enhanced visual appeal and other sensory qualities. For instance, to create nutrient-rich 3D-printed foods for children, materials can be sourced from fruits, vegetables, legumes, or others. With 3D printing, foods with high nutritional value can score better in sensorial acceptance among children, which can be attributed to their appealing structures (Escalante-Aburto et al., 2021). Additionally, the 3D printing approach can also be considered in functional food development, such as, prebiotic/postbiotic incorporated products (Sadeghi et al., 2024; Yoha et al., 2021), fiber-enriched foods (Keerthana et al., 2020), protein-fortified foods (Zhang et al., 2022b), and low sugar foods (Ukkunda et al., 2025).

The selection of materials and the tuning of printing settings are frequently tedious steps in 3D food printing. Despite the printable nature of the material, retaining its structure is a challenging endeavor. The printing procedures are also influenced by the proximate composition of the material supplies. The popular extrusion-based technique relies heavily on the moisture and nutritional composition of the material sources. For instance, fibers present in the food are not conducive to printing; one major reason is the chances of nozzle clogging during the material extrusion process. In an interesting work, a 3D-printed snack was developed using a different approach; the binder jet technique was used for printing protein-rich material (Zhu et al., 2022). However, it is certain that food printing can support unmatched levels of customization in different parameters, including the nutritional content of the final products. Foods can also be structured with food 3D printing and/or other hybrid approaches, such as integration with electrospinning to develop products that have superior health benefits (Ejiohuo, 2023).

Studies have also documented the fortification of minerals and vitamins with the 3D printing approach (Eswaran et al., 2023; Varvara et al., 2021). Similarly, nutritional-rich foods can be directly added to the printing material supplies. Such processes can contribute to both printability and postprinting quality. For instance, incorporating pearl millet flour into idli batter was found to increase the nutritional profiles after fermentation (Raja et al., 2023a). Studies also reported that 3D printing of composite materials could facilitate the development of products with superior nutritional profiles (Kewuyemi et al., 2023; Krishnaraj et al., 2019; Varvara et al., 2021). In the concept of personalized food development, 3D printing can help develop meals with the nutritional requirements needed by the human body based on predictive models available in the database. This application presents futuristic concepts for developing low GI meals, dysphagia diets, and lower fat and higher protein products; however, the approach depends on the person's mentality, health conditions, and nutritional needs. Typically, individuals diagnosed with diabetes will have to share their blood glucose data to a cloud or web-based platform, and subsequently, a computer processes the nutrition-balanced recipe for the upcoming meal using an internal software connected with the 3D printer that can generate the personalized plan for the diabetic patient's food (Chen et al., 2024). Similarly, the 3D printing approach can also be used in developing medical foods that supply specific nutrients to the patient.

2.3.5 Foods for Special Needs

There are revolutionary opportunities to create personalized meals for those with special requirements through 3D food printing. This creative method makes it possible to precisely regulate the nutritional content of meals, guaranteeing that they satisfy particular dietary needs and can supply vital nutrients. For instance, 3D printing can modify food textures to make them easier to handle for people who have trouble chewing or swallowing (such as youngsters, newborns, and patients with dysphagia) (Lorenz et al., 2022; Pereira et al., 2021). 3D-printed foods can find new and futuristic applications by altering the oral processing behavior. In their study, Pattarapon et al. (2022) found that 3D-printed food strengthens patients' immune systems and decreases the duration of hospitalization and the severity of the illness based on studies considering 3D-printed and conventional "unappetizing" foods. Patients prefer 3D-printed food over conventional foods because of its potential to deliver customized texture, individualized nutrients, and appearance. Food printing has also found applications in sports, space, and military environments, called "special foods" (Lee, 2021). In an interesting approach, a NASA long-duration mission used 3D printers to provide meals at astronauts' request. In collaboration with a company, the latest demonstration successfully created a zero-gravity 3D printer capable of producing meals specifically designed for astronauts (Santhoshkumar et al., 2023).

Recently, the United States' Army has been investigating the potential of 3D food printing technology for military applications, citing its advantages such as immediate meal creation, customized nutrition for soldiers' body requirements, and prolonged shelf life achieved by preserving components in their raw state (Caulier et al., 2020). Researchers have also conducted trials with ultrasonic agglomeration to produce a wide variety of meals. They are developing a small 3D compact unit to transform forage plant materials like tree bark and berries into food products (Liu et al., 2017). The capability of 3D printing technology extends to creating personalized meals for athletes and pregnant women. This can be performed reasonably easily by eliminating, lowering, or changing the protein, fat, fiber, cellulose, and vitamin levels, among other modifications, and by introducing functional substances like anthocyanins and carotenoids to deliver the desired functions (Zhao et al., 2021). Several reports have documented the scope for developing special foods using 3D food printing (Caulier et al., 2020; Enfield et al., 2022; Liu et al., 2017; Santhoshkumar et al., 2023; Varvara et al., 2021; Zhao et al., 2021).

2.3.6 Consumer Requirements: Conveniences and Variety

Often, the primary factors influencing consumer food selection are taste, cost, and convenience. Each of these aspects comes along with its own multidisciplinary research approaches, criteria, and measurements. The psychological domain of taste and smell, as well as the neurobiological mechanisms of reward, includes the sensory components of food and its associated reward value. Food economics encompasses the interrelationships among food prices, diet quality, and diet cost. The affordability of food is determined by the ratio of food prices to household earnings. Convenience refers to the level of convenience regarding the interactions between the customer and the culinary environment. It is well known that the modern world looks toward convenience foods whilst providing the best of palatability, satiety, less preparation time, and sufficient nutritional profile for the person's needs (Brunner et al., 2010; Drewnowski & Monsivais, 2020). Now, more prominent ways have been introduced; 3D food printing can transform the culinary industry by enabling the precise and creative production of personalized meals on demand. The technology has the capability to produce complex food shapes and structures that pose challenges that are unattainable using conventional cooking techniques. Consumers gain from the option to modify their meals to satisfy dietary demands (Derossi et al., 2020), esthetic preferences (Califano & Spence, 2024), or nutritional goals, all while decreasing food waste through perfect portioning. Through the facilitation of a diverse range of textures, flavors, and forms, 3D food printing provides an unmatched degree of personalization and variation, so enabling the accessibility and customization of gourmet experiences to suit individual preferences with minimum exertion (Anandharamakrishnan et al., 2022). Copious research reports published in the past decade using 3D food printing focus on consumers' personalized diets and the customization of nutrients (Abedini et al., 2024; Burke-Shyne et al., 2020; Dankar et al., 2018; Escalante-Aburto et al., 2021; Eswaran et al., 2023; Varvara et al., 2021) explain the production of animal-based nutrient-rich foods, alternative foods, digital production, and automation in the food business. Importantly, consumers demand variety in daily and special diets. Most often, healthy food options and most foods that are consumed on a regular basis are associated with boredom. With 3D food printing, this limitation can be overcome, and foods with improved variety can be conveniently prepared, even up to the level of customization on a daily basis or meal-to-meal and snack-to-snack basis.

2.3.7 Toward Waste Valorization and Cleaner Production

Regarding sustainability, as an AM approach, 3D food printing supports reducing waste generation during fabrication. Additionally, 3D food printing enables on-demand production, simplifying supply chains and reducing waste (Rogers & Srivastava, 2021). Unlike the food sector, this aspect is among its most notable benefits in the manufacturing sector. 3D printing reduces the fabrication and material cost of particular items. Food wastage arises at every level of agricultural production, industrial manufacturing and processing, retail and domestic consumption. Other than unpredictable waste, most waste is generated by humans due to satiation, dislike of product taste, color, etc. 3D food printing can give the exact quantity and quality selected by the consumer with sufficient nutrients. Subsequently, the uneatable portion is avoided for cooking/consumption and considered a waste in the food industry though it contains rich fiber or essential nutrients. This portion can utilize 3D food printing to develop a value-added product from discards. For instance, products such as the development of potato peel-based noodles (Muthurajan et al., 2021), grape pomace-based functional cookies (Jagadiswaran et al., 2021), jackfruit seed-based snacks (Hooi Chuan Wong et al., 2022), and watermelon rind-based gummies (Keerthy et al., 2025) are examples. Similarly, fish-sourced gelatin can be used to enhance the printability of foods (Carvajal-Mena et al., 2022), and this logic can extend to different other food and food discard sources. Recently, a study reported using a 3D food printing approach along with food waste for the development of a controlled drug delivery system (Wang et al., 2024).

Another facet is the development of customized nonedible forms from food and agro-processing waste streams. For instance, food packaging casings have been successfully developed by researchers from sugarcane baggase, banana peel, and rice husk (Nida et al., 2021a, b, 2022, 2023). These were reported to be biodegradable and potential alternatives to single-use plastics. Accordingly, recent reports extensively document the potential of using agricultural waste streams for 3D food printing applications (Pant et al., 2023; Tyupova & Harasym, 2024; Yoha & Moses, 2023). Many countries worldwide have taken a step forward to reach sustainable goals. “Circular economy” and “zero waste economy” are not only 21st-century concepts but are progressively becoming integral components of the mainstream narrative to promote an ethical and environmentally aware lifestyle that benefits not only humans but also the entire ecosystem. Novel strategies are being aimed to accomplish this objective, one of which is the conversion of food waste into goods with added value. Within the realm of 3D food printing, waste reduction and circular economy represent distinctive applications that have the potential to dominate the food sector in the future.

2.3.8 Improving Food Packaging

3D food printing is transforming the food packaging industry by introducing customization, sustainability, and innovation. This technology allows for highly personalized packaging designs tailored to fit specific food items, enhancing functionality and esthetics (Nachal et al., 2019). It promotes the development of sustainable packaging alternatives, such as biodegradable and edible materials, to lessen environmental effects. The ability to create sophisticated and distinct designs improves the visual attractiveness of packaging and helps food preservation by adding enhanced barrier qualities (Sonkaria & Khare, 2024). As reported in the previous section, 3D printing can also be used to develop casings from agro-processing wastes that can be customized to different shapes and forms and used for packaging cakes, chocolates, and specialty foods. Recently, edible tags have been fabricated using 3D food printing, with excellent reception among consumers. Such edible tags carry information about the product and can be used as a live tracking tool (Santhoshkumar et al., 2024). Quick response (QR) codes can also be fabricated by modifying the infill of the 3D food printing system (Punpongsanon et al., 2022). Similarly, 3D food printing systems have been used for intelligent packaging applications to monitor the quality of food or to act as a time, temperature, or pH indicator (Tracey et al., 2022). Also, sensor fabrications can involve 3D printing approaches, including biosensors and nanosensors for food quality sensing applications. In particular, this can be used for smart food packaging applications (Cruz et al., 2024; dos Santos et al., 2022; Osmólska et al., 2022). Interactive packaging and the potential integration of food and packaging further highlight the technology's versatility. Despite current constraints in terms of materials, initial costs and process speeds, 3D food printing is poised to revolutionize the industry by offering more efficient, customized, and ecofriendly packaging solutions. In the future, an appealing food product combined with attractive packaging will certainly strengthen marketing strategies.

2.4 Conclusion

This chapter has comprehensively addressed the potential of 3D food printing to transform the food processing industry. Effective use of 3D printing in the food industry requires understanding its potential benefits, especially in improving material properties and increasing product diversity. Extrusion-based technology, which is thought to be the most promising approach, is the main driver propelling the development of 3D food printing in the food processing sector. Research and development activities are required in the calibration techniques for 3D-printed food ingredients, the design and dimensional accuracy of printed products, and the optimization of printing configurations. Moreover, there is a desperate need for advanced computational models and awareness initiatives to inform stakeholders and mitigate neophobia associated with this nascent technology. Advancements in postprocessing techniques for texturing, glazing, and surface finishing have significantly improved the esthetic value, practicality, and quality of 3D-printed foods, effectively merging technology with culinary arts. 3D food printing has significant potential to revolutionize commercial and culinary practices by facilitating personalization, scalability, and creativity. Overall, this technology has promising prospects for the food industry.

References

Abedini, A., Sohrabvandi, S., Sadighara, P., Hosseini, H., Farhoodi, M., Assadpour, E., Alizadeh Sani, M., Zhang, F., Seyyedi-Mansour, S., & Jafari, S. M. (2024). Personalized nutrition with 3D-printed foods: A systematic review on the impact of different additives. *Advances in Colloid and Interface Science*, *328*, 103181. doi:10.1016/j.cis.2024.103181

Ali, M. H., Mir-Nasiri, N., & Ko, W. L. (2016). Multi-nozzle extrusion system for 3D printer and its control mechanism. *The International Journal of Advanced Manufacturing Technology*, *86*(1–4), 999–1010. doi:10.1007/s00170-015-8205-9

Anandharamakrishnan, C., Moses, J. A., & Anukiruthika, T. (2022). *3D printing of foods*. John Wiley & Sons Ltd.

Beg, S., Almalki, W. H., Malik, A., Farhan, M., Aatif, M., Rahman, Z., Alruwaili, N. K., Alrobaian, M., Tarique, M., & Rahman, M. (2020). 3D printing for drug delivery and biomedical applications. *Drug Discovery Today*, *25*(9), 1668–1681. doi:10.1016/j.drudis.2020.07.007

Brunner, T. A., van der Horst, K., & Siegrist, M. (2010). Convenience food products. Drivers for consumption. *Appetite*, *55*(3), 498–506. doi:10.1016/j.appet.2010.08.017

Burke-Shyne, S., Gallegos, D., & Williams, T. (2020). 3D food printing: Nutrition opportunities and challenges. *British Food Journal*, *123*(2), 649–663. doi:10.1108/BFJ-05-2020-0441

Califano, G., & Spence, C. (2024). Consumer preference and willingness to pay for 3D-printed chocolates: A discrete choice experiment. *Future Foods*, *9*, 100378. doi:10.1016/j.fufo.2024.100378

Caron, E., de Walle, D., Van, D., & K., & Marchesini, F. H. (2024). State of the art, challenges, and future prospects for the multi-material 3D printing of plant-based meat. *Food Research International*, *114712*. doi:10.1016/j.foodres.2024.114712

Carvajal-Mena, N., Tabilo-Munizaga, G., Pérez-Won, M., & Lemus-Mondaca, R. (2022). Valorization of salmon industry by-products: Evaluation of salmon skin gelatin as a biomaterial suitable for 3D food printing. *LWT*, *155*, 112931. doi:10.1016/j.lwt.2021.112931

Caulier, S., Doets, E., & Noort, M. (2020). An exploratory consumer study of 3D printed food perception in a real-life military setting. *Food Quality and Preference*, *86*, 104001. doi:10.1016/j.foodqual.2020.104001

Chao, C., Nam, H. K., Park, H. J., & Kim, H. W. (2024). Potentials of 3D printing in nutritional and textural customization of personalized food for elderly with dysphagia. *Applied Biological Chemistry*, *67*(1), 25. doi:10.1186/s13765-023-00854-7

Chen, J., Teng, X., Zhang, M., Bhandari, B., Adhikari, B., & Yu, D. (2023). 5D food printing with color change induced by probiotic growth in a starch-protein-based gel system. *Food and Bioprocess Technology*, *16*(10), 2304–2314. doi:10.1007/s11947-023-03064-7

Chen, Y., Bi, S., Gu, J., Che, Q., Liu, R., Li, W., ... Zhang, Y. (2024). Achieving personalized nutrition for patients with diabetic complications via 3D food printing. *International Journal of Bioprinting*, *10*(2), 1862. doi:10.36922/ijb.1862

Cruz, R. M. S., Albertos, I., Romero, J., Agriopoulou, S., & Varzakas, T. (2024). Innovations in food packaging for a sustainable and circular economy. *Advances in Food and Nutrition Research*, *108*, 135–177. doi:10.1016/bs.afnr.2023.10.003

Dankar, I., Haddarah, A., Omar, F. E. L., Sepulcre, F., & Pujolà, M. (2018). 3D printing technology: The new era for food customization and elaboration. *Trends in Food Science & Technology*, *75*, 231–242. doi:10.1016/j.tifs.2018.03.018

Derossi, A., Husain, A., Caporizzi, R., & Severini, C. (2020). Manufacturing personalized food for people uniqueness. An overview from traditional to emerging technologies. *Critical Reviews in Food Science and Nutrition*, *60*(7), 1141–1159. doi:10.1080/10408398.2018.1559796

Derossi, A., Spence, C., Corradini, M. G., Jekle, M., Fahmy, A. R., Caporizzi, R., Devahastin, S., Moses, J. A., Zhou, W., Zhang, M., Bhandari, B., & Severini, C. (2024). Personalized, digitally designed 3D printed food towards the reshaping of food manufacturing and consumption. *Npj Science of Food*, *8*(1), 1–10. doi:10.1038/s41538-024-00296-5

dos Santos, D. M., Cardoso, R. M., Migliorini, F. L., Facure, M. H. M., Mercante, L. A., Mattoso, L. H. C., & Correa, D. S. (2022). Advances in 3D printed sensors for food analysis. *TrAC Trends in Analytical Chemistry*, *154*, 116672. doi:10.1016/j.trac.2022.116672

Drewnowski, A., & Monsivais, P. (2020). Taste, cost, convenience, and food choices. In B. P. Marriott, D. F. Birt, V. A. Stallings, & A. A. Yates (Eds.), *Present knowledge in nutrition* (pp. 185–200). Elsevier. https://doi.org/10.1016/B978.0-12-818460-8.00010.1

Ejiohuo, O. (2023). A perspective on the synergistic use of 3D printing and electrospinning to improve nanomaterials for biomedical applications. *Nano Trends*, *4*, 100025. doi:10.1016/j.nwnano.2023.100025

Enfield, R. E., Pandya, J. K., Lu, J., McClements, D. J., & Kinchla, A. J. (2022). The future of 3D food printing: Opportunities for space applications. *Critical Reviews in Food Science and Nutrition*, *63*. doi:10.1080/10408398.2022.2077299

Escalante-Aburto, A., Trujillo-de Santiago, G., Álvarez, M. M., & Chuck-Hernández, C. (2021). Advances and prospective applications of 3D food printing for health improvement and personalized nutrition. *Comprehensive Reviews in Food Science and Food Safety*, *20*(6), 5722–5741. doi:10.1111/1541-4337.12849

Eswaran, H., Ponnuswamy, R. D., & Kannapan, R. P. (2023). Perspective approaches of 3D printed stuffs for personalized nutrition: A comprehensive review. *Annals of 3D Printed Medicine*, *12*, 100125. doi:10.1016/j.stlm.2023.100125

Fujiwara, K., Suzuki, Y., Toba, K., Ogawa, J., Furukawa, H., Hashizume, M., Noji, T., Teratani, K., & Ito, N. (2025). Multifunctional 3D food printer with quad-screw nozzle for four-color printing and dual ink mixing. *Journal of Food Engineering*, *388*, 112384. doi:10.1016/j.jfoodeng.2024.112384

Gao, Y., Wang, Z., Xue, C., & Wei, Z. (2022). Modulation of fabrication and nutraceutical delivery performance of ovalbumin-stabilized oleogel-based nanoemulsions via complexation with gum arabic. *Food*, *11*(13), 1859. doi:10.3390/foods11131859

Gibson, I., Rosen, D., Stucker, B., & Khorasani, M. (2021). *Additive manufacturing technologies*. doi:10.1007/978-3-030-56127-7/COVER

Goole, J., & Amighi, K. (2016). 3D printing in pharmaceutics: A new tool for designing customized drug delivery systems. *International Journal of Pharmaceutics*, *499*(1–2), 376–394. doi:10.1016/j.ijpharm.2015.12.071

Hooi Chuan Wong, G., Pant, A., Zhang, Y., Kai Chua, C., Hashimoto, M., Huei Leo, C., & Tan, U.-X. (2022). 3D food printing z – sustainability through food waste upcycling. *Materials Today Proceedings*, *70*, 627–630. doi:10.1016/j.matpr.2022.08.565

Jagadiswaran, B., Alagarasan, V., Palanivelu, P., Theagarajan, R., Moses, J. A., & Anandharamakrishnan, C. (2021). Valorization of food industry waste and by-products using 3D printing: A study on the development of value-added functional cookies. *Future Foods*, *4*, 100036. doi:10.1016/j.fufo.2021.100036

Jang, J., Park, J. Y., Gao, G., & Cho, D.-W. (2018). Biomaterials-based 3D cell printing for next-generation therapeutics and diagnostics. *Biomaterials*, *156*, 88–106. doi:10.1016/j.biomaterials.2017.11.030

Jeon, W. Y., Yu, J. Y., Kim, H. W., & Park, H. J. (2021). Production of customized food through the insertion of a formulated nanoemulsion using coaxial 3D food printing. *Journal of Food Engineering*, *311*, 110689. doi:10.1016/j.jfoodeng.2021.110689

Karavasili, C., Gkaragkounis, A., Moschakis, T., Ritzoulis, C., & Fatouros, D. G. (2020). Pediatric-friendly chocolate-based dosage forms for the oral administration of both hydrophilic and lipophilic drugs fabricated with extrusion-based 3D printing. *European Journal of Pharmaceutical Sciences*, *147*, 105291. doi:10.1016/j.ejps.2020.105291

Kavimughil, M., Leena, M. M., Moses, J. A., & Anandharamakrishnan, C. (2022). 3D printed MCT oleogel as a co-delivery carrier for curcumin and resveratrol. *Biomaterials*, *287*, 121616. doi:10.1016/j.biomaterials.2022.121616

Kavimughil, M., Mahalakshmi, L., Moses, J. A., & Anandharamakrishnan, C. (2023). Delivery of nutraceutical ingredients through three-dimensional food printing. In C. Anandharamakrishnan & P. Subramanian (Eds.), *Industrial application of functional foods, ingredients and nutraceuticals* (pp. 253–278). Elsevier. https://doi.org/10.1016/B978-0-12-824312-1.00008-X

Keerthana, K., Anukiruthika, T., Moses, J. A., & Anandharamakrishnan, C. (2020). Development of fiber-enriched 3D printed snacks from alternative foods: A study on button mushroom. *Journal of Food Engineering*, *287*, 110116. doi:10.1016/j.jfoodeng.2020.110116

Keerthy, M. A., Dhanaselvam, K. R., Santhoshkumar, P., Ravikrishnan, V., & Moses, J. A. (2025). Improving the printability of watermelon rind using pectin: 3D printing optimization for the development of gummies. *Food Biomacromolecules*, *2*(2), 143–156. doi:10.1002/fob2.70008

Kewuyemi, Y. O., Kesa, H., Meijboom, R., Alimi, O. A., & Adebo, O. A. (2023). Comparison of nutritional quality, phenolic compounds, and antioxidant activity of conventional and 3D printed biscuits from wholegrain and multigrain flours. *Innovative Food Science & Emerging Technologies*, *83*, 103243. doi:10.1016/j.ifset.2022.103243

Kokane, S. B., Anjaly, P., Thangalakshmi, S., & Arora, V. K. (2024). Current trends in additive manufacturing based 4D food printing technology: A review. *Future Foods*, *10*, 100450. doi:10.1016/j.fufo.2024.100450

Kong, D., Zhang, M., Mujumdar, A. S., Fan, D., & Li, C. (2025). Novel coaxial 3D printing for future foods: Multi-material structuring for functional and sustainable resource innovation. *Innovative Food Science & Emerging Technologies*, *104*, 104087. doi:10.1016/j.ifset.2025.104087

Krishnaraj, P., Anukiruthika, T., Choudhary, P., Moses, J. A., & Anandharamakrishnan, C. (2019). 3D Extrusion printing and post-processing of fibre-rich snack from indigenous composite flour. *Food and Bioprocess Technology*, *12*(10), 1776–1786. doi:10.1007/s11947-019-02336-5

Lanz, M., Hartmann, C., Egan, P., & Siegrist, M. (2024). Consumer acceptance of cultured, plant-based, 3D-printed meat and fish alternatives. *Future Foods*, *9*, 100297. doi:10.1016/j.fufo.2024.100297

Lee, C. P., Ng, M. J. Y., Chian, N. M. Y., & Hashimoto, M. (2024). Multi-material direct ink writing 3D food printing using multi-channel nozzle. *Future Foods*, *10*, 100376. doi:10.1016/j.fufo.2024.100376

Lee, J. (2021). A 3D food printing process for the new normal era: A review. *Processes*, *9*(9), 1495. doi:10.3390/pr9091495

Lee, S. H., Kim, H. W., & Park, H. J. (2023). Integrated design of micro-fibrous food with multi-materials fabricated by uniaxial 3D printing. *Food Research International*, *165*, 112529. doi:10.1016/j.foodres.2023.112529

Leena, M. M., Anukiruthika, T., Moses, J. A., & Anandharamakrishnan, C. (2022). Co-delivery of curcumin and resveratrol through electrosprayed core-shell nanoparticles in 3D printed hydrogel. *Food Hydrocolloids*, *124*, 107200. doi:10.1016/J.FOODHYD.2021.107200

Liu, Z., Zhang, M., Bhandari, B., & Wang, Y. (2017). 3D printing: Printing precision and application in food sector. *Trends in Food Science and Technology*, *69*. doi:10.1016/j.tifs.2017.08.018

Lorenz, T., Iskandar, M. M., Baeghbali, V., Ngadi, M. O., & Kubow, S. (2022). 3D food printing applications related to dysphagia: A narrative review. *Food*, *11*(12), 1789. doi:10.3390/foods11121789

Marques, A. M., Oliveira, S. M., Martins, A. J., Bourbon, A. I., Rousta, N., Taherzadeh, M. J., Teixeira, J. A., Cerqueira, M. A., & Pastrana, L. M. (2025). Use of a coaxial nozzle-assisted 3D printing and ionic crosslinking to control the texture of a mycoprotein-based food product. *Future Foods*, *11*, 100530. doi:10.1016/j.fufo.2024.100530

Muthurajan, M., Veeramani, A., Rahul, T., Gupta, R. K., Anukiruthika, T., Moses, J. A., & Anandharamakrishnan, C. (2021). Valorization of food industry waste streams using 3D food printing: A study on noodles prepared from potato peel waste. *Food and Bioprocess Technology*, *14*(10), 1817–1834. doi:10.1007/s11947-021-02675-2

Nachal, N., Moses, J. A., Karthik, P., & Anandharamakrishnan, C. (2019). Applications of 3D printing in food processing. *Food Engineering Reviews*, *11*(3), 123–141. doi:10.1007/s12393-019-09199-8

Nida, S., Anukiruthika, T., Moses, J. A., & Anandharamakrishnan, C. (2021a). 3D printing of grinding and milling fractions of rice husk. *Waste and Biomass Valorization*, *12*(1), 81–90. doi:10.1007/s12649-020-01000-w

Nida, S., Moses, J. A., & Anandharamakrishnan, C. (2021b). 3D printed food package casings from sugarcane bagasse: A waste valorization study. *Biomass Conversion and Biorefinery*, *15*. doi:10.1007/s13399-021-01982-0

Nida, S., Moses, J. A., & Anandharamakrishnan, C. (2022). 3D extrusion printability of sugarcane bagasse blended with banana peel for prospective food packaging applications. *Sugar Tech*, *24*(3), 764–778. doi:10.1007/s12355-021-01095-y

Nida, S., Moses, J. A., & Anandharamakrishnan, C. (2023). Converting fruit waste to 3D printed food package casings: The case of banana peel. *Circular Economy*, *2*(1), 100023. doi:10.1016/j.cec.2022.100023

Okonkwo, C. E., Ofoedu, C. E., Hussain, S. Z., Adeyanju, A. A., Naseer, B., Inyinbor, A. A., Olaniran, A. F., & Kamal-Eldin, A. (2022). Application of biogels for bioactives delivery: Recent developments and future research insights. *Applied Food Research*, *2*(2), 100238. doi:10.1016/j.afres.2022.100238

Osmólska, E., Stoma, M., & Starek-Wójcicka, A. (2022). Application of biosensors, sensors, and tags in intelligent packaging used for food products—a review. *Sensors*, *22*(24), 9956. doi:10.3390/s22249956

Ottolenghi, C., D'Amico, S., & Iasevoli, G. (2024). Packaging customization in the Italian food industry. When is it really worth it? *British Food Journal*, *126*(7), 2646–2667. doi:10.1108/BFJ-02-2023-0131

Pant, A., Ni Leam, P. X., Chua, C. K., & Tan, U.-X. (2023). Valorisation of vegetable food waste utilising three-dimensional food printing. *Virtual and Physical Prototyping*, *18*(1). doi:10.1080/17452759.2022.2146593

Pattarapon, P., Zhang, M., & Mujumdar, A. S. (2022). Application potential of 3D food printing to improve the oral intake for immunocompromised patients: A review. *Food Research International*, *160*, 111616. doi:10.1016/j.foodres.2022.111616

Pereira, T., Barroso, S., & Gil, M. M. (2021). Food texture design by 3D printing: A review. *Food*, *10*(2), 320. doi:10.3390/foods10020320

Prasad, L. K., & Smyth, H. (2016). 3D Printing technologies for drug delivery: A review. *Drug Development and Industrial Pharmacy*, *42*(7), 1019–1031. doi:10.3109/03639045.2015.1120743

Punpongsanon, P., Miyatake, Y., Iwai, D., & Sato, K. (2022). *Demonstration of interiqr: unobtrusive edible tags using food 3D printing*. Adjunct proceedings of the 35th annual ACM symposium on user interface software and technology, 1–3. https://doi.org/10.1145/3526114.3558630

Qiu, L., Zhang, M., Bhandari, B., Chitrakar, B., & Chang, L. (2023). Investigation of 3D printing of apple and edible rose blends as a dysphagia food. *Food Hydrocolloids*, *135*, 108184. doi:10.1016/j.foodhyd.2022.108184

Raheem, D., Carrascosa, C., Ramos, F., Saraiva, A., & Raposo, A. (2021). Texture-modified food for dysphagic patients: A comprehensive review. *International Journal of Environmental Research and Public Health*, *18*(10), 5125. doi:10.3390/ijerph18105125

Raja, V., Moses, J. A., & Anandharamakrishnan, C. (2023a). Effect of 3D printing conditions and post-printing fermentation on pearl millet fortified idli. *Journal of the Science of Food and Agriculture*, *103*(5), 2401–2412. doi:10.1002/jsfa.12410

Raja, V., Moses, J. A., & Anandharamakrishnan, C. (2023b). Effect of 3D printing conditions and post-printing fermentation on pearl millet fortified idli. *Journal of the Science of Food and Agriculture*, *103*(5), 2401–2412. doi:10.1002/jsfa.12410

Reuben, S. J., Chandra, S., Praveena, B., Santhoshkumar, P., & Moses, J. A. (2025). 3D printing of alternative proteins: Approaches, challenges and advances. *Sustainable Materials and Technologies*, *43*, e01285. doi:10.1016/j.susmat.2025.e01285

Rogers, H., & Srivastava, M. (2021). Emerging sustainable supply chain models for 3D food printing. *Sustainability*, *13*(21), 12085. doi:10.3390/su132112085

Sadeghi, A., Can Karaca, A., Ebrahimi, M., Assadpour, E., & Jafari, S. M. (2024). The 3D printed probiotic products; an emerging category of the functional foods for the next-generations. *Trends in Food Science & Technology*, *148*, 104526. doi:10.1016/j.tifs.2024.104526

Santhoshkumar, P., Negi, A., & Moses, J. A. (2023). 3D printing for space food applications: Advancements, challenges, and prospects. *Life Sciences in Space Research*, *40*. doi:10.1016/j.lssr.2023.08.002

Santhoshkumar, P., Ramu, D., Mahalakshmi, L., & Moses, J. A. (2024). 3D printed edible electronics: Components, fabrication approaches and applications. *Biosensors and Bioelectronics*, *272*, 117059. doi:10.1016/j.bios.2024.117059

Schüler, K., Marques, D. M. C., Gusmão, A., Jabouille, M., Leite, M., Cabral, J. M. S., Sanjuan-Alberte, P., & Ferreira, F. C. (2024). 3D printing of plant-based fat inks towards manufacturing complex cellular agriculture products with fatty structures. *Food Hydrocolloids*, *157*, 110369. doi:10.1016/j.foodhyd.2024.110369

Shang, W., Sun, Y., Song, J., Zhang, P., Hou, Y., Wang, H., & Tan, M. (2023). Novel high internal phase oleogels-in-water pickering emulsions stabilized solely by whey protein isolate for 3D printing and fucoxanthin delivery. *Food Hydrocolloids*, *140*, 108609. doi:10.1016/j.foodhyd.2023.108609

Sharma, R., Chandra Nath, P., Kumar Hazarika, T., Ojha, A., Kumar Nayak, P., & Sridhar, K. (2024). Recent advances in 3D printing properties of natural food gels: Application of innovative food additives. *Food Chemistry*, *432*, 137196. doi:10.1016/j.foodchem.2023.137196

Silva, F., Pereira, T., Mendes, S., Gordo, L., & Gil, M. M. (2024). Consumer's perceptions and motivations on the consumption of fortified foods and 3D food printing. *Future Foods*, *10*, 100423. doi:10.1016/j.fufo.2024.100423

Sonkaria, S., & Khare, V. (2024). 3D printing of antimicrobial agents for food packaging. In A. K. Jaiswal & S. Shankar (Eds.), *Food packaging and preservation* (pp. 153–168). Elsevier. https://doi.org/10.1016/B978-0-323-90044-7.00009-4

Subhasri, D., Dutta, S., Leena, M. M., Moses, J. A., & Anandharamakrishnan, C. (2022). Gastronomy: An extended platform for customized nutrition. *Future Foods*, *5*, 100147. doi:10.1016/j.fufo.2022.100147

Swathika, B. S., Santhoshkumar, P., Moses, J. A., & Sinija, V. R. (2025). Novel Three-Dimensional-Printed ι-Carrageenan–Carnauba Wax Bigel Formulation for Codelivery Applications. *ACS Food Science & Technology*, *5*(3), 1183–1192. doi:10.1021/acsfoodscitech.5c00004

Tang, T., Zhang, M., Bhandari, B., & Li, C. (2024). Personalized transformation of 3D printing for traditional multi-material food with stuffing: A review. *Food Bioscience*, *59*, 104112. doi:10.1016/j.fbio.2024.104112

Tian, Z., Zhong, Q., Zhang, H., Yin, T., Zhao, J., Liu, G., Zhao, Y., Li, H. N., & He, Y. (2024). Optimising printing fidelity of the single-nozzle based multimaterial direct ink writing for 3D food printing. *Virtual and Physical Prototyping*, *19*(1). doi:10.1080/17452759.2024.2352075

Tracey, C. T., Predeina, A. L., Krivoshapkina, E. F., & Kumacheva, E. (2022). A 3D printing approach to intelligent food packaging. *Trends in Food Science & Technology*, *127*, 87–98. doi:10.1016/j.tifs.2022.05.003

Tyupova, A., & Harasym, J. (2024). Valorization of fruit and vegetables industry by-streams for 3D printing—a review. *Food*, *13*(14), 2186. doi:10.3390/foods13142186

Ukkunda, N. S., Santhoshkumar, P., & Moses, J. A. (2025). Impact of different natural sweeteners on 3D printability and post-printing quality of marzipan. *Future Foods*, *11*, 100557. doi:10.1016/j.fufo.2025.100557

Varvara, R.-A., Szabo, K., & Vodnar, D. C. (2021). 3D food printing: Principles of obtaining digitally-designed nourishment. *Nutrients*, *13*(10), 3617. doi:10.3390/nu13103617

Wang, L., Yong, L. X., & Loo, S. C. J. (2024). Utilizing food waste in 3D-printed PLA formulations to achieve sustainable and customizable controlled delivery systems. *ACS Omega*, *9*(31), 34140–34150. doi:10.1021/acsomega.4c05155

Wen, Y., Chao, C., Che, Q. T., Kim, H. W., & Park, H. J. (2023). Development of plant-based meat analogs using 3D printing: Status and opportunities. *Trends in Food Science & Technology*, *132*, 76–92. doi:10.1016/j.tifs.2022.12.010

Wu, X., Shi, W., Liu, X., & Gu, Z. (2024). Recent advances in 3D-printing-based organ-on-a-chip. *EngMedicine*, *1*(1), 100003. doi:10.1016/j.engmed.2024.100003

Yang, G., Tao, Y., Wang, P., Xu, X., & Zhu, X. (2022). Optimizing 3D printing of chicken meat by response surface methodology and genetic algorithm: Feasibility study of 3D printed chicken product. *LWT*, *154*, 112693. doi:10.1016/j.lwt.2021.112693

Yoha, K. S., Anukiruthika, T., Anila, W., Moses, J. A., & Anandharamakrishnan, C. (2021). 3D printing of encapsulated probiotics: Effect of different post-processing methods on the stability of Lactiplantibacillus plantarum (NCIM 2083) under static in vitro digestion conditions and during storage. *LWT*, *146*, 111461. doi:10.1016/j.lwt.2021.111461

Yoha, K. S., & Moses, J. A. (2023). 3D printing approach to valorization of agri-food processing waste streams. *Food*, *12*(1), 212. doi:10.3390/foods12010212

Zhang, C., Wang, C.-S., Girard, M., Therriault, D., & Heuzey, M.-C. (2024). 3D printed protein/polysaccharide food simulant for dysphagia diet: Impact of cellulose nanocrystals. *Food Hydrocolloids*, *148*, 109455. doi:10.1016/j.foodhyd.2023.109455

Zhang, L., Zaky, A. A., Zhou, C., Chen, Y., Su, W., Wang, H., Abd El-Aty, A. M., & Tan, M. (2022a). High internal phase Pickering emulsion stabilized by sea bass protein microgel particles: Food 3D printing application. *Food Hydrocolloids*, *131*, 107744. doi:10.1016/j.foodhyd.2022.107744

Zhang, Y., Lee, A. Y., Pojchanun, K., Lee, C. P., Zhou, A., An, J., Hashimoto, M., Tan, U.-X., Leo, C. H., Wong, G., Chua, C. K., & Pant, A. (2022b). Systematic Engineering approach for optimization of multi-component alternative protein-fortified 3D printing food Ink. *Food Hydrocolloids*, *131*, 107803. doi:10.1016/j.foodhyd.2022.107803

Zhao, L., Zhang, M., Chitrakar, B., & Adhikari, B. (2021). Recent advances in functional 3D printing of foods: A review of functions of ingredients and internal structures. *Critical Reviews in Food Science and Nutrition*, *61*(21), 3489–3503. doi:10.1080/10408398.2020.1799327

Zhu, S., Vazquez Ramos, P., Heckert, O. R., Stieger, M., van der Goot, A. J., & Schutyser, M. (2022). Creating protein-rich snack foods using binder jet 3D printing. *Journal of Food Engineering*, *332*, 111124. doi:10.1016/J.jfoodeng.2022.111124

3

Postprinting and Postprocessing

Need, Methods, and Quality Assessment

CHAPTER MENU

3.1 Introduction

3D food printing is a digitally controlled food manufacturing process involving the progressive layering of food materials. To enhance the structural integrity, esthetics, palatability, and overall applicability of printed products, postprocessing stands as an integral component in 3D food printing. This applies to both ready-to-cook (RTC) and ready-to-eat (RTE) food products (Alghamdy et al., 2024; Varvara et al., 2021). To reinforce the structure and maintain the shape of the 3D-printed foods, available postprocessing techniques such as baking, frying, drying, or freezing are typically conventional food processing unit operations (Kewuyemi et al., 2022). These methods can significantly alter (positively or negatively) sensory qualities, such as texture, flavor, and mouthfeel, which are crucial for consumer acceptability (Abedini et al., 2024; Chirico Scheele et al., 2022). In the case of RTC foods, the postprocessing stage must not negatively affect its quality, up to the level of significantly varying from its intended shape or quality (Waseem et al., 2024). From a different perspective, postprocessing can enable the enhancement of the esthetic appeal of printed foods through uniform cooking and surface finishing (Schiffer et al., 2024). It also

4D Printing of Foods: Fundamentals, Applications and Prospects, First Edition.
C. Anandharamakrishnan, Jeyan A. Moses, and P. Santhoshkumar.

guarantees that specific foods adhere to safety regulations, certifying that the final product is safe for consumption. In this background, postprocessing is a critical phase in integrating the technological capabilities of 3D food printing with consumers' functional and culinary expectations.

A typical 3D food printing process follows three integral stages: preprocessing, printing, and postprocessing. Herein, the transformation of material into a printable format is referred to as material preprocessing. Subsequently, the processed material is sent for printing. The printed construct is then postprocessed into a consumable form, and this is mandatory for cold extrusion or 3D printing at room temperature (Theagarajan et al., 2021). The quality and stability of printed products heavily rely on the postprocessing technique employed. Conversely, a food printed with hot-extrusion printing undergoes the required changes during the printing stage itself owing to the heat applied. The preceding chapters presented the fundamentals of 3D food printing, the requisite characteristics of materials, and trends in research and development in this field. This chapter focuses on postprocessing components and their effects on product quality and sensory attributes, contributing to the overall acceptability of the developed 3D-printed food.

3.2 Postprinting Stability and Applicability

Cold extrusion in 3D food printing is employed to create RTC foods, whilst hot extrusion printing is utilized for RTE food products. In addition to conventional extrusion-based printing, gel-forming extrusion combined with a heating element can facilitate ink or gel-based 3D printing (Zhang et al., 2022a). However, ensuring the quality and durability of printed constructs is a challenging endeavor. For instance, a starch-based product developed using hot extrusion 3D printing can maintain its structural stability after heating (Santhoshkumar et al., 2025). This can be attributed to the gelatinization and retrogradation processes that happen during the printing stage. Similarly, fat is crucial in chocolate-based printing; however, high levels of fat in the printing material supply may result in poor postprinting stability. This explains the significance of material composition on printing, postprocessing, overall postprinting quality and highlights the need for extensive research in this area. For instance, although chocolate is a promising material in food 3D printing and has several applications, limited research endeavors have documented the impact of hot extrusion in chocolate printing (VA et al., 2023; Santhoshkumar et al., 2025).

Postprocessing techniques can also improve the flavor and texture of the product while simultaneously reducing the moisture content of the printed sample (Santhoshkumar et al., 2024b). Additives employed in 3D food printing, including hydrocolloids, can improve the printability of the material and maintain the shape of the printed structures during postprocessing treatments. A study on 3D-printed cookies utilized 0.5% xanthan gum to improve material printability, resulting in enhanced postbaking shape accuracy (Hussain et al., 2022). Similarly, carrageenan and ethyl cellulose were used as additives in oleogel-based 3D food printing (Kavimughil et al., 2022a, 2022b). For hot extrusion 3D printing, konjac gum was used to improve the printability of cake gel (Santhoshkumar & Moses, 2025), and guar gum used for blueberry gel (Santhoshkumar et al., 2025). Likewise, studies have reported the shape fidelity, structural recoverability, and alterations in the structure/morphology of materials following postprocessing treatments (Kim et al., 2019; Pulatsu et al., 2020; Tay et al., 2023; Theagarajan et al., 2021; Yatmaz, 2024). Particularly, postprinting methods are known to affect the internal structure of the printed sample, and these are evident when materials are subjected to high- or low-temperature processes (Jo et al., 2021).

It is indispensable that the material of interest must be thoroughly characterized. To achieve this, several techniques are available, and a judicious selection of methods and approaches can support a robust 3D printing optimization process. A dynamic rheology test facilitates the prediction of form recoverability following printing and postprocessing. Particularly, the storage and loss modulus values quantify the materials' elasticity and viscosity under shear stress. Also, the three-interval thixotropy test (3ITT) and creep recovery tests assess the material's recoverability of postprinting stage (Raja & Moses, 2022; Santhoshkumar et al., 2024a). Differential scanning calorimetry (DSC) and thermogravimetric analysis (TGA) have been employed to ascertain the thermal properties of printing material supplies (Liu et al., 2020). Based on the thermal behavior of the materials, it is possible to identify and select appropriate postprocessing treatments and their conditions (e.g., temperature, air flow rate, or others). In this regard, the temperature chosen for hot extrusion-based 3D food printing is determined by considering the results of thermal behavior analyses. For instance, starch-based samples undergoing hot extrusion printing must consider the temperature-dependent effects of material gelatinization and retrogradation (Theagarajan et al., 2021). Further, entropy, enthalpy, and glass transition temperatures of the materials are other critical parameters (Rando & Ramaioli, 2021). However, the choice of parameter and test conditions will have to be decided on a case-by-case basis.

A printed product's "strong" structure can be achieved through increased infill density and appropriate selection of infill pattern/design. Figure 3.1 depicts the unique infill pattern and density for a 3D structure and printed potato samples. Increased infill density correlates with enhanced stability as compared to lower infill. Further, the infill pattern can relate to product esthetics; nevertheless, the structural integrity of the printed construct must be kept under consideration, particularly for geometries with greater heights (Zhao et al., 2021). The stability of the printed structure heavily depends on the wall perimeter and layer height (Mazlan et al., 2023). The printing parameters, particularly nozzle diameter, significantly affect postprinting stability. Also, during postprocessing, constructs with lower infill density have higher void areas and, therefore, show varying effects as compared with constructs having higher infill density (Anukiruthika et al., 2020). Further, lower infill density supports homogeneous cooking of the product and, often, relatively lower volumetric changes as compared to counterparts with higher infill density (Santhoshkumar et al., 2024b).

The influence of postprocessing techniques, such as microwave vacuum drying, on varying infill densities has been documented. It was observed that the 100% infill density construct resembled a cast-molded sample after the postprocessing stage (Liu & Zhang, 2021). Additionally, in another study of 3D food printing of fortified products and fermented-based food preparation, it was observed that different infill levels significantly influence the shape changes of the product (Raja & Moses 2022). Although varying infill patterns and density can provide several other benefits, such as in the development of texture-modified foods, it is important to strike a balance between such applications and the impact of infill levels on postprinting stability and postprocessing stability.

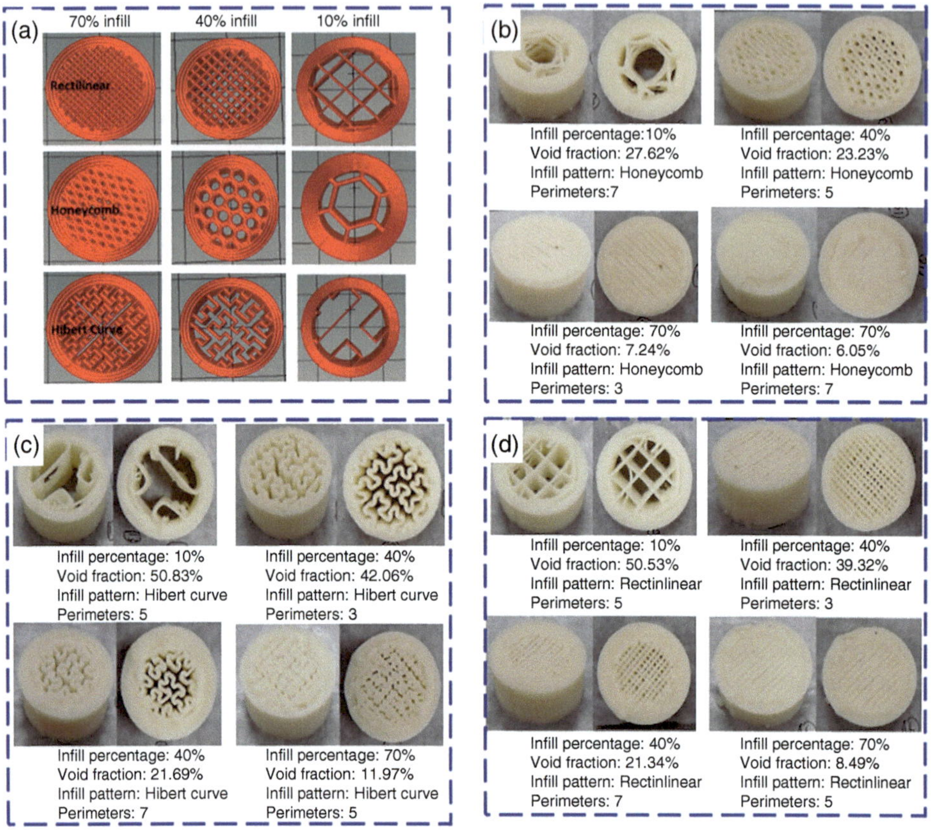

Figure 3.1 Different infill density and patterned 3D-printed potato samples with varying infill density and patterns. (a) 3D model, (b) honeycomb, (c) hibert curve, and (d) rectilinear. *Source:* Liu et al. (2018) / with permission of Elsevier.

3.3 Postprocessing Methods and Impact on Quality

Postprocessing is an essential phase of 3D food printing and can significantly alter the shape, structure, flavor, and overall quality of the printed product. A variety of postprocessing techniques have been employed for extrusion-based 3D-printed foods. Essentially, a postprocessing method is necessary when the material is fabricated through cold-extrusion printing approaches. Further, in contrast to hot melt extrusion-based printing, selective laser sintering, binder jetting, and inkjet printing methods, which necessitate temperature application during printing, do not require distinct postprocessing methods.

As research advances, many studies have changed their focus on the effects of postprocessing on the nutritional and phytochemical alterations in 3D-printed products (Krishnaraj et al., 2019; Raja & Moses, 2022; Santhoshkumar et al., 2024b; Theagarajan et al., 2021). Wang et al. (2024) provided a comprehensive analysis of the postprocessing effects on quality and internal structures. In addition to conventional cooking methods, current innovations in food technology, including laser, ultraviolet, infrared cooking, and enhanced or modified drying techniques, have been documented for 3D-printed food products (Blutinger et al., 2021; Deng et al., 2023a; Shen et al., 2023a; Yang et al., 2019). Nonetheless, aside from physical measures, the most significant alterations in quality were noted in the nutritional profile and phytochemical attributes during postprocessing. Primarily, postprocessing affects dimensions, resulting in either expanded or contracted geometries, often due to thermal effects, closely associated with the ingredients present in the material. For example, steaming, baking, and frying processes can significantly alter a product's dimensions (Santhoshkumar et al., 2024b; Theagarajan et al., 2021). Additionally, Waseem et al. (2024) recent review presents the most prevalent postprocessing strategies in 3D food printing, highlighting their impact on quality alterations, associated challenges, and the advantages of specific processing methods.

Postprocessing treatments substantially influence the dimensions and microstructure of processed samples. Theagarajan et al. (2021) examined various treatments for 3D-printed rice starch samples. Methods such as blanching, steaming, microwaving, roasting, and deep and shallow frying were considered; significant dimensional, microstructural, and quality alterations were observed. For this product, microwave treatment was found to result in a greater deformation of the product's shape as compared to other postprocessing treatments. Likewise, Santhoshkumar et al. (2024b) documented the preparation of a pearl millet-based 3D-printed snack employing steaming and baking techniques, concluding that sensory panelists preferred the baked samples. Similarly, studies have reported the impact of different approaches on the quality of mushrooms (Keerthana et al., 2020), chicken nuggets (Wilson et al., 2020), fortified fermented products (e.g., idli) (Raja & Moses, 2022), and fiber-rich snacks (Krishnaraj et al., 2019), demonstrating distinct effects owing to the choice of the postprocessing approach and the process parameters involved.

As in extrusion-based 3D food printing, starch and sugar are popular in other printing approaches and have been used in creating intricate shapes and structures (Wu et al., 2023). With differences in printing techniques used, the kind of parameters that must be studied and established vary, and these are mainly associated with the underlying mechanisms involved in such printing techniques. In the context of postprocessing, approaches may employ nonthermal processing techniques either alone or combined with conventional thermal processing procedures. Such innovations can resolve challenges encountered in existing postprocessing methods, particularly in terms of beneficial effects on the nutritional and sensorial quality of the final 3D-printed products. From a different perspective, it is important to note that recent studies have explored the scope of using nonthermal pretreatments to modify the printability characteristics of materials, and they may also be used as a first stage of postprocessing (He et al., 2020; Shen et al., 2023a,b; Zhou et al., 2024; Liu et al., 2024). However, the commercial viability and scalability of such approaches require further investigation and detailed documentation.

With moisture removal in drying, the resulting products exhibit enhanced hardness and brittleness. Infrared, microwave, and microwave-assisted drying methods significantly exhibit different changes, including color, textural, and nutritional quality (Lv et al., 2024; Shen et al., 2023a; Tang et al., 2024). As expected, freeze drying as the postprocessing approach can facilitate better retention of shape, color, and nutritional compounds more effectively (Deng et al., 2023b; Swathika et al., 2025). Freeze drying has also been used as a postprocessing method to avoid denaturation of bioactive compounds (Kavimughil et al., 2022a), and the approach holds great scope for nutraceutical delivery applications.

Accordingly, innovations in drying, such as novel techniques or assisted methods, can address challenges encountered in traditional drying techniques. For example, in a work on 3D printing probiotic-infused snacks, conductive hydro drying was employed as a post-processing method, and results indicated that cell viability was maintained with results comparable to freeze-dried counterparts (Yoha et al., 2021).

Postprocessing heat treatments can also result in the degradation of the nutritional profile of 3D-printed products. For instance, intense heat treatment can lead to a reduction in proteins, carbohydrates, and fats due to denaturation, the formation of new compounds, and other effects applicable to both micronutrients and macronutrients (Pérez et al., 2019). In the context of visual quality, the mechanical stability and shape fidelity of the printed and postprocessed products are of importance. Basic dimensional deviation measurements have often been considered a fundamental test to identify a product's deviations in shape/size along the process. Wen et al. (2024) recently documented the impact of postprocessing on the dimensional changes of printed products. For a more detailed understanding and to minimize ambiguity in process and variable selection, simulation tools can be used to predict and explain postprinting stability and postprocessing effects on food products (Figure 3.2). Similarly, in meat printing, Dick et al. (2019) reported the development of beef-based printed products and assessed the impact of postprocessing using simulations. In the future, it can be merged with artificial intelligence (AI) based algorithm to predict the post-processing impact of the printed products.

At present, individuals view postprocessing after printing as a laborious endeavor. A recent study showcased an innovative 3D food printer that simultaneously extrudes material from the nozzle and applies heat to the food layer, resulting in a cooked product (Lv et al., 2023). In a different approach, earlier Blutinger et al. (2021) demonstrated an approach involving the application of heat to each printed layer. In the future, as systems evolve to commercialization, the time, resource, space, and energy requirements of such alterations in processes must be established. Certainly, these would have slight variation focuses when the application is considered for industry-scale production or pilot or domestic-level usage. Overall, postprocessing considerably adds to major modifications after printing in terms of sensory,

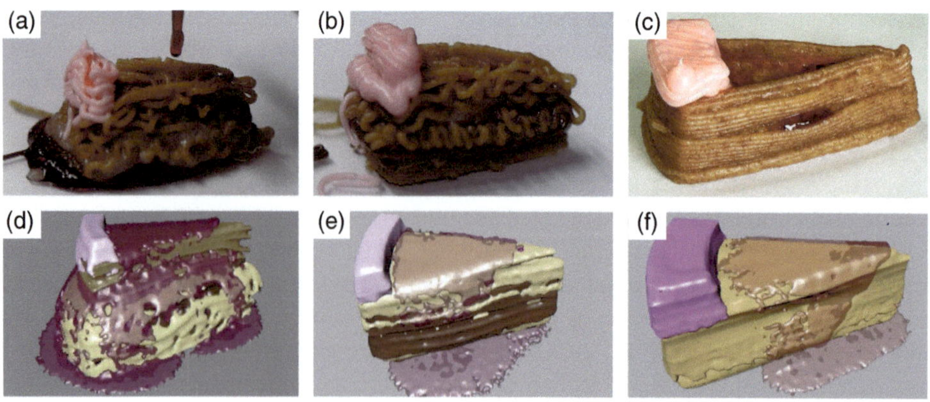

Figure 3.2 Actual and simulated views of three different cheesecake prints with different viscosities (a, b, c – actual samples, d, e, f – simulated samples). *Source:* Goldfinger et al. (2024) / Elsevier / CC BY 4.0.

structure, and shape of the printed constructs and nutritional ingredients. Therefore, this has been a topic of emerging research interest in the field of 3D food printing.

3.4 Approaches for Quality Assessment

A well-printed food product that loses its quality after postprocessing will certainly not be preferred by consumers. To assess the effects of postprinting and postprocessing on product quality and other parameters, different approaches, both subjective and objective, have been explored. While typical sensory evaluation studies are subjective, there has been a rising interest in different instrumental methods, with both destructive and nondestructive quality assessment approaches. However, it is important to note that most of these techniques were not specifically designed for 3D food printing applications and are fundamental wet lab procedures, computer-assisted techniques, spectroscopic evaluation, or assessments of engineering properties. This section focuses on common quality assessment approaches for postprocessed 3D-printed foods.

3.4.1 Texture and Color Assessment

Textural profile analysis is employed to identify the textural modifications due to printing and postprocessing. Certain postprocessing methods remove the moisture from the product, thus increasing the hardness of the product. For instance, drying and frying result in higher hardness than the steaming approach (Theagarajan et al., 2021). Conversely, steaming results in higher adhesiveness and gumminess values (Santhoshkumar et al., 2024b). However, these textural properties also depend on the nutritional compounds present in the printed sample (Nachal et al., 2019). For special health needs, when 3D food printing is used, the criteria and focus quality parameters may be specific to the requirement. For example, dysphagic diets prepared through 3D printing should have the appropriate texture for subsequent oral processing, bolus formation, and swallowing (Molimi et al., 2025). Therefore, the choice of postprocessing method will also play a crucial role. Typically, a higher infill increases the product's hardness due to the higher material deposition during printing, and a lower infill would result in lower hardness values (Derossi et al., 2021; Oral et al., 2021). Nevertheless, the choice of post-processing methods is significantly affecting the color values of the printed products. Copious research studies have reported the impact of postprocessing on the color and textural properties of foods (Keerthana et al., 2020; Santhoshkumar et al., 2024b; Raja et al., 2023; Keerthy et al., 2025). In terms of color, except freeze drying and certain nonthermal processing techniques, most postprocessing methods result in significant color changes. This logic is similar to the case of conventional non-3D-printed foods as well.

3.4.2 Dimensional Assessment

To evaluate the influence of postprocessing on the quality of the printed structures, many assessment techniques have been employed. An adaptation of the fundamental dimension deviation approach was developed to quantify the accuracy of shapes after printing and the effects of postprocessing, namely, expansion and shrinkage effects. An investigation of the printing of mushroom-based snacks revealed dimensional alterations following post-processing (Keerthana et al., 2020). Ideally, comparisons are made between the actual 3D CAD geometry, the printed construct's dimensions, and the dimensions of the post-processed 3D-printed food. Similarly, Theagarajan et al. (2021) reported the physical dimensional changes in rice starch constructs subjected to a range of postprocessing techniques, highlighting significant variations between the methods. In some cases, products may undergo severe structural changes and damage, breakage, or otherwise.

The following sections outline the quality evaluation methods used for assessing post-printing stability and postprocessing shape fidelity. For example, to calculate a square-shaped 3D-printed product, the postprocessing dimensional deviation can be calculated using the following equations (3.1–3.6) (where l is the length, w is the width, and h is the height):

$$\text{Surface area}\left(\text{mm}^2\right) = 2lw + 2lh + 2hw \tag{3.1}$$

$$\text{Volume}\left(\text{mm}^3\right) = lwh \tag{3.2}$$

$$\text{Porosity}\left(\%\right) = \frac{\text{Volume of the actual geometry}\left(\text{cm}^3\right) - \text{Volume of estimated geometery}\left(\text{cm}^3\right)}{\text{Volume of the actual geometry}\left(\text{cm}^3\right)} \times 1 \tag{3.3}$$

$$\text{Shrinkage}\left(\%\right) = \frac{\left(\text{Raw}\,l - \text{Cooked}\,l\right) + \left(\text{Raw}\,w - \text{Cooked}\,w\right) + \left(\text{Raw}\,h - \text{Cooked}\,h\right)\left(\text{mm}\right)}{\text{Raw}\,l + \text{Raw}\,w + \text{Raw}\,h\left(\text{mm}\right)} \times 100 \tag{3.4}$$

$$\text{Shape fidelity}\left(\%\right) = \frac{\text{Measured dimension}\left(\text{mm}\right)}{\text{Theoretical dimension}\left(\text{mm}\right)} \times 100 \tag{3.5}$$

$$\text{Cooking loss}\left(\%\right) = \frac{\text{Initial weight}\left(\text{g}\right) - \text{Final weight}\left(\text{g}\right)}{\text{Initial weight}\left(\text{g}\right)} \times 100 \tag{3.6}$$

3.4.3 Sensory Evaluation

Sensory assessments are often performed using a hedonic scale rated from 1 to 9, and panelists are asked to indicate their preferences. Specifically, in 3D food printing studies, two types of sensory assessment have been reported: visual sensory and product sensory. Herein, the former focuses on parameters such as appearance, dimensional stability, shape, layer definitions, texture and finishing, binding property, and thread quality. These are defined below:

- *Appearance*: Pertains to the entire visual impression, which encompasses the absence of visual defects in the printed food, vibrancy, and consistency.

- *Dimensional stability*: Preserving the structure of the printed foods' intended dimensions and proportions without deformation, warping, or contraction.
- *Shape*: The precision of the printed food aligns with the planned design or geometry.
- *Layer definition*: The finishing and uniformity of discrete layers within the printed composition.
- *Texture and finishing*: The printed food's physical quality and visual presentation.
- *Binding property*: The capacity of the printed material to cohesively retain its shapes and ensure correct adhesion between layers.
- *Thread quality*: Uniformity of the material during the extrusion.

Accordingly, panelists are briefed about these terminologies, and then standard sensory evaluation protocols are followed. Overall, the visual sensory quality is the first decider of the 3D-printed product's quality. Further, it is well accepted that a 3D-printed food would have a better esthetic appeal than a conventional product, and therefore, visual sensory quality parameters are given significant attention in 3D printing studies (Santhoshkumar et al., 2024a; Theagarajan et al., 2020).

Similarly, the next type, product sensory quality, covers parameters such as the color, flavor, texture, taste, aftertaste, and overall acceptance of postprocessed products. Many 3D food printing studies reported this aspect (Santhoshkumar et al., 2024b; Theagarajan et al., 2020; Wilson et al., 2020). Interestingly, these results can correlate with instrumental results, such as results of texture or color analyses or those from E-nose and E-tongue.

3.4.4 Image Analysis

Numerous research studies use visual sensory analysis to detect printing inaccuracies and categorize instances of under-extrusion and over-extrusion (Anukiruthika et al., 2020; Santhoshkumar et al., 2024a; Theagarajan et al., 2020). Nevertheless, to prevent human errors and biases, image processing techniques can be developed to quantify the changes in shape caused by printing and subsequent processing. A recent study by Santhoshkumar et al. (2024b) involved measuring shape changes following printing and postprocessing. Sample images were processed using ImageJ software to compare the outer dimensions of the actual geometry, printed, and postprocessed samples. All printed and postprocessed samples differed in shape from the desired (actual) geometry, with levels varying based on printing and postprocessing conditions (Figure 3.3). Earlier, Raja et al. (2023) documented the impact of fermentation on shape fidelity. The fundamental workflow of the image processing system is represented in Figure 3.4. It has to be kept in mind that most conventional shape fidelity analyses solely focus on the edge (outer) image of the product and do not include any evaluation of inside modifications.

In a recent investigation, researchers used an automated image analysis technique and compared the digital measures with human evaluations to determine the corresponding error tolerance in both tests. Figure 3.5 shows the quantification of the accuracy of 3D-printed food using layer-wise image analytics. It was reported that human subjective assessment may be unable to identify an under-extrusion process; hence, the image analysis technique enables its detection more accurately. In contrast, in the case of over-extrusion, both approaches were found to be feasible (Ma et al., 2023b).

Sample	Image	Edge image	Surface area (mm^2)	Change in surface area (%)
Control			2146.4	–
Raw 1			2366.9	10.3
Raw 2			2471.2	15.2
Raw 3			2416.2	12.6
Steam 1			2399.3	11.8
Steam 2			2486.7	15.8
Steam 3			2383.2	11.0
Oven 1			2453.1	14.3
Oven 2			2458.1	14.5
Oven 3			2392.4	11.5

Figure 3.3 Shape fidelity study for printed and postprocessed millet-based 3D-printed samples. *Source:* Santhoshkumar et al. (2024b) / with permission of Elsevier.

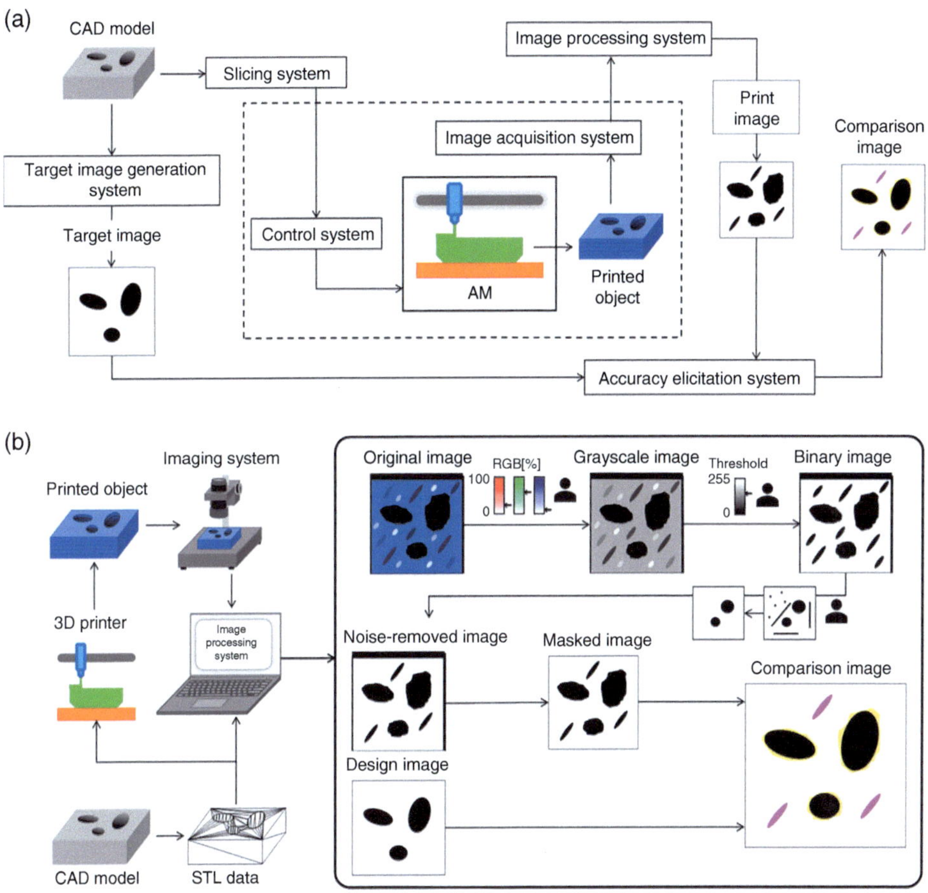

Figure 3.4 Image processing setup for 3D-printed objects. (a) Workflow for image processing and (b) working principle of the image processing system. *Source:* Okamoto & Ura (2024) / MDPI / CC BY 4.0.

3.4.5 Computer-Assisted Quality Evaluation

In industry, computer-assisted quality evaluation has emerged as a powerful and accurate tool, providing huge savings in time and resources. By utilizing computer vision techniques, researchers could control a 3D food printing nozzle's feedforward velocity (Ma et al., 2023a). Through modeling and simulation techniques, one can visualize realistic changes using suitable software (Ma et al., 2021; Raja et al., 2023). The use of machine learning (ML) programs and AI has been increasingly popular in recent years to forecast the printability of a product and its stability after printing. A recent report details the potential of AI applications in 3D printing (Bedoya et al., 2022). Such tools can support material selection, formulations, printability assessment, process conditions and optimization, postprinting quality assessment, selection and standardization of postprocessing methods and conditions, and assessment of the impact of postprocessing on 3D-printed foods. Interesting applications have been documented in several recent studies (Lu et al., 2023; Outrequin et al., 2024). For example, a recent work reports the possibility of combining ML with computational fluid dynamics (CFD) (Chen et al., 2024). CFD, as a powerful tool, has several applications in the

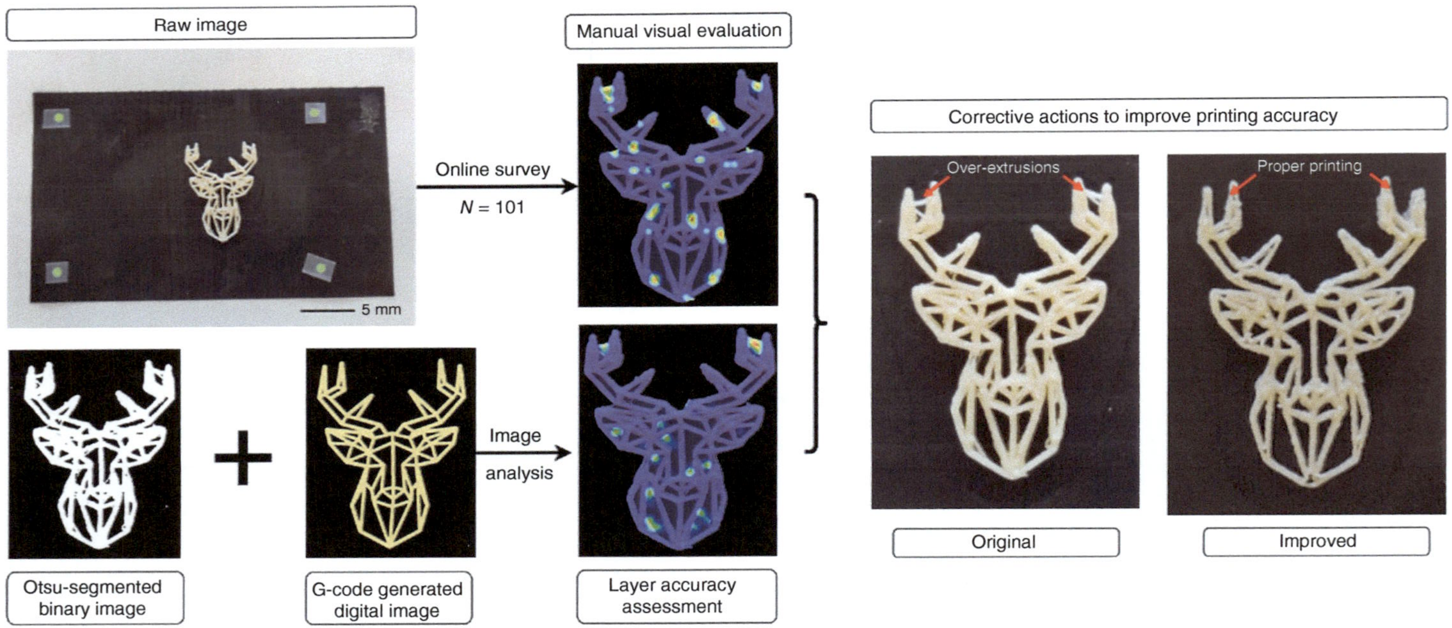

Figure 3.5 Contextualizing automated image analysis with human evaluations for 3D food-printed cookies. *Source:* Ma et al. (2023b) / Elsevier / CC BY 4.0 / Elsevier / CC BY 4.0.

food industry, including newer concepts emerging in the case of 3D food printing. These researchers integrated artificial neural networks (ANN), ML, fuzzy systems, and CFD with a focus on enhancing production quality by developing an automated decision-control model to optimize the system's performance. Although this is an emerging area of research, there are plenty of avenues for development and improvement in this segment.

3.4.6 Microstructure Imaging

Microstructural imaging can explore how internal structures change to printing, postprinting, and postprocessing. Studies have reported various microscopic approaches, such as X-ray (Okamoto & Ura, 2024), computed tomography (CT) (Derossi et al., 2022), scanning electron microscopy (SEM), and confocal laser scanning microscopy (CLSM), to assess changes in surface morphology and internal geometries. For instance, Figure 3.6 shows

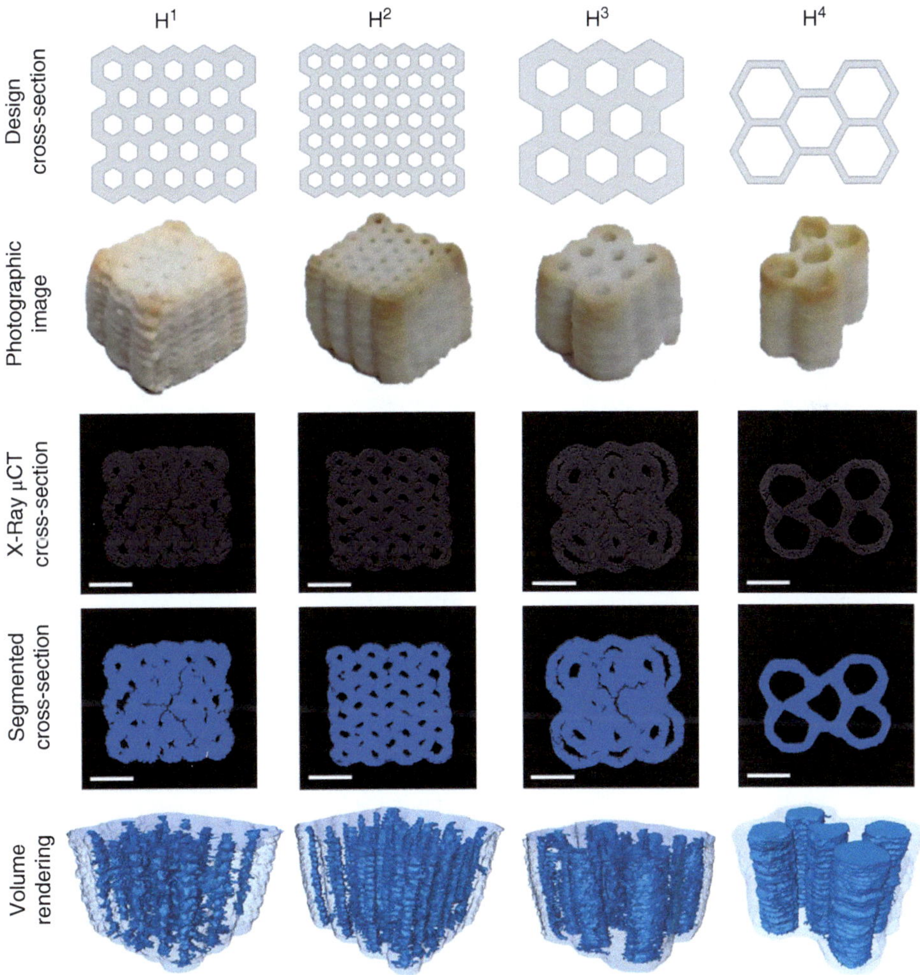

Figure 3.6 Micro-CT images of 3D-printed postprocessed cookies. *Source:* Piovesan et al. (2020) / MDPI / CC BY 4.0.

micro-CT images revealing the effects of different infill patterns and postprocessing on microstructures (Piovesan et al., 2020). Likewise, Dick et al. (2021) reported the microstructure of 3D-printed beef pastes and their corresponding cooked samples (Figure 3.7a, b). Recently, Xie et al. (2025) reported SEM-energy dispersive X-ray spectroscopy analysis to confirm the strong interaction between the protein-based formulations.

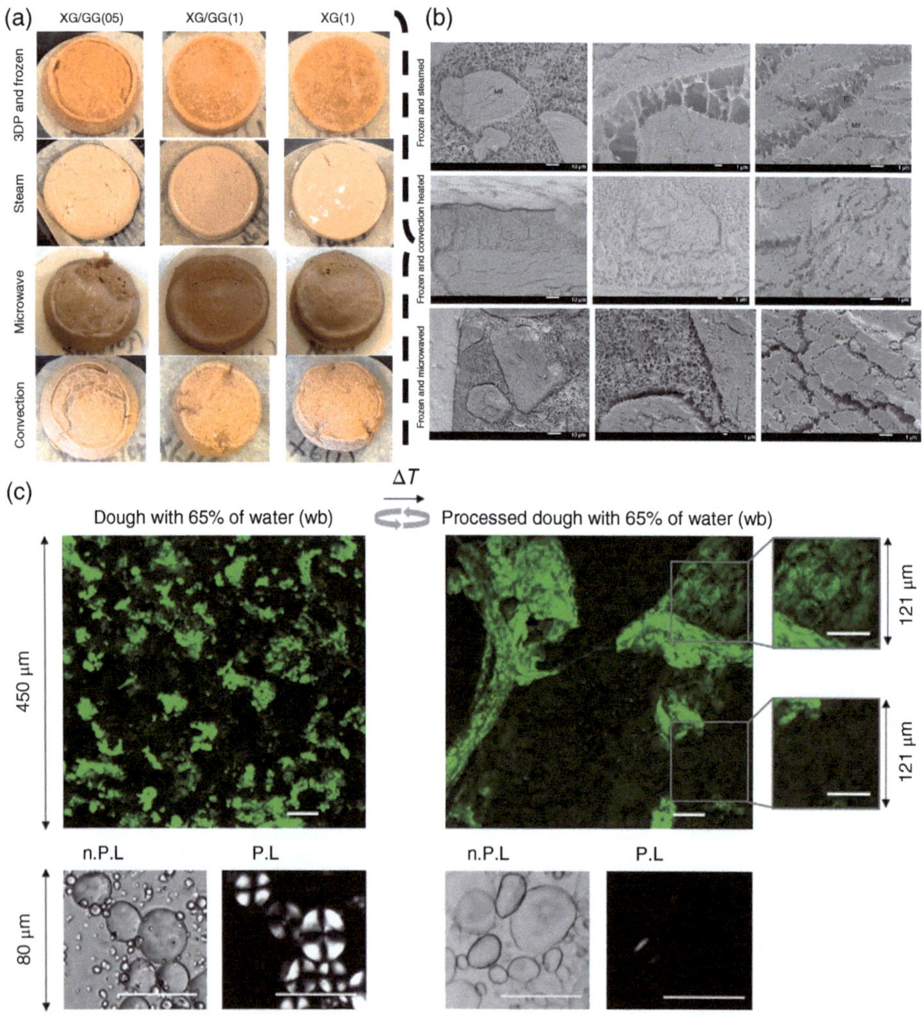

Figure 3.7 (a) Photos of 3D-printed cooked beef pastes in the frozen state and after steam, microwave, and convection heating (formulations XG/GG(05) and XG/GG(1) = 1:1 blend of xanthan gum and guar gum in concentrations of 0.5 g/100 g and 1 g/100 g, XG(1) = 1 g/100 g added xanthan gum), (b) cryogenic scanning electron microscopy images. Microstructure of XG (1) 3D-printed cooked beef pastes as affected by the reheating method (steaming, convection, and microwave heating). Left: 1000×; center: 3000× for frozen and steamed, 5000× for frozen and convection heated and frozen and microwaved; right: 10000× (MF: muscle fibers, M: matrix, MY: myofibrils, IC: intermyofibrillar cavities, F: fat) *Source:* Dick et al. (2021) / with permission of Elsevier, (c) CLSM images of protein distribution in wheat dough and heat-processed wheat dough (fast green) (magnification ×10; proteins are stained green, P.L. - polarized light; magnification ×50, n.P.L. - nonpolarized light) *Source:* Masbernat et al. (2021) / with permission of Elsevier.

Furthermore, CLSM enables the observation of microstructural variations and alterations in protein neural networks. For instance, studies on 3D-printed wheat dough indicate that CLSM images can quantify starch distribution through polarized light microscopy and protein distribution through CLSM imagery (Figure 3.7c) (Masbernat et al., 2021). Similarly, optical microscopic images were used to confirm the bioactive compounds after printing (Leena et al., 2022). Additionally, polarized and fluorescence microscopes have been mostly used in quality assessment in pharmaceutical, nutraceutical and bioprinting applications. By examining the microstructural and protein distribution data, researchers may refine the relevant parameters of the respective procedures to improve their quality.

3.4.7 Structural Stability Assessment

Considering the postprinting stability aspects, the printed products' stability was assessed by keeping the product at room temperature or refrigeration conditions for short or long durations. For instance, Ukkunda et al. (2025) assessed the postprinting stability of 3D-printed marzipan with different natural sweeteners. A relatively complex shape (chalice) was chosen for the study, and printed samples were left for observation over a period of 36 hours at room temperature (~25°C). It was concluded that a higher infill ratio supported better structural stability than a lower infill counterpart (Figure 3.8). Such assessments can be used to understand the structure or duration of shape retention "stability" after printing.

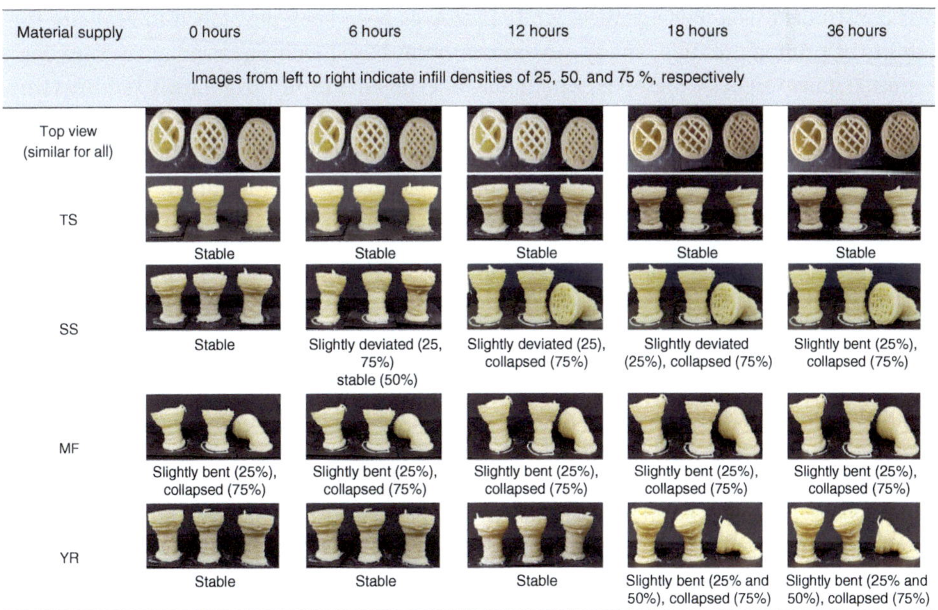

Figure 3.8 Postprinting stability assessment for 3D-printed Marzipan at different infills. *Source:* Ukkunda et al. (2025) / Elsevier / CC BY-NC 4.0.

3.5 Challenges and Future Needs

Current postprinting and postprocessing quality assessment techniques have both benefits and drawbacks contingent upon the materials, printing setups, and chosen postprocessing processes and conditions. Typically, in postprinting processes, shape retention is viewed to be challenging, especially for objects with complex geometries and greater heights; accuracy and printing precision are crucial for achieving the desired form (Raja et al., 2023). AI models and simulation methods have also been developed to address such issues. However, issues in terms of precision, consideration of overall features, etc., remain challenges (Hassan et al., 2024). Similarly, although such advanced systems and technologies surpass human labor, they necessitate further training and optimization.

In terms of postprocessing methods, currently, many heat-based technologies are being explored to facilitate material cooking during the printing or layer deposition processes. In recent days, microwaves, lasers, ultraviolet, and infrared technologies have been utilized for postprocessing setups (Zhang et al., 2022b). Such advances have shown benefits such as decreasing postprocessing duration while concurrently enhancing product quality as compared to traditional postprocessing techniques. In most cases, postprocessing exposure time varies based on product composition, size, infill percentage, and patterns. Each product necessitates distinct optimization and characterization studies. Simulation systems can support anticipating the heat distribution path of the product (Raja et al., 2023). However, such endeavors necessitate additional criteria, for instance, if the product composition alters, along with the need for technical human resources and time-intensive processes.

Globally, various research groups in 3D food printing have reported significant findings and adopted diverse quality evaluation methods tailored to their materials and specific needs. However, no universally accepted standard for assessing the quality of 3D-printed food constructs exists. For instance, while protocols have been developed to evaluate the stability of printed products, many rely on computer-based techniques adapted from conventional image analysis methods, which may not consider infill variations. To address this gap, it is essential to develop standardized protocols to enhance robust quality assessment techniques in 3D food printing. Moreover, there is a noticeable deficiency in efficient postprocessing methods that maintain the printed product's shape and nutritional value. To improve these aspects, advancements in CFD and AI-driven modeling systems are crucial for refining postprinting and postprocessing workflows.

3.6 Scope for Four-Dimensional (4D) Printing

3D printing incorporating an additional dimension is called fourth-dimensional (4D) printing; this may pertain to color or shape, influenced by temporal changes (Anandharamakrishnan et al., 2022). The color and shape fluctuate in response to external or internal stimuli, such as heat, pH, light, magnetism, electricity, and solvents (Wan et al., 2024). 4D printing has numerous uses across various industries, including polymers, biomedicine, medication delivery, and bioprinting (Ahmed et al., 2021). The emergence of 4D printing in gastronomy signifies a transformative advancement in culinary and food technology. In food printing, 3D technology is essential; subsequently, alterations in shape and color occur in response to stimuli, referred to as 4D printing (Navaf et al., 2022). Currently, in documented studies about food, the sources of stimuli are referred to as postprocessing. For instance, heat sources serve as external stimuli from various origins (e.g., microwave, infrared and so on) to alter the configuration of the printed items (Chen et al., 2021; Lv et al., 2024). Likewise, contact with solvents is a postprocessing technique employed to alter the color of printed materials (Shanthamma et al., 2021). Internal stimuli elicit responses based on the composition of various materials, whereby cellular traction over time induces alterations in the color of the products. Overall, postprocessing holds a prominent place in the 4D printing of foods.

This capability gives exciting potential for customized nutrition, which enables the personalization of foods to match individuals' dietary requirements and preferences, thereby delivering specific nutrients at the perfect intervals. In addition, the use of 4D printing has the potential to transform dining experiences by creating dynamic food items that change as they are consumed, thereby enhancing flavor and texture (Derossi et al., 2024; Fahmy et al., 2023). Nevertheless, previous to the widespread implementation of 4D printing in the food industry, several challenges need to be addressed and handled. These challenges include technological limitations in material innovation, increased costs, and regulatory concerns. The potential of 4D printing in the culinary industry provides significant advancements in the design, preparation, and consumption of our meals despite the challenges that have been presented. In the following chapters, the mechanism that underlines the stimulus–response relationship of food-smart materials is explained in detail.

3.7 Conclusion

The stability and integrity of the printed structures are significantly impacted during the postprocessing stage, an essential step in 3D food printing. Therefore, apart from the focus on material printability, an equal focus on the postprinting stability and postprocessing behavior of food materials remains crucial. The chapter delves into the pros and cons of different postprocessing procedures, emphasizing quality parameter assessment before, during, and after these processes. To support research and commercial endeavors, several methods have been explored to assess the quality of postprocessed 3D-printed foods. Further, to minimize experimental trials and to better support process optimization, a range of modeling and simulation tools are being considered. Importantly, with the emergence of computer-based solutions, AI and ML are finding increasing applications in this segment, supporting food printing works in selection, design, process optimization, error identification, quality assessment, and overall process control. However, there is still room for improvement in terms of accuracy and precision. With such advances, as an interesting area with prominent consumer interest, 4D printing is emerging, with current applications majorly focusing on enhanced consumer involvement and food preferencing.

References

Abedini, A., Sohrabvandi, S., Sadighara, P., Hosseini, H., Farhoodi, M., Assadpour, E., Alizadeh Sani, M., Zhang, F., Seyyedi-Mansour, S., & Jafari, S. M. (2024). Personalized nutrition with

3D-printed foods: A systematic review on the impact of different additives. *Advances in Colloid and Interface Science*, *328*, 103181. doi:10.1016/j.cis.2024.103181

Ahmed, A., Arya, S., Gupta, V., Furukawa, H., & Khosla, A. (2021). 4D printing: Fundamentals, materials, applications and challenges. *Polymer*, *228*, 123926. doi:10.1016/j.polymer.2021.123926

Alghamdy, M., He, I., Satsangee, G. R., Keramati, H., & Ahmad, R. (2024). Enhancing printability through design feature analysis for 3D food printing process optimization. *Applied Sciences*, *14*(20), 9587. doi:10.3390/app14209587

Anandharamakrishnan, C., Moses, J. A., & Anukiruthika, T. (2022). 4D printing technology. In *3D printing of foods* (pp. 310–354). Wiley. https://doi.org/10.1002/9781119669838.ch10

Anukiruthika, T., Moses, J. A., & Anandharamakrishnan, C. (2020). 3D printing of egg yolk and white with rice flour blends. *Journal of Food Engineering*, *265*. doi:10.1016/j.jfoodeng.2019.109691

Bedoya, M. G., Montoya, D. R., Tabilo-Munizaga, G., Pérez-Won, M., & Lemus-Mondaca, R. (2022). Promising perspectives on novel protein food sources combining artificial intelligence and 3D food printing for food industry. *Trends in Food Science & Technology*, *128*, 38–52. doi:10.1016/j.tifs.2022.05.013

Blutinger, J. D., Tsai, A., Storvick, E., Seymour, G., Liu, E., Samarelli, N., Karthik, S., Meijers, Y., & Lipson, H. (2021). Precision cooking for printed foods via multiwavelength lasers. *Npj Science of Food*, *5*(1), 24. doi:10.1038/s41538-021-00107-1

Chen, C., Zhang, M., Guo, C., & Chen, H. (2021). 4D printing of lotus root powder gel: Color change induced by microwave. *Innovative Food Science & Emerging Technologies*, *68*, 102605. doi:10.1016/j.ifset.2021.102605

Chen, J., Kong, Y., & Huang, Q. (2024). Analysis of surimi extrusion behavior during 3D printing by modified computational fluid dynamics (CFD) and quick prediction of printability using machine learning based on texture data. *Innovative Food Science & Emerging Technologies*, *94*, 103698. doi:10.1016/j.ifset.2024.103698

Chirico Scheele, S., Hartmann, C., Siegrist, M., Binks, M., & Egan, P. F. (2022). Consumer assessment of 3D-printed food shape, taste, and fidelity using chocolate and marzipan materials. *3D Printing and Additive Manufacturing*, *9*(6), 473–482. doi:10.1089/3dp.2020.0271

Deng, Y., Cheng, K., Shao, X., Dong, X., Jiang, H., & Xiao, G. (2023). Effects of drying method on the stability and quality of post-processing of 3D-printed processed cheese. *Drying Technology*, *41*(6), 1060–1067. doi:10.1080/07373937.2022.2126494

Derossi, A., Caporizzi, R., Paolillo, M., & Severini, C. (2021). Programmable texture properties of cereal-based snack mediated by 3D printing technology. *Journal of Food Engineering*, *289*, 110160. doi:10.1016/j.jfoodeng.2020.110160

Derossi, A., Paolillo, M., Verboven, P., Nicolai, B., & Severini, C. (2022). Extending 3D food printing application: Apple tissue microstructure as a digital model to create innovative cereal-based snacks. *Journal of Food Engineering*, *316*, 110845. doi:10.1016/j.jfoodeng.2021.110845

Derossi, A., Spence, C., Corradini, M. G., Jekle, M., Fahmy, A. R., Caporizzi, R., Devahastin, S., Moses, J. A., Le-Bail, A., Zhou, W., Zhang, M., Bhandari, B., & Severini, C. (2024). Personalized, digitally designed 3D printed food towards the reshaping of food manufacturing and consumption. *Npj Science of Food*, *8*(1), 54. doi:10.1038/s41538-024-00296-5

Dick, A., Bhandari, B., & Prakash, S. (2019). Post-processing feasibility of composite-layer 3D printed beef. *Meat Science*, *153*, 9–18. doi:10.1016/j.meatsci.2019.02.024

Dick, A., Bhandari, B., & Prakash, S. (2021). Effect of reheating method on the post-processing characterisation of 3D printed meat products for dysphagia patients. *LWT*, *150*, 111915. doi:10.1016/J.LWT.2021.111915

Fahmy, A. R., Derossi, A., & Jekle, M. (2023). Four-dimensional (4D) printing of dynamic foods—definitions, considerations, and current scientific status. *Food*, *12*(18), 3410. doi:10.3390/foods12183410

Goldfinger, S., Lipson, H., & Blutinger, J. (2024). Dynamic simulation of 3D-printed foods. *Future Foods*, *9*, 100375. doi:10.1016/j.fufo.2024.100375

Hassan, M., Misra, M., Taylor, G. W., & Mohanty, A. K. (2024). A review of AI for optimization of 3D printing of sustainable polymers and composites. *Composites Part C: Open Access*, *15*, 100513. doi:10.1016/j.jcomc.2024.100513

He, C., Zhang, M., & Fang, Z. (2020). 3D printing of food: Pretreatment and post-treatment of materials. *Critical Reviews in Food Science and Nutrition*, *60*(14), 2379–2392. doi:10.1080/10408398.2019.1641065

Hussain, S., Malakar, S., & Arora, V. K. (2022). Extrusion-based 3D food printing: Technological approaches, material characteristics, printing stability, and post-processing. *Food Engineering Reviews*, *14*(1), 100–119. doi:10.1007/s12393-021-09293-w

Jo, G. H., Lim, W. S., Kim, H. W., & Park, H. J. (2021). Post-processing and printability evaluation of red ginseng snacks for three-dimensional (3D) printing. *Food Bioscience*, *42*, 101094. doi:10.1016/j.fbio.2021.101094

Kavimughil, M., Leena, M. M., Moses, J. A., & Anandharamakrishnan, C. (2022a). 3D printed MCT oleogel as a co-delivery carrier for curcumin and resveratrol. *Biomaterials*, *287*, 121616. doi:10.1016/j.biomaterials.2022.121616

Kavimughil, M., Leena, M. M., Moses, J. A., & Anandharamakrishnan, C. (2022b). Effect of material composition and 3D printing temperature on hot-melt extrusion of ethyl cellulose based medium chain triglyceride oleogel. *Journal of Food Engineering*, *329*, 111055. doi:10.1016/j.jfoodeng.2022.111055

Keerthana, K., Anukiruthika, T., Moses, J. A., & Anandharamakrishnan, C. (2020). Development of fiber-enriched 3D printed snacks from alternative foods: A study on button mushroom. *Journal of Food Engineering*, *287*, 110116. doi:10.1016/j.jfoodeng.2020.110116

Keerthy, M. A., Dhanaselvam, K. R., Santhoshkumar, P., Ravikrishnan, V., & Moses, J. A. (2025). Improving the printability of watermelon rind using pectin: 3D printing optimization for the development of gummies. *Food Biomacromolecules*, *2*(2), 143–156. doi:10.1002/fob2.70008

Kewuyemi, Y. O., Kesa, H., & Adebo, O. A. (2022). Trends in functional food development with three-dimensional (3D) food printing technology: Prospects for value-added traditionally processed food products. *Critical Reviews in Food Science and Nutrition*, *62*(28), 7866–7904. doi:10.1080/10408398.2021.1920569

Kim, H. W., Lee, I. J., Park, S. M., Lee, J. H., Nguyen, M.-H., & Park, H. J. (2019). Effect of hydrocolloid addition on dimensional stability in post-processing of 3D printable cookie dough. *LWT*, *101*, 69–75. doi:10.1016/j.lwt.2018.11.019

Krishnaraj, P., Anukiruthika, T., Choudhary, P., Moses, J. A., & Anandharamakrishnan, C. (2019). 3D extrusion printing and post-processing of fibre rich snack from indigenous composite flour. *Food and Bioprocess Technology*, *12*(10), 1776–1786. doi:10.1007/s11947-019-02336-5

Leena, M. M., Anukiruthika, T., Moses, J., & Anandharamakrishnan, C. (2022). Co-delivery of curcumin and resveratrol through electrosprayed core-shell nanoparticles in 3D printed hydrogel. *Food Hydrocolloids*, *124*, 107200. doi:10.1016/j.foodhyd.2021.107200

Liu, Y., Sun, J., Wen, Z., Wang, J., Roopesh, M., Pan, D., & Du, L. (2024). Functionality enhancement of pea protein isolate through cold plasma modification for 3D printing application. *Food Research International*, *197*, 115267. doi:10.1016/j.foodres.2024.115267

Liu, Y., Tang, T., Duan, S., Qin, Z., Zhao, H., Wang, M., Li, C., Zhang, Z., Liu, A., Han, G., Wu, D., Chen, H., Shen, L., Wu, W., & Dong, Y. (2020). Applicability of rice doughs as promising food materials in extrusion-based 3D printing. *Food and Bioprocess Technology*, *13*(3), 548–563. doi:10.1007/s11947-020-02415-y

Liu, Z., Bhandari, B., Prakash, S., & Zhang, M. (2018). Creation of internal structure of mashed potato construct by 3D printing and its textural properties. *Food Research International*, *111*, 534–543. doi:10.1016/j.foodres.2018.05.075

Liu, Z., & Zhang, M. (2021). Texture properties of microwave post-processed 3D printed potato snack with different ingredients and infill structure. *Future Foods*, *3*, 100017. doi:10.1016/j.fufo.2021.100017

Lu, Y., Rai, R., & Nitin, N. (2023). Image-based assessment and machine learning-enabled prediction of printability of polysaccharides-based food ink for 3D printing. *Food Research International*, *173*, 113384. doi:10.1016/j.foodres.2023.113384

Lv, Y., Lv, W., Li, G., & Zhong, Y. (2023). The research progress of physical regulation techniques in 3D food printing. *Trends in Food Science & Technology*, *133*, 231–243. doi:10.1016/j.tifs.2023.02.004

Lv, Y., Wang, B., Cheng, Y., Lv, W., Zeng, S., & Xiao, H. (2024). Printing characteristics and microwave infrared-induced 4D printing of chestnut powder composite paste. *Journal of Food Engineering*, *382*, 112197. doi:10.1016/j.jfoodeng.2024.112197

Ma, Y., Potappel, J., Chauhan, A., Schutyser, M. A. I., Boom, R. M., & Zhang, L. (2023a). Improving 3D food printing performance using computer vision and feedforward nozzle motion control. *Journal of Food Engineering*, *339*, 111277. doi:10.1016/j.jfoodeng.2022.111277

Ma, Y., Potappel, J., Schutyser, M. A. I., Boom, R. M., & Zhang, L. (2023b). Quantitative analysis of 3D food printing layer extrusion accuracy: Contextualizing automated image analysis with human evaluations. *Current Research in Food Science*, *6*, 100511. doi:10.1016/j.crfs.2023.100511

Ma, Y., Schutyser, M. A. I., Boom, R. M., & Zhang, L. (2021). Predicting the extrudability of complex food materials during 3D printing based on image analysis and gray-box data-driven modelling. *Innovative Food Science & Emerging Technologies*, *73*, 102764. doi:10.1016/j.ifset.2021.102764

Masbernat, L., Berland, S., Leverrier, C., Moulin, G., Michon, C., & Almeida, G. (2021). Structuring wheat dough using a thermomechanical process, from liquid food to 3D-printable food material. *Journal of Food Engineering*, *310*, 110696. doi:10.1016/j.jfoodeng.2021.110696

Mazlan, M. A., Anas, M. A., Nor Izmin, N. A., & Abdullah, A. H. (2023). Effects of infill density, wall perimeter and layer height in fabricating 3D printing products. *Materials*, *16*(2), 695. doi:10.3390/ma16020695

Molimi, M. B., Egan, P., & Adebo, O. A. (2025). Progress in three-dimensional (3D) printed foods for dysphagia patients: Food sources, processing techniques, printability, nutrition, acceptability, and safety aspects. *Food Research International*, *202*, 115629. doi:10.1016/j.foodres.2024.115629

Nachal, N., Moses, J. A., Karthik, P., & Anandharamakrishnan, C. (2019). Applications of 3D printing in food processing. *Food Engineering Reviews*, *11*(3), 123–141. doi:10.1007/s12393-019-09199-8

Navaf, M., Sunooj, K. V., Aaliya, B., Akhila, P. P., Sudheesh, C., Mir, S. A., & George, J. (2022). 4D printing: A new approach for food printing; effect of various stimuli on 4D printed food properties. A comprehensive review. *Applied Food Research*, *2*(2), 100150. doi:10.1016/j.afres.2022.100150

Okamoto, T., & Ura, S. (2024). Verifying the accuracy of 3D-printed objects using an image processing system. *Journal of Manufacturing and Materials Processing*, *8*(3), 94. doi:10.3390/jmmp8030094

Oral, M. O., Derossi, A., Caporizzi, R., & Severini, C. (2021). Analyzing the most promising innovations in food printing. Programmable food texture and 4D foods. *Future Foods*, *4*, 100093. doi:10.1016/j.fufo.2021.100093

Outrequin, T. C. R., Gamonpilas, C., Sreearunothai, P., Deepaisarn, S., & Siriwatwechakul, W. (2024). Machine learning assisted evaluation of the filament spreading during extrusion-based 3D food printing: Impact of the rheological and printing parameters. *Journal of Food Engineering*, *381*, 112166. doi:10.1016/j.jfoodeng.2024.112166

Pérez, B., Nykvist, H., Brøgger, A. F., Larsen, M. B., & Falkeborg, M. F. (2019). Impact of macronutrients printability and 3D-printer parameters on 3D-food printing: A review. *Food Chemistry*, *287*, 249–257. doi:10.1016/j.foodchem.2019.02.090

Piovesan, A., Vancauwenberghe, V., Aregawi, W., Delele, M. A., Bongaers, E., de Schipper, M., van Bommel, K., Noort, M., Verboven, P., & Nicolai, B. (2020). Designing mechanical properties of 3D printed cookies through computer aided engineering. *Food*, *9*(12), 1804. doi:10.3390/foods9121804

Pulatsu, E., Su, J.-W., Lin, J., & Lin, M. (2020). Factors affecting 3D printing and post-processing capacity of cookie dough. *Innovative Food Science & Emerging Technologies*, *61*, 102316. doi:10.1016/j.ifset.2020.102316

Raja, V., & Moses, J. A. (2022). Effect of 3D printing conditions and post-printing fermentation on pearl millet fortified idli. *Journal of the Science of Food and Agriculture*, *103*(5), 2401–2412. doi:10.1002/jsfa.12410

Raja, V., Nimbkar, S., Moses, J. A., Ramachandran Nair, S. V., & Anandharamakrishnan, C. (2023). Modeling and simulation of 3D food printing systems—scope, advances, and challenges. *Food*, *12*(18), 3412. doi:10.3390/foods12183412

Rando, P., & Ramaioli, M. (2021). Food 3D printing: Effect of heat transfer on print stability of chocolate. *Journal of Food Engineering*, *294*, 110415. doi:10.1016/j.jfoodeng.2020.110415

Santhoshkumar, P., & Moses, J. A. (2025). Impact of konjac glucomannan on hot extrusion 3D printability of cake gel. *International Journal of Biological Macromolecules*, *305*, 141162. doi:10.1016/j.ijbiomac.2025.141162

Santhoshkumar, P., Raja, V., Priyadarshini, S. R., & Moses, J. A. (2024a). Evaluating the 3D printability of pearl millet flour with banana pulp blends. *Journal of the Science of Food and Agriculture*. doi:10.1002/jsfa.13389

Santhoshkumar, P., Raja, V., Priyadarshini, S. R., & Moses, J. A. (2024b). Impact of baking and steaming of 3D printed pearl millet-Banana pulp blends on physicochemical and sensory attributes. *International Journal of Gastronomy and Food Science*, *37*, 100962. doi:10.1016/j.ijgfs.2024.100962

Santhoshkumar, P., Swathika, B. S., Negi, A., & Moses, J. A. (2025). Improving gel performance and hot extrusion three-dimensional printability of blueberry gel with guar gum. *Food Biophysics*, *20*(2), 1–15. doi:10.1007/s11483-025-09964-9

Schiffer, A., Sahai, N., & Ratnavel, S. (2024). *3D printing technologies and automation in food processing industry: An innovative marketing approach for health and digital gastronomy perspective*. 10.26434/chemrxiv-2024-q37xr

Shanthamma, S., Preethi, R., Moses, J. A., & Anandharamakrishnan, C. (2021). 4D printing of sago starch with turmeric blends: A study on pH-triggered spontaneous color transformation. *ACS Food Science & Technology*, *1*(4), 669–679. doi:10.1021/acsfoodscitech.0c00151

Shen, C., Chen, W., Li, C., Chen, X., Cui, H., & Lin, L. (2023a). 4D printing system stimulated by curcumin/whey protein isolate nanoparticles: A comparative study of sensitive color change and post-processing. *Journal of Food Engineering*, *342*, 111357. doi:10.1016/j.jfoodeng.2022.111357

Shen, D., Zhang, M., Mujumdar, A. S., & Li, J. (2023b). Advances and application of efficient physical fields in extrusion based 3D food printing technology. *Trends in Food Science & Technology*, *131*, 104–117. doi:10.1016/j.tifs.2022.11.017

Swathika, B. S., Santhoshkumar, P., Moses, J. A., & Sinija, V. R. (2025). Novel Three-Dimensional-Printed ι-Carrageenan–Carnauba Wax Bigel Formulation for Codelivery Applications. *ACS Food Science & Technology*, *5*(3), 1183–1192. doi:10.1021/acsfoodscitech.5c00004

Tang, T., Zhang, M., Bhandari, B., & Li, C. (2024). Personalized transformation of 3D printing for traditional multi-material food with stuffing: A review. *Food Bioscience*, *59*, 104112. doi:10.1016/j.fbio.2024.104112

Tay, J. U., Zhou, C., Lee, H. W., Lu, Y., & Huang, D. (2023). 3D printing of salmon fillet mimic: Imparting printability via high-pressure homogenization and post-printing texturization via transglutaminase. *Food Hydrocolloids*, *140*, 108564. doi:10.1016/j.foodhyd.2023.108564

Theagarajan, R., Moses, J. A., & Anandharamakrishnan, C. (2020). 3D extrusion printability of rice starch and optimization of process variables. *Food and Bioprocess Technology*, *13*(6), 1048–1062. doi:10.1007/s11947-020-02453-6

Theagarajan, R., Nimbkar, S., Moses, J. A., & Anandharamakrishnan, C. (2021). Effect of post-processing treatments on the quality of three-dimensional printed rice starch constructs. *Journal of Food Process Engineering*, *44*(9), e13772. doi:10.1111/jfpe.13772

Ukkunda, N. S., Santhoshkumar, P., & Moses, J. A. (2025). Impact of different natural sweeteners on 3D printability and post-printing quality of marzipan. *Future Foods*, *11*, 100557. doi:10.1016/j.fufo.2025.100557

Va, A., Udayarajan, C. T., Goksen, G., Brennan, C. S., & P, N. (2023). A brief review on 3D printing of chocolate. *International Journal of Food Science & Technology*, *58*(6), 2811–2828. doi:10.1111/ijfs.16415

Varvara, R.-A., Szabo, K., & Vodnar, D. C. (2021). 3D food printing: Principles of obtaining digitally-designed nourishment. *Nutrients*, *13*(10), 3617. doi:10.3390/nu13103617

Wan, X., Xiao, Z., Tian, Y., Chen, M., Liu, F., Wang, D., Liu, Y., Bartolo, P. J. D. S., Yan, C., Shi, Y., Zhao, R. R., Qi, H. J., & Zhou, K. (2024). Recent advances in 4D printing of advanced materials and structures for functional applications. *Advanced Materials*, *36*(34). doi:10.1002/adma.202312263

Wang, X., Zhang, M., Phuhongsung, P., & Mujumdar, A. S. (2024). Impact of internal structural design on quality and nutritional properties of 3D printed food products during post-printing: A critical review. *Critical Reviews in Food Science and Nutrition*, *64*(12), 3713–3724. doi:10.1080/10408398.2022.2134979

Waseem, M., Tahir, A. U., & Majeed, Y. (2024). Printing the future of food: The physics perspective on 3D food printing. *Food Physics*, *1*, 100003. doi:10.1016/j.foodp.2023.100003

Wen, Y., Che, Q. T., Wang, S., Park, H. J., & Kim, H. W. (2024). Elaboration of dimensional quality in 3D-printed food: Key factors in process steps. *Comprehensive Reviews in Food Science and Food Safety*, *23*(1). doi:10.1111/1541-4337.13267

Wilson, A., Anukiruthika, T., Moses, J. A., & Anandharamakrishnan, C. (2020). Customized shapes for chicken meat–based products: Feasibility study on 3D-printed nuggets. *Food and Bioprocess Technology*, *13*(11), 1968–1983. doi:10.1007/s11947-020-02537-3

Wu, H., Sang, S., Weng, P., Pan, D., Wu, Z., Yang, J., Liu, L., Farag, M. A., Xiao, J., & Liu, L. (2023). Structural, rheological, and gelling characteristics of starch-based materials in context to 3D food printing applications in precision nutrition. *Comprehensive Reviews in Food Science and Food Safety*, *22*(6), 4217–4241. doi:10.1111/1541-4337.13217

Xie, J., Bi, J., Liu, X., Blecker, C., Jacquet, N., & Lyu, J. (2025). Development of soy protein isolate-chelator soluble pectin composite gels as extrusion-based 3D food printing inks: Effects of mingling strategy. *Food Hydrocolloids*, *162*, 110904. doi:10.1016/j.foodhyd.2024.110904

Yang, F., Zhang, M., & Liu, Y. (2019). Effect of post-treatment microwave vacuum drying on the quality of 3D-printed mango juice gel. *Drying Technology*, *37*(14), 1757–1765. doi:10.1080/07373937.2018.1536884

Yatmaz, H. A. (2024). 3D-printed meat paste using minimal additive: Assessment of rheological and printing behavior with post-processing stability. *Food Biophysics*, *19*(3), 503–516. doi:10.1007/s11483-024-09866-2

Yoha, K. S., Anukiruthika, T., Anila, W., Moses, J. A., & Anandharamakrishnan, C. (2021). 3D printing of encapsulated probiotics: Effect of different post-processing methods on the stability of Lactiplantibacillus plantarum (NCIM 2083) under static in vitro digestion conditions and during storage. *LWT*, *146*, 111461. doi:10.1016/j.lwt.2021.111461

Zhang, J., Li, Y., Cai, Y., Ahmad, I., Zhang, A., Ding, Y., Qiu, Y., Zhang, G., Tang, W., & Lyu, F. (2022a). Hot extrusion 3D printing technologies based on starchy food: A review. *Carbohydrate Polymers*, *294*, 119763. doi:10.1016/j.carbpol.2022.119763

Zhang, J. Y., Pandya, J. K., McClements, D. J., Lu, J., & Kinchla, A. J. (2022b). Advancements in 3D food printing: A comprehensive overview of properties and opportunities. *Critical Reviews in Food Science and Nutrition*, *62*(17), 4752–4768. doi:10.1080/10408398.2021.1878103

Zhao, L., Zhang, M., Chitrakar, B., & Adhikari, B. (2021). Recent advances in functional 3D printing of foods: A review of functions of ingredients and internal structures. *Critical Reviews in Food Science and Nutrition*, *61*(21), 3489–3503. doi:10.1080/10408398.2020.1799327

Zhou, S., Chen, W., & Fan, K. (2024). Recent advances in combined ultrasound and microwave treatment for improving food processing efficiency and quality: A review. *Food Bioscience*, *58*, 103683. doi:10.1016/j.fbio.2024.103683

4

Fundamentals of 4D Printing Technology

CHAPTER MENU

4.1 Introduction

The growth of additive manufacturing is remarkably faster than anticipated thus far, and it has introduced a new dimension to the engineering world. The extension of 3D printing into the temporal dimension is referred to as "4D printing" and arises from the interaction of environmental elements with 3D-printed structures. Figure 4.1 summarizes the progression of 4D printing from 3D printing. The term 4D printing was initially introduced by Tibbits in a TED talk in 2013, coinciding with the publication of the first research article in this domain using multimaterials (Kuang et al. 2019; Momeni and Ni 2020). Since then, temporal alterations in shape, devoid of motors and wires, have gained prominent attention across numerous domains, including bioprinting, biomedical,

4D Printing of Foods: Fundamentals, Applications and Prospects, First Edition.
C. Anandharamakrishnan, Jeyan A. Moses, and P. Santhoshkumar.

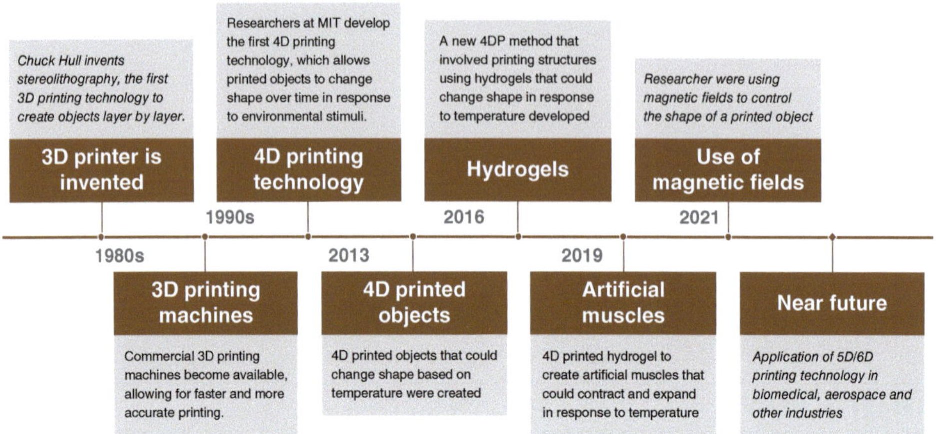

Figure 4.1 Historical development in additive manufacturing. *Source:* Manshor et al. (2023) / with permission of Elsevier.

pharmaceutical, microfluidics, soft robotics, polymer, aerospace (Chu et al. 2020), textile (Hu et al. 2012), and the food industry (Shanthamma et al. 2021). In contemporary fields such as quantum computing, soft robotics, human augmentation, and deep learning, 4D printing has gained significant interest within scientific communities in the past decade. 4D printing is based on 3D printing, while the material adaptability and technology applicability of 4D printing often differ from those of 3D printing. Consequently, the printer types for 4D printing resemble those of 3D printing; nevertheless, the material selection for this method may differ. The diverse applications of this technology engage with different fields, and prominent concepts involving shape shifting, color and functionality changing mechanisms, and the associated challenges are detailed in this chapter.

4.2 Fundamental Concepts and Definitions

Tibbits originally defined 4D printing using the formula "4D printing = 3D printing + time" (Sydney Gladman et al. 2016; Tibbits 2013). However, currently, its definition differs across various industry applications although the underlying principle remains consistent. For instance, alterations resulting from external or internal stimuli can result in changes to the shape or color or other parameters of printed materials over time, and in this case, time may be considered as the fourth dimension. In other words, 4D printing is contingent on time and independent of the printer, enabling the attainment of the desired shape and functionality. Self-assembly and repair, together with multifunctionality, occur when temporal alterations and stimulus factors affect the products (Momeni et al. 2017). Kuang et al. (2019) provided an alternative definition of 4D printing is the evolution of the shape and function of 3D-printed objects. To understand this better, Figure 4.2 illustrates the distinction between 3D and 4D printing.

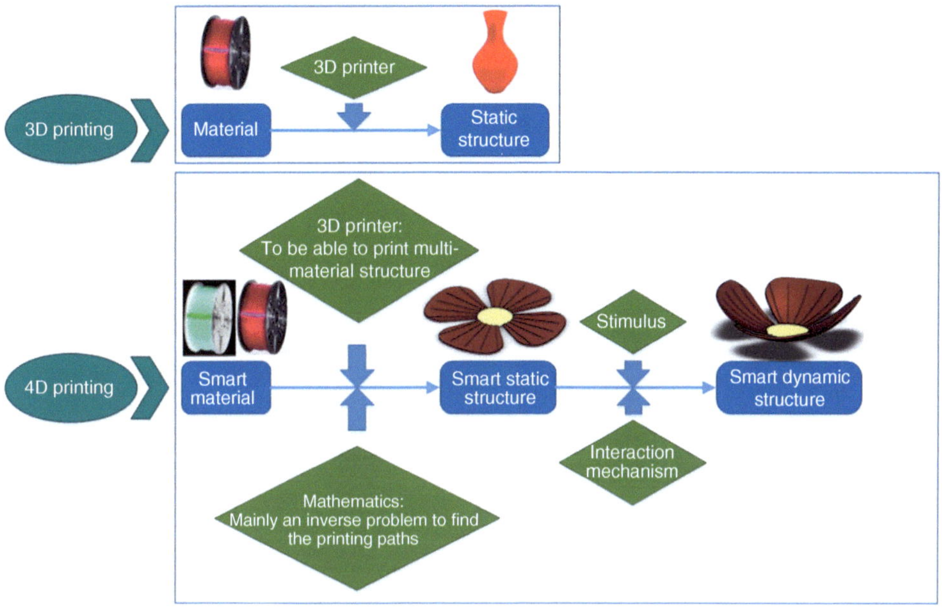

Figure 4.2 Fundamental difference between 3D and 4D printing. *Source:* Momeni et al. (2017) / with permission of Elsevier.

4.2.1 Stimulus and Material Design

The key aspects of 4D printing encompass 3D printing techniques, stimuli and stimuli-responsive materials, interaction or shape-shifting mechanisms, and concepts of simulation. Initially, 3D printing techniques are employed to fabricate a 3D product and utilize distinct printing methodologies regardless of their purposes. Then, the stimulus may consist of water, light, pH, heat, or humidity, either alone or in combination, applied to printed constructs. Subsequently, the stimulus-responsive materials, referred to as smart materials, react to these internal or external stimuli. In 4D printing, stimuli are generally categorized as follows: thermo-responsive (heat-dependent), mechano-responsive (pressure/stress), electro-responsive (current/voltage), magneto-responsive (magnetic field), chemo-responsive (pH/moisture/solvent), and photo-responsive (light) materials (Sun et al. 2012). These represent the predominant external stimuli utilized in 4D printing applications. Moreover, such intelligent materials may possess self-sensing, adaptability, repair capabilities, responsiveness, shape memory, and multifunctionality capabilities (Roy et al. 2010).

Often, the interaction mechanism defines the fourth dimension of the 3D structures, representing an intrinsic response between the stimuli and stimulus-responsive materials. This explains activation in shape, functionality, or properties of the materials. In this context, existing categories of stimulus-responsive materials are outlined in Figure 4.3a. The interaction mechanism depends on the concentration/effect of the stimulus or the sequences shown for a specified duration. A popular interaction mechanism identified in 4D printing methodologies is thermo-mechanics for shape transformation. This results in structural deformation due to the applied temperature and external load; thereafter,

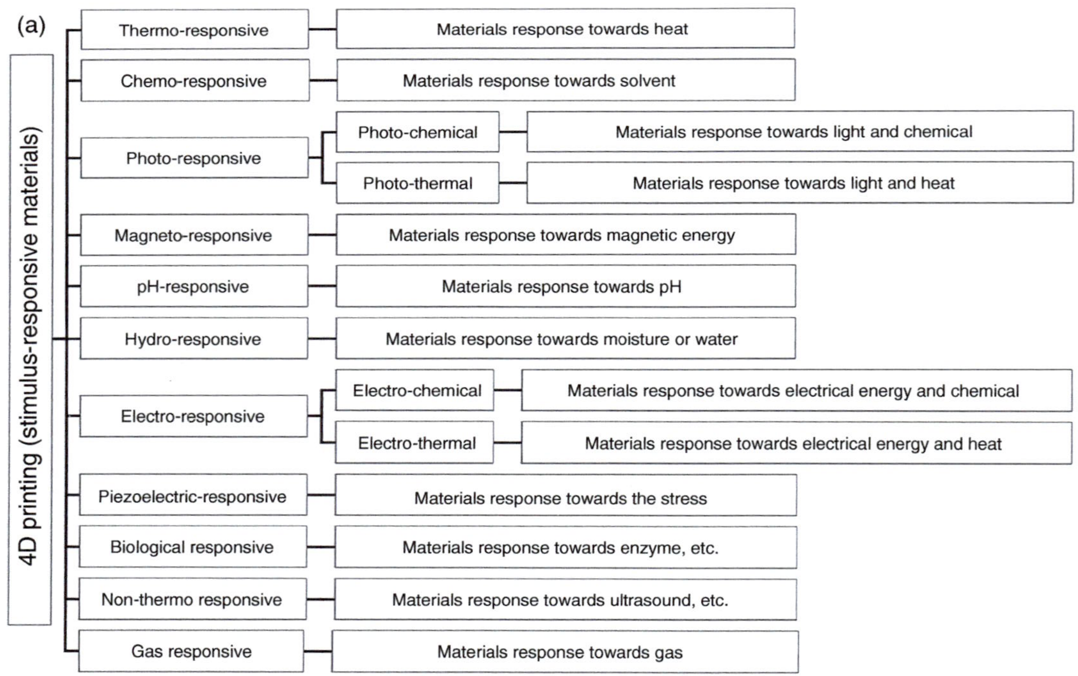

Figure 4.3 (a) Classification of stimulus-responsive materials, (b) smart materials and stimulus-response mechanisms. *Source:* Rastogi and Kandasubramanian (2019) / with permission of Elsevier.

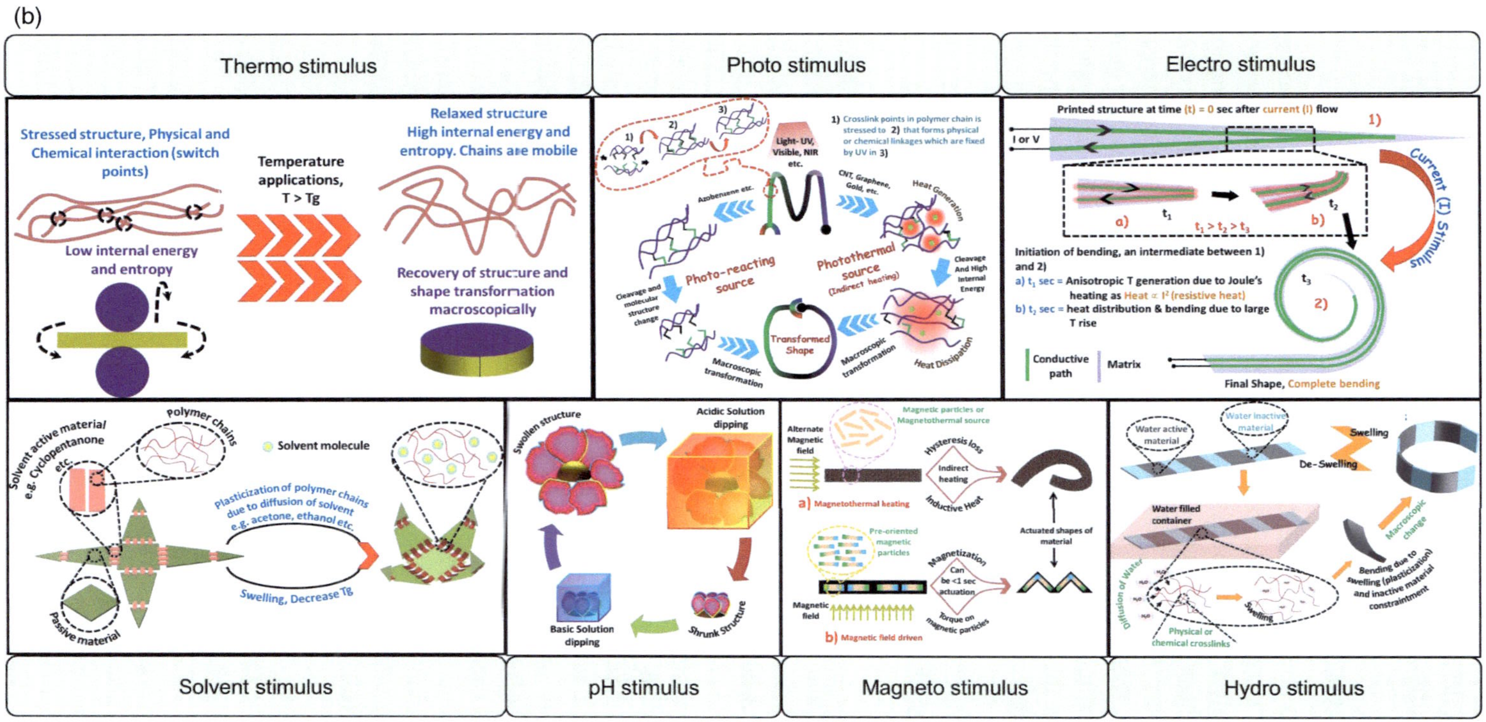

Figure 4.3 (Continued)

reformation occurs with the removal of the external load and the application of reheating temperature (Slavkovic et al. 2023). Similarly, different stimulus-response mechanisms are shown in Figure 4.3b.

4.2.2 Modeling and Simulation

Very often, 4D printing of materials requires modeling applications to anticipate the fourth dimension of the structures in terms of changes in functionality, shape, or properties with the application of stimuli to the 3D-printed constructs. Therefore, several mathematical models have been developed by identifying four fundamental components: material properties, stimulus properties, material structure, and stimulus characteristics (Biswas et al., 2021a). These simulation tools enhance printing accuracy, product quality, and the capability of simulating the postprinted product behavior. Additionally, modeling and simulation reduce the time, which is required for the experimental trails of specific materials. The upcoming sections detail the fundamental phenomena of 4D printing approaches.

4.3 Laws of 4D Printing

4D printing is the integration of intelligent/smart (active, stimuli-responsive) materials, mathematics, and multimaterial additive manufacturing. These three elements produce an encoded, multimaterial (multicomponent), single-piece smart construct that exhibits predictable/programmable and dynamic, intelligent behaviors in response to appropriate stimuli via an interaction mechanism. The science behind the fourth dimension in printing has been explained using three laws of 4D printing. These laws enhance the comprehension of the physics underlying the shape-shifting capability of 4D-printed structures. Few studies have demonstrated the idea of using single-material 4D-printed constructs. Practically, the future lies in multimaterial structures and printing. The logic of multimaterials consider active and passive layers; active or smart materials can be used, enabling self-actuating operation. Figure 4.4 summarizes the laws underlying the concept of multimaterial 4D printing. As mentioned in the introductory paragraph, 4D printing of multimaterials was the first research work published in the field of 4D printing (Tibbits 2014). A single-material process is considered as an active operation execution; in contrast, a multimaterial process is considered as an active and passive operation (Momeni and Ni 2020). These researchers presented the equation (Figure 4.4) and explained the laws behind 4D printing using multimaterials, listed as follows,

- *First law: "all shape-morphing behavior, such as twisting, coiling, curling, etc, of multi-material 4D printing relies on the fundamental phenomenon of 'relative expansion' between the active and passive materials."*
- *Second law: "covers the four basic physical factors behind shape-changing: mass diffusion, thermal expansion, molecular transformation, and organic growth, and all these factors lead to the relative expansion between the active and passive materials, which results in shape-changing under the stimulus."*
- *Third law: "time-dependent shape-morphing behavior of nearly all multi-material 4D printed structures is governed by two different types of time constants."*

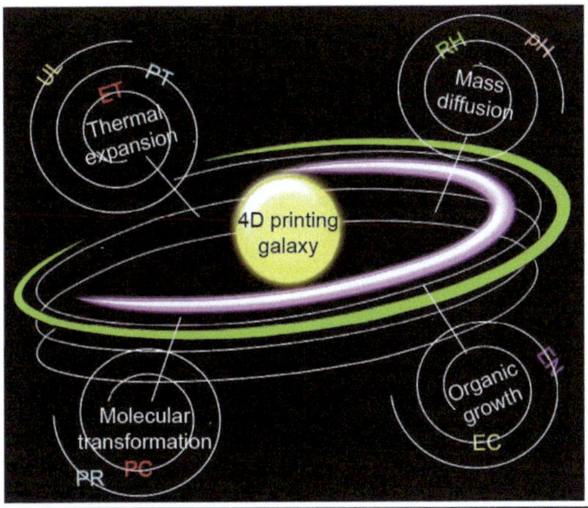

1st law of 4D printing:
Almost all shape-morphing behaviors of multi-material 4D structures originate from "relative expansion" between active and passive materials

2nd law of 4D printing:
The shape-morphing behaviors of almost all multi-material 4D structures have four types of physics (4D printing galaxy)

Unified formula of the 2nd law:

C_j depends on different factors
τ_j depends on different factors

$$\left.\begin{cases} \varepsilon_{\text{mass diffusion}}(t) = C_1[1-\exp(-t/\tau_1)] \\ \varepsilon_{\text{thermal expansion/contraction}}(t) = C_2[1-\exp(-t/\tau_2)] \\ \varepsilon_{\text{molecular transformation}}(t) = C_3[1-\exp(-t/\tau_3)] \\ \varepsilon_{\text{organic growth}}(t) = C_4[1-\exp(-t/\tau_4)] \end{cases}\right\} \Rightarrow C_j\left[1-\exp\left(-\frac{t}{\tau_j}\right)\right]$$

Formula of the 3rd law:

$$\kappa(t) = \frac{H_{\text{I}}\left(1-e^{\frac{-t}{\tau_{\text{I}}}}\right)+H_{\text{II}}\left(1-e^{\frac{-t}{\tau_{\text{II}}}}\right)}{\frac{h}{2}+\frac{2(E_1I_1+E_2I_2)}{h}\left(\frac{1}{E_1a_1}+\frac{1}{E_2a_2}\right)}$$

3rd law of 4D printing:
Time-dependent shape-morphing behavior of almost all multi-material 4D structures is governed by two "types" of time constants

Figure 4.4 Three laws of 4D printing (EC, electrochemical; EN, enzyme; ET, electrothermal; PC, photochemical; PR, pressure; PT, photothermal; RH, relative humidity; UL, ultrasound). *Source:* Momeni and Ni (2020) / with permission of Elsevier / CC BY-NC-ND 4.0.

Overall, Momeni and Ni (2020) highlighted the physical science behind 4D printing by explaining changes in the shape of 3D-printed constructs over time. In addition, the different models used in shape shifting for 4D printing, governing equations, and their limitations are listed in Table 4.1, and detailed explanations have been presented by these researchers.

4.4 Nature-Inspired Concepts in 4D Printing

Typically, predominant shape transformations in 4D printing have been derived from natural phenomena. Accordingly, the approach has been termed as bioinspired shape transformation. For instance, fundamental morphological alterations such as folding, curling, and bending typically occur in plants. Figure 4.5 illustrates several natural qualities attainable through 4D printing. The mechanisms of nature-driven and synthetic actuation are fundamentally distinct. With 4D printing, it is possible to execute distinctive modifications through meticulous observation of the behaviors of plants and animals. Phytomimetic research, centered on plant-inspired innovations, has gained prominence in recent decades.

Table 4.1 Summary of different models in 4D printing for shape-shifting behaviors.

Equation[a]	Limitations
$C = \frac{1}{r}$ C – bending and unbending curvatures r – radius of the imaginary circle drawn around the actuated object (bend radius)	• Simple geometry • Time-independent model • Not suitable for various stimuli as well as materials
$r = \frac{\left[(1/\varepsilon_R)-1\right]t}{2}$ $\varepsilon_R = \frac{L_R - L_0}{L_0}$ $L_R = R \times \theta$ r – bend radius (inner radius) ε_R – strain at the (outer) surface t – thickness θ – bending angle L_0 – original length L_R – bending length at the outer surface R – outer radius	• Depends on geometry • Time-independent model • Not suitable for various stimuli as well as materials
$\varepsilon = \frac{l_0 - l_c}{l_c} = \frac{(R + y_b)\theta - R\theta}{R\theta} = \frac{y_b}{R} = Ky_b$ ε – bending strain l_0 – swelled length l_c – free length R – curvature radius θ – bending angle K – bending curvature y_b – distance between the swelled layer and reference plane (e.g., half of the thickness)	• Similar to the previous model, but with different curvature radii • Depends on geometry • Time-independent model • Not suitable for various stimuli as well as materials
$\alpha \approx \frac{Nd}{r}, l \approx \alpha\left(r + \frac{1}{2}D\right) \approx \left(\alpha\frac{Nd}{\alpha} + \frac{1}{2}D\right) = Nd + \frac{1}{2}\alpha D$ α – folding angle r – folding radius N – number of disks d – disk thickness D – disk diameter l – total length	• Depends on geometry • Customized model (N disks) • Time-independent model • Not suitable for various stimuli as well as materials

Table 4.1 (Continued)

Equation[a]	Limitations
$\alpha = K\sqrt{\frac{L^2h}{d^3}}, K = \sqrt{\frac{24\times(1-\nu^2)\mu\rho q}{E^T}}$ α – bending angle K – coefficient related to the temperature and material properties h – thickness of the active layer d – thickness of the passive layer L – length of the bilayer ρ – density of the active material q – unit releasing energy of the active layer μ – mass shrinking ratio of the active layer ν – poisson ration of the active layer E^T – flexural modulus of the passive layer cured temperature T	• The energy generated in the contraction of the active layer will lead to the deformation of the passive layer was assumed to obtain this equation • Time-independent model • Not suitable for various stimuli as well as materials
$\varepsilon(z) = \varepsilon^0 + kz, M = \int_A \sigma z dA = \int_A zE(\varepsilon^0 + kz)dA$ $= E\varepsilon^0 \int_A z dA + kE\int_A z^2 dA \Rightarrow M = kEI = \frac{EI}{r} = \frac{\theta EI}{L}$ $\Leftrightarrow \theta = \frac{ML}{EI}$ ε – strain ε^0 – strain at the mid-axis σ – stress z – third axis in the cartesian coordinate system A – cross-sectional area of the beam k – curvature θ – bending angle r – curvature radius E – Young's modulus I – second moment of area M – moment L – length	• Classic Euler-Bernoulli beam theory model • Time-independent model • Not suitable for various stimuli as well as materials
$\frac{1}{\rho} = \frac{6(\varepsilon_2 - \varepsilon_1)(1+m^2)}{h\left[3\times(1+m^2)+(1+mm)\left(m^2+\frac{1}{mn}\right)\right]}$ ρ – radius of the curvature ε_1, ε_2 – thermal expansion coefficients of the two layers m – the ratio of the thickness of the two layers n – ratio of Young's modulus of the two layers h – thickness of the bilayer	• Classical Timoshenko bimetal model • Time-independent model • Not suitable for various stimuli as well as materials

(Continued)

Table 4.1 (Continued)

Equation[a]	Limitations
$\theta(x,t') = \dfrac{6\varepsilon_r\left(1-e^{\frac{-t'}{\tau_f}}\right)E'_{pm}h_{pm}b_{pm}\left(h_{pm}+1\right)x}{\left(E'_{pm}h_{pm}+1\right)\left(E'_{pm}b_{pm}h_{pm}^3+1\right)h_m}$ θ – bending angle $h_{pm}=h_p/h_m$ – thickness of the two layers $b_{pm}=b_p/b_m$ – widths ratio of the two layers $E'_{pm}=E_{pe}/E_m$ – Young's (elastic) moduli ratio of the two layers (E_{pe} is the elastic modulus of the active layer above its glass transition temperature) X – any value between zero and l (i.e., $x \in [0,l]$) l – length of the composites t' – time starting from the point of the glass transition temperature (i.e., $t' = t - t_0$) t – total time t_0 – time at the glass transition temperature τ_f – relaxation time ε_r – stored internal strain p – printed polymer (active) m – membrane of paper (passive)	• Extension of classical Timoshenko bimetal model to an exponential model • Not suitable for various stimuli as well as materials
$H = c_1\dfrac{\alpha_\parallel-\alpha_\perp}{h}\dfrac{\sin^2(\theta)}{c_2-c_3\cos(2\theta)+m^4\cos(4\theta)}$, $K = -c_4\dfrac{\left(\alpha_\parallel-\alpha_\perp\right)^2}{h^2}\dfrac{\sin^2(\theta)}{c_5-c_6\cos(2\theta)+m^4\cos(4\theta)}$ $H = ½$ *trace (k)* – mean curvature K – curvature tensor $K = det\ (k)$ – gaussian curvature $\alpha_\parallel$, $\alpha_\perp$ – longitudinal and transverse swelling (expansion) strains θ – angle between the two layers in the bilayer h – sum of the layer thickness m – ratio of the layer thickness c_i – functions of 'm' and the elastic moduli	• Extension of classical Timoshenko bimetal model to 2D when an angle *θ exists between the two layers* • Mean and Gaussian curvatures are present in 2D (plane) • Time-independent model • Not suitable for various stimuli as well as materials • In the limit that θ approaches 0°, the classical Timoshenko bimetal model is recovered

Source: Adapted from Momeni and Ni (2020).
[a] Symbols are relatively independent.

In pursuit of expanding avenues in 4D printing, researchers have focused on diverse natural materials to illustrate phytomimetic actuation (Arsuffi et al., 2025). For example, in nature, the movements of plants are stimulated by sunlight, water, touch stimuli, and gravity. These facilitate many biological actions at distinct scales, for instance, macroscopically, the flowering of plants and the curling of tendrils, and microscopically, the expansion or

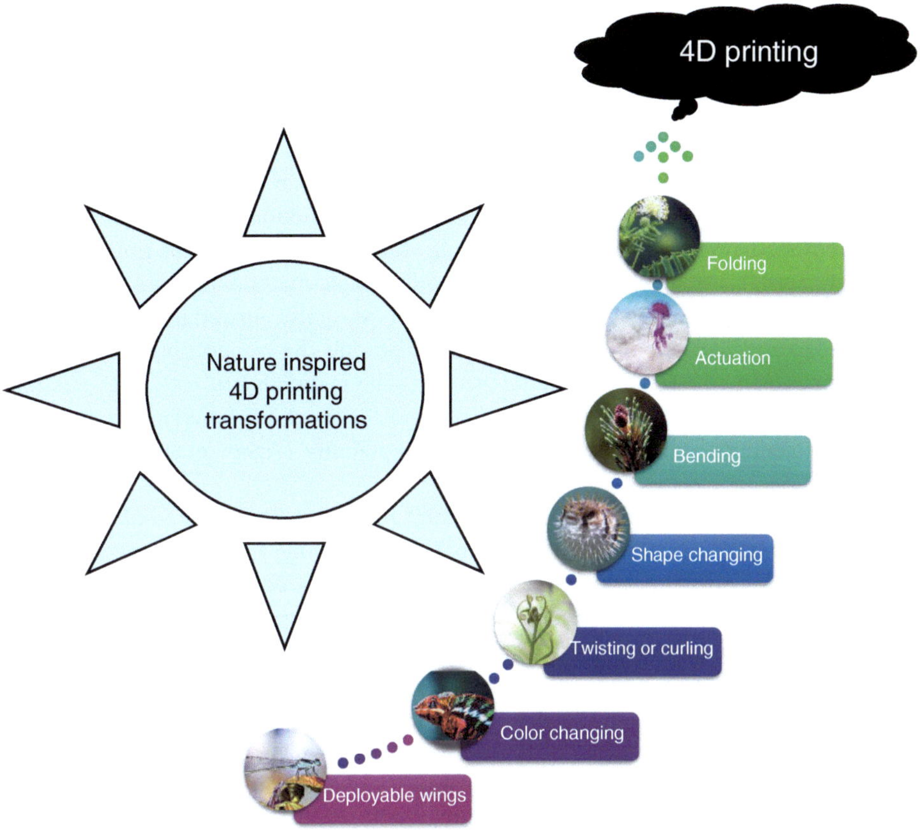

Figure 4.5 Nature-inspired 4D printing transformations. *Source:* Barnes Dr Thomas G, USFWS/ pixnio.com; Emma Birdsey / Wikimedia Commons / CC BY-SA 4.0; USDA-NRCS PLANTS Database / Herman, D.E. et al. 1996. North Dakota tree handbook; Saspotato/flickr.com; fififum/Pixabay; George Lebada / Pexels; Laitche / Wikimedia Commons / Public domain.

contraction of cells resulting from osmosis. Although natural shape-morphing procedures differ from artificial actuations, alternatives, including employing temperature gradients and comprehending material qualities, can be utilized to replicate such complex transitions.

Likewise, animals have also showcased the practicality of 4D printing across numerous biological actions and mechanisms. Unlike plants, animal mobility entails intricate muscular tension and contractions. For instance, organisms with hydrostatic skeletons execute intricate movements by spatially regulating the expansion and contraction of their soft tissues. Consequently, rather than completely replicating animal locomotion, basic movements are more attainable with the existing 4D printing technology. Ongoing progress in 4D printing will certainly enable the realization of intricate nature-inspired actuation (Ayushi et al. 2021; Magalhães and Almeida 2023; Tee and Tran 2021).

4.5 Materials in 4D Printing

A 4D-printed structure can be seen as the progeny of a 3D printer and intelligent materials, transforming a static form into a dynamic one. These intelligent materials are categorized according to their respective stimuli, such as thermo-responsive, magneto-responsive, or electro-responsive types. Further, the fundamental characteristics and requirements for 4D printing are classified into two categories, each presenting a range of behaviors and concepts to be considered: (i) smartness: single/dual/multiresponsiveness, self-decision/sensing/healing, response speed, self-actuating; and (ii) printability: rheology and viscosity modifier, sacrificial agent, cross-linking agent.

Typically, smart materials possess the capability to traverse in reaction to external stimuli via an interaction mechanism, and they enhance their locomotion efficiency through the application of mathematical concepts (Chen et al., 2024). Overall, materials utilized in 4D printing are classified as active and passive materials. Single-material printing and reversible materials printing employ substances known as active materials (Sadraei and Naghib 2025). Combinations of active and passive materials can be performed with multi-material printing and irreversible material printing. The materials utilized in 4D printing are referred to as smart or intelligent materials, and these exhibit functionality through intelligent capabilities, multifunctionality, and adaptability (Azher et al., 2025; Ding et al., 2025).

Previously, materials were characterized by their responses to environmental changes over time; recently, this concept has been expanded to include materials that provide a beneficial reversible effect upon encountering external stimuli. This outcome swiftly alters the color, shape, volume, stress/strain, or refraction, and such modifications, alone or collectively, influence the ultimate 3D-printed structures, converting them into 4D forms.

Furthermore, 4D printing smart materials may exhibit capabilities of self-healing, self-sensing, self-adaptability, self-actuation, self-diagnosis, self-assembly, and shape memory effects (SMEs). For instance, materials can revert to their previous shapes after deformation due to external stimuli, a phenomenon known as the SME, and such materials are referred to as shape memory materials (SMMs). Additionally, these materials are categorized according to their applications and presented in Figure 4.6. The SME and SMM, along with their applications contingent upon certain stimuli and situations, were comprehensively reported by Zafar and Zhao (2020). Numerous research conflates a focus on SMEs with shape change effects (SCEs). The primary distinction between SME and SCE is that prior to the application of the appropriate stimulus, shape recovery in SMMs is constrained if it exists at all; however, in SCE, reversible shape alteration (either instantaneously or progressively) relies on the application of the correct stimulus (Li et al. 2017).

The SME is additionally categorized into three primary phases: (i) one-way, (ii) two-way, and (iii) three-way SME. In the one-way SME, deformation into a transient shape is irreversible and becomes permanent following interaction with a stimulus. The two-way SME denotes the ability of some shape memory alloys (SMAs) to revert to a specified shape when heated over the transition temperature and thereafter return to a different configuration upon cooling. The three-way/multiple-way SME exhibits an intermediate configuration between its original and transient states, achievable by incorporating multiple two-way shape memory polymers (SMPs) with differing glass transition temperatures (Mathur et al. 2023; Pei and Loh 2018). Further, SMMs are classified according to their SME outcomes. For instance, SMPs are predominantly documented in irreversible effects or one-way SMEs (Pisani et al. 2022). Figure 4.7 illustrates SMPs, representing three types of SME processes. The advantages offered can propel interest in this topic across several sectors. The subsequent sections provide details regarding 4D printing materials.

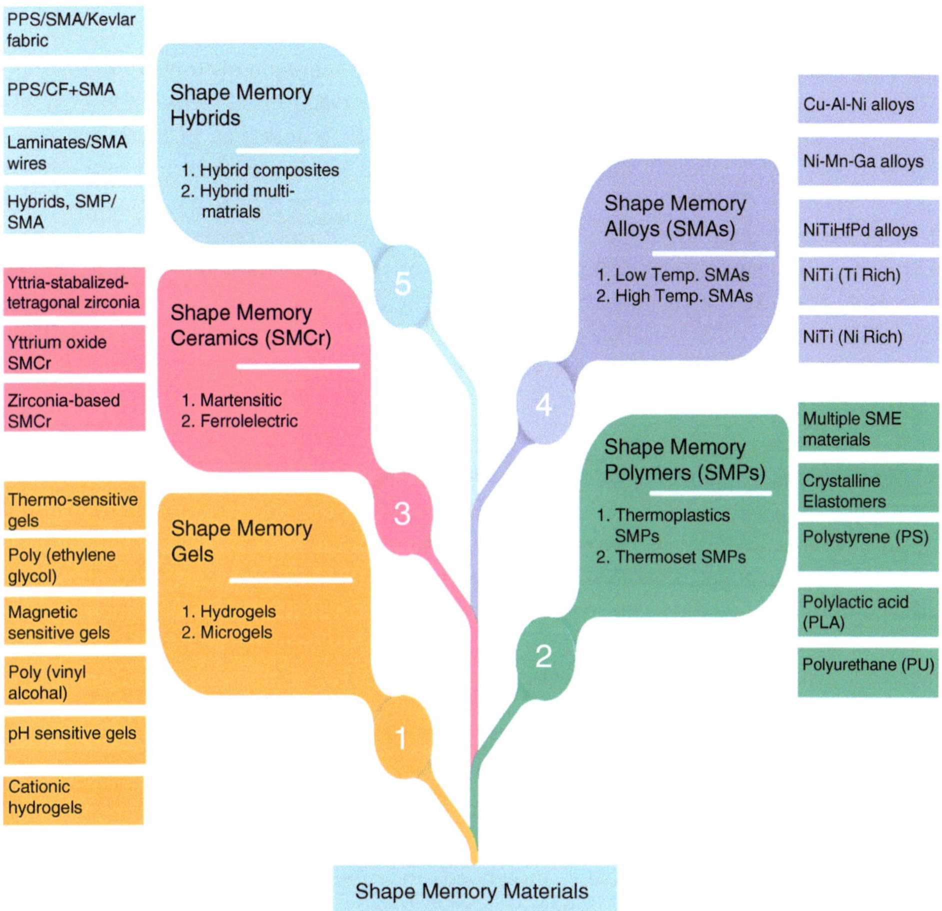

Figure 4.6 4D printing – shape memory materials. *Source:* Khalid et al. (2022) / with permission of Elsevier.

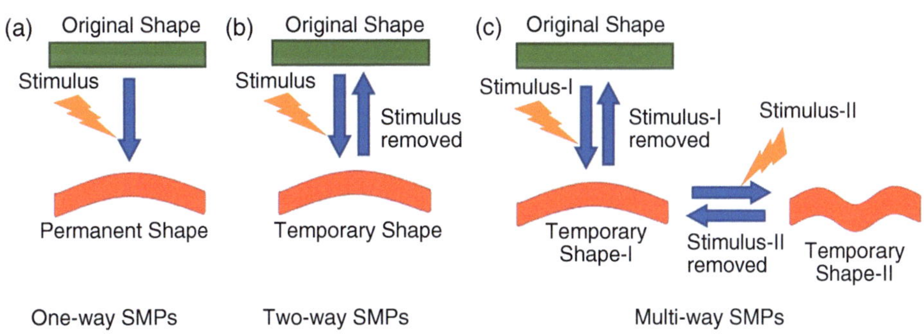

Figure 4.7 Shape memory effects (SME) of shape memory polymers (SMPs): (a) one-way, (b) two-way, and (c) multiway. *Source:* Khalid et al. (2022) / with permission of Elsevier.

4.5.1 Single Material 4D Printing

The predominant smart material employed in 3D printing subsequently transitions to 4D by external stimuli; this method utilizes SMP or liquid crystal elastomers (LCEs). In 4D printing, a singular SMP can accomplish a one-way SME, whereas an LCE can achieve a two-way SME. As previously stated, these materials fall under the broad category termed "active" materials. This active substance serves as the core of the printed construct, enabling it to detect and react to certain stimuli. Furthermore, it may possess many qualities, manifested through color or a singular transformation with lasting consequences. SMAs have been employed in methodologies described by Subeshan et al. (2021). SME faucets produced using single-material 4D printing are categorized and examined in the subsequent sections. The basic shape-shifting mechanisms for single material prints are presented in Figure 4.8.

4.5.1.1 Stimulus Responsiveness by Single SME

The single material utilized for this printing possesses the ability to react to external stimuli, hence extending its fourth dimension. The unique SME properties pertain to a material referred to as single SMM, such as SMAs and SMP. SMPs are predominantly utilized for individual SMEs because of their intrinsic covalent crosslinks, which restrict material stretchability, although enhanced stretchability may be attainable when the material surpasses its transient temperature (Ge et al. 2016). Nonetheless, the capacity for shape-shifting and material extensibility is contingent upon the thermal properties of the material, which encompass transient temperature, melting temperature, and glass-transition temperature, particularly when the stimulus is thermal in nature. Thermally activated SME can be induced by direct or indirect heating via joule heating, photothermal effects, hysteresis, and magnetic fields (Kuang et al. 2019). The unidirectional and bidirectional SME is extensively integrated into this methodology, necessitating programming to revert to its original configuration. The SMEs of single material performance, simulated by a simulation method, yield substantial predictions of experimental results, as depicted in Figure 4.9. Additionally, they illustrate the material's recoverability, utilizing stimuli for various forms produced via 4D printing techniques. Additionally, aside from thermal stimuli, research has also indicated the utilization of light (Ahmed et al. 2021) and magnetic fields (Kim et al. 2018). The predominant materials employed in these singular materials and single SMEs are SMP and shape memory gels (SMGs) rather than SMA (Ge et al. 2016).

4.5.1.2 Stimulus Responsiveness by Multiple SME

The raw materials employed in this printing method are single-material/SMM, exhibiting multi-SME features. Using this material concept, studies have documented automatic shape-changing behavior without the need for programming. This notion encompasses numerous investigations, regardless of their applications (Khalid et al. 2022; Kuang et al. 2016; Wan et al. 2022; Yu et al. 2015). In addition to the various shape alterations, multicolor transformations have also been reported in 4D printing research utilizing inks derived from SMPs (Zhang et al. 2021) and are regarded as one of the distinctive SMEs.

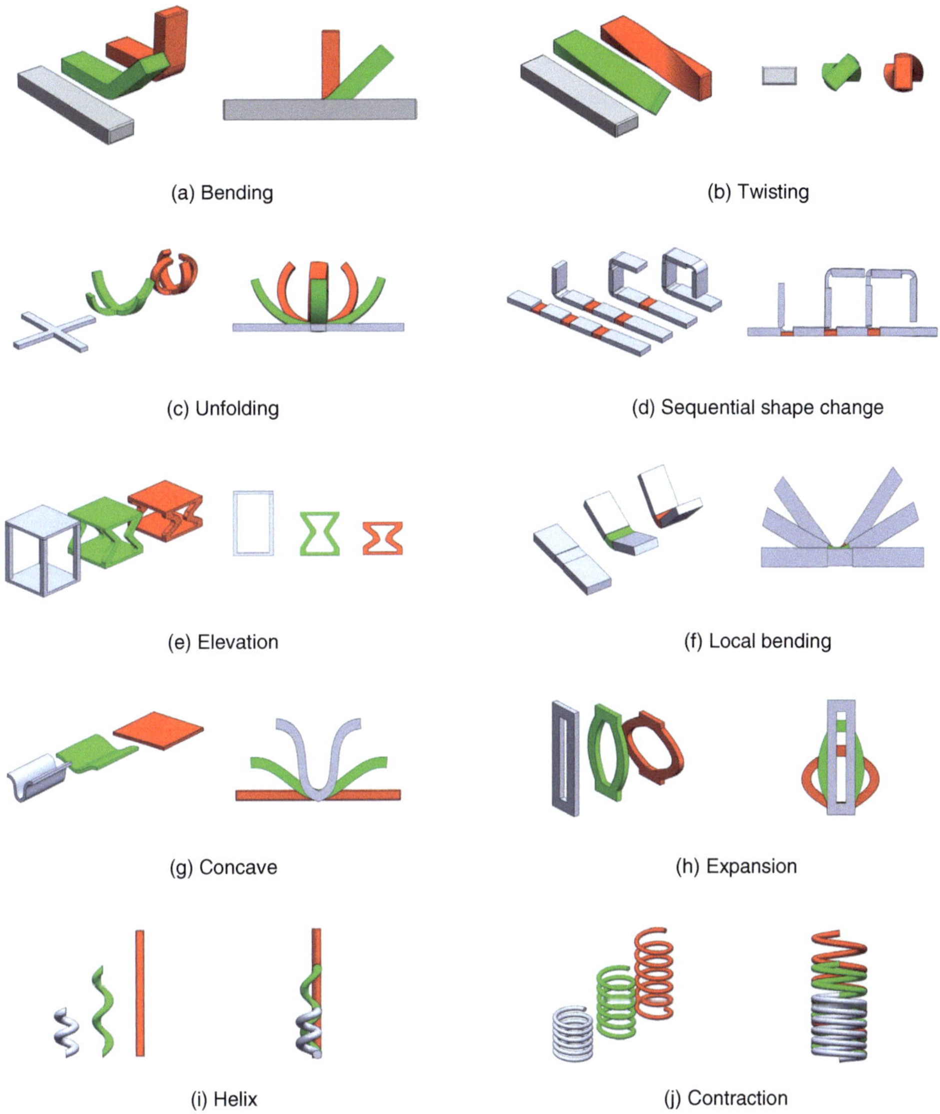

Figure 4.8 Frontal views of shape programming for single material prints (active material: colors reflex strain induced by the object; red – full strain; green – mid-range strain; grey – no strain). *Source:* Alshebly et al. (2021) / with permission of Elsevier.

Furthermore, these numerous SMEs depend on all three methods of shape shifting in nature, in contrast to single SMEs that utilize only one or two methods. This approach leverages the intrinsic properties and capabilities of smart materials, which have numerous applications in this domain. The production of lighter things can have impacts on their role in operations. Figure 4.10 illustrates the process of generating SMP-based 4D printing, resulting in a lightweight structure that undergoes many folds over time to form a complex

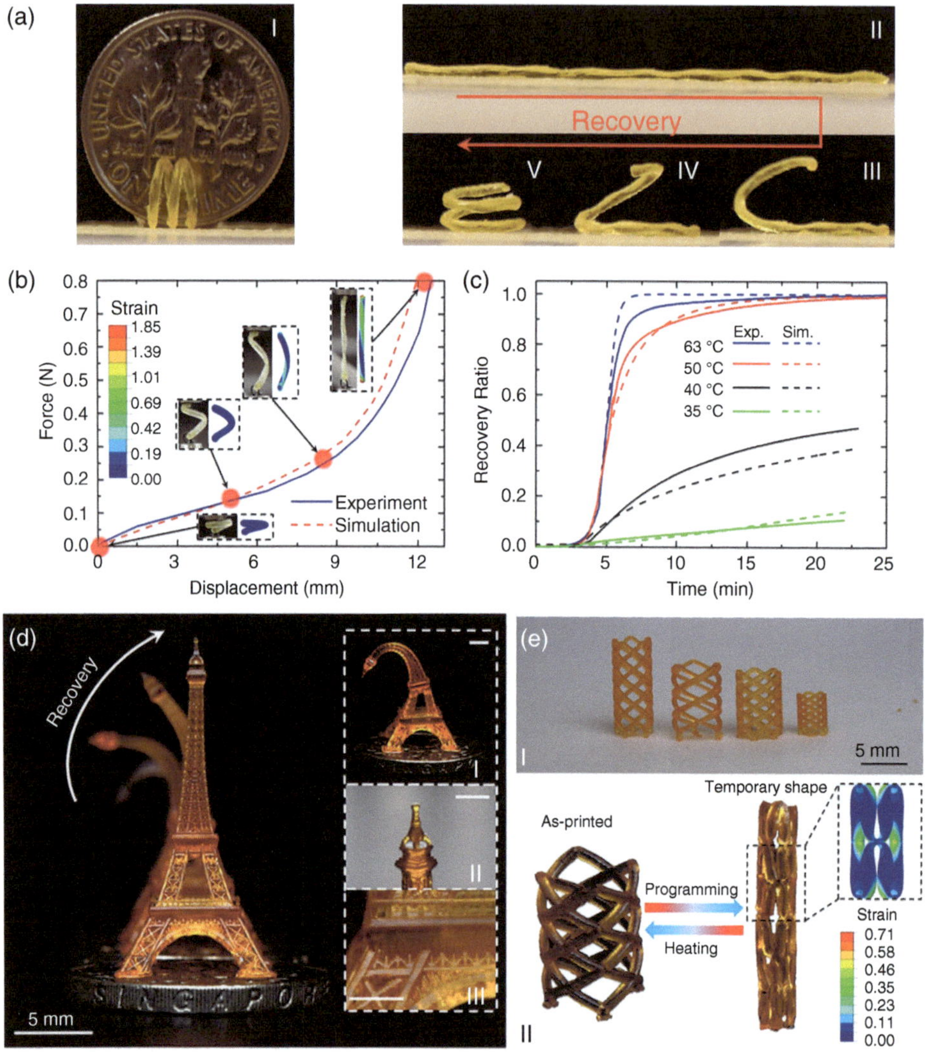

Figure 4.9 Single SME by single material in 4D printing. (a) Spring (I) was programmed to a straight strand temporary configuration (II) and returned to its original shape upon heating (III–V), (b) simulation responses, (c) comparison of experiments and simulations results at different temperatures, (d) eiffel tower, and (e) stents. *Source:* Ge et al. (2016) / Springer Nature / CC BY 4.0.

shape in response to stimuli. Likewise, triple SMEs are utilized in truss structures (Yu et al. 2015), molecular switches (Kuang et al. 2016), SMAs (Akbar et al. 2022), and SMP-based (Kuang et al. 2019) applications. This technique enables sequential sheet folding with respect to time and space by manipulating light wavelength or other stimuli. Such advanced shape-shifting techniques enable self-folding origami, instability-driven pop-up mechanisms, and "sequential" shape-shifting, markedly improving the functionality of 4D printing with a single material.

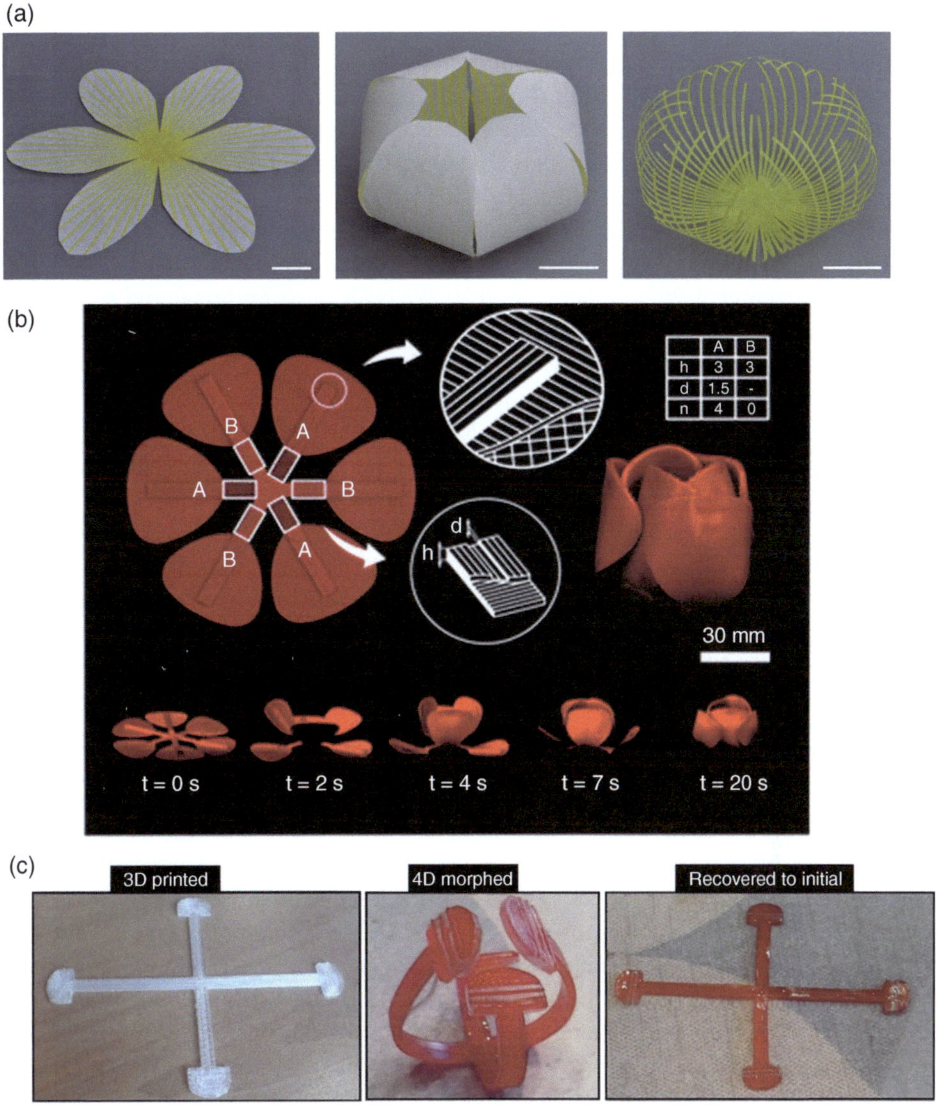

Figure 4.10 Multiple-SME by single material in 4D printing. (a) Planar shape sheet self-folded into the final flower shape after a stimulus is applied (heating and cooling) (scale bars = 2 cm). *Source:* Zhang et al. 2016 / Springer Nature / CC BY 4.0, (b) a two-step folding to create a tulip flower-like 3D shape from the initially flat petals. *Source:* van Manen et al. 2017 / The Royal Society of Chemistry / CC BY 3.0, and (c) SMPs – color and shape transformations. *Source:* Akbar et al. 2022 / with permission of Elsevier.

Nonetheless, a single material exhibits constrained applications in 4D printing functionality. Numerous investigations in this domain about the utilization of a single material have reported alterations in color and shape. Therefore, from a commercial standpoint, such issues must be addressed through substantial advancements in research within this domain. Consequent to the obstacles and constraints inherent in single-material 4D printing, it is surpassed by the advantages of utilizing multimaterials.

4.5.2 Multimaterial 4D Printing

The first study documented that 4D printing applications employed multimaterials to achieve multifunctionality. The utilization of diverse materials within a singular printed structure can confer benefits to intricate patterns that possess enhanced attributes as compared to those made from a single material. The utilization of various materials typically enhances strain, increases actuation angles, accelerates shape recovery, improves shape fidelity, and allows for more design complexity. Differential strain levels between two materials are typically employed to produce shape-shifting programming in a printed structure (Alshebly et al. 2021; Rafiee et al. 2020; Wu et al. 2016). Multimaterial 4D-printed structures comprise active and passive materials to enable self-actuation after the sensing or decision-making processes of the active materials.

Actuators are a type of 4D-printed multimaterial component advocated for enhancing the functionality of soft robots (Jamróz et al. 2018). The stress in nonactive or passive materials is inherently present. Internal stress is imparted to the nonactive materials during the printing or programming process. Typically, this internal tension is mitigated during the application of nonactive materials, as it influences the product's performance and stability. However, if this stress is harnessed positively and systematically, it can become advantageous and facilitate the attainment of shape-altering geometries with passive materials (Ahmed et al. 2021). The fundamental shape-shifting presents the multimaterial prints is presented in Figure 4.11.

In brief, the integration of various materials within a singular design effectively produces dynamic 4D-printed structures. The form morphing capability of these multimaterial systems typically results from their varied responses to applied stimuli. This disparity in response can be utilized to engineer intricate shape transformations, for instance, by employing one active material and one passive material. Furthermore, the use of various stimuli-responsive materials with distinct activation thresholds, such as varying transient temperature ranges, can facilitate successive deformations. Another method for fabricating multimaterial structures involves the combination of materials that respond differently to a singular stimulus or a multimaterial system in which each material reacts to distinct stimuli (McLellan et al. 2022). This passive material is used to achieve a strength of the constructs (for example, carbon nanotube and nanoparticles), enhancing biodegradability through the addition of plant oil polymer composition (e.g., epoxidized soyabean oil incorporation in polycaprolactone) (Tsujimoto et al. 2015). Notably, this will facilitate superior SME recovery behavior and is highly reusable, rendering it suitable for shape-shifting smart devices (Joshi et al. 2020).

Self-progress materials provide another example of 4D-printed multimaterials. Water exposure causes these materials to transition into their predetermined form, as they are composed of hydrophilic polymers that absorb water and can expand to potentially double their initial volume (Subeshan et al. 2021). Similarly, active origami exemplifies 4D-printed multimaterials. These materials facilitate the integration of large components into compact volumes, such as airbags in vehicles and photovoltaic solar cells in panels, owing to their capacity for form alteration (Teoh et al. 2018). Recently, significant advancements have been made in the medical industry using 4D printing, frequently associated with 4D bio-printing techniques (Osouli-Bostanabad et al. 2022). Integration of diverse stimuli

Figure 4.11 Schematic representation of shape shifting in multimaterial prints. (Active material: colors reflect strain induced by the object, red – full strain, green – mid-range strain, grey – no strain; passive material (yellow)). *Source:* Alshebly et al. (2021) / with permission of Elsevier.

elements may support various functions simultaneously. Nonetheless, incorporating the diversity of materials in this composition concurrently elevates the complexities related to handling and printing execution. These multimaterials encompass multiple small and medium enterprises in 4D printing applications. The fundamental distinction between single and multimaterial prints is summarized in Table 4.2. Especially for multimaterial 4D printing, the material distribution over the single constructs can be visualized through the simulation software called digital materials (Momeni et al. 2017). The comprehensive view

Table 4.2 Shape programming terminologies in single and multimaterial prints.

	Description		
Terminology	Single material	Multimaterials	References
Bending	Restoration to original form following a strain-induced deformation	Strain variation between two materials when the SME of one material is activated poststrain	Alshebly et al. (2021)
Twisting	The recovery of the printed spring configuration following strain-induced straightening	Printing of shape memory materials in a helical pattern around a support structure	
Folding	—	Flexing of the hinge segment to facilitate the folding of the remaining structure	
Unfolding	Restored to form following the folding of edges	—	
Sequential shape change	Activation of consecutive hinges by varying section thicknesses	Activation of consecutive hinges through graded mixing	
Elevation	Printing of auxetic patterns that expand upon activation	A multitude of hinges are employed to elevate a base upon activation	
Rotation	—	Manufacturing of a centered base with connectors utilizing a bi-stable design composed of hard and elastomeric materials	
Local bending	Activation of specific portions of the structure using photothermal heating or another stimulus	—	
Concave	Concentric printing pattern of SMM, adjusting the print temperature to regulate induced printing strain	—	
Expand to fit a mold	Hexagonal patterns restore the completely expanded form	—	
Helix	Restoration to the printed helical configuration	Printing a pattern of a substance in a diagonal orientation	
Contract	Strain is generated throughout the printing process, and the original form is restored upon activation	—	
Uniform patterns	—	Recovery of SMM patterns imprinted on the fabric	
Nonuniform patterns	Modulation of polymer cross-linking by grayscale values	Nonuniform graded morphological alteration predicated on varying proportions of two materials for hydration-induced expansion	
Gripper	—	The curvature of numerous segments toward the center	
Surface expansion	—	Origami designs utilizing dynamic materials	

of 3D and 4D printing of multimaterials across numerous industries, their diverse material sources, the varied types of printers, and the associated challenges are detailed in a recent report by Rafiee et al. (2020).

Based on their applications and their functional usage, the following sections consider three major categories: hydrogels, composite materials, and multifunctional materials.

4.5.2.1 Materials Used in Multimaterial 4D Prints

Essentially, hydrogels form gels when in contact with water/moisture; that is, they expand when in contact with water. Hydrogels are widely used in 4D printing applications in various industries. Importantly, the first 4D printing study utilized these materials for multimaterial 4D print applications using a Stratasys Connex 500 printer (Tibbits 2014). Compared with SMPs, many studies have been reported with SMGs, including in biomedical and healthcare applications, especially in the field of bioprinting (Agarwal et al. 2023). These materials are apt for active as well as passive functions, thus resulting in single and multimaterial prints in 4D printing applications. In terms of stimulus responses, water, and heat-responsive applications are widely explored. Due to the inherent nature of being moistened, powder-based printing methods are not suitable for this concept. To some extent, by applying cooling stimuli after the deformation of the shape, it becomes a rigid structure; further, to return to its original shape, heat is required as an external stimulus. However, although it has wider applications in this field, there are many challenges and limitations. Champeau et al. (2020) documented the scope and potential of different smart/responsive materials and nonresponsive materials, particularly hydrogels, in 4D printing applications.

Composite materials in 4D printing are majorly reported for applications in SMP-based constructs. In addition to the SMP-based material, fiber-reinforced active composites have been used to strengthen constructs. Additionally, microscale composites, nanocomposites, nanoparticles, bilayer composites, and multimaterial desolvation are widely used in 4D printing irrespective of their applications (Alshebly et al. 2021; Biswas et al., 2021b).

Finally, multifunctional materials are those representing unique properties of 4D-printed constructs. Particularly, they have demonstrated capabilities in executing multifunctionalities. This execution takes place while applying different stimuli respective to their concentration. Especially, these multifunctional materials are utilized for bioprinting and self-healing operations in 4D printing. As a field with growing research and commercial interest, detailed documentation regarding multimaterial sources and their function in 4D printing approaches have been reported by different researchers (Ali et al. 2019; González-Henríquez et al. 2019; Shraim et al. 2021).

4.6 Printers in 4D Printing

A 3D printer is a fundamental component of the 4D printing process, functioning in a layer-by-layer manner according to computer instructions to fabricate objects from materials such as polymers, ceramics, metals, and others. 4D-printed components are produced by additive manufacturing with one or a combination of multiple smart/intelligent materials that can transform over a designated period. The variations in material properties, such as shape memory, thermal expansion, contraction, and compression, result in the shape-shifting process of the entire structure (Nam and Pei 2019). A suitable 3D printing approach can be the foundation for the subsequent 4D printing component. The fundamental ASTM classification of additive manufacturing (as 7 categories) is elaborated in Chapter 1.

4.6.1 Printer Configurations

The previous chapters explained the principles, material selections, and applications of additive manufacturing processes. Additionally, Table 4.3 summarizes the additive manufacturing process. This section covers the possible printing techniques in 4D printing approaches for different materials.

In 4D printing fabrication of smart materials, the most widely adapted printing techniques include fusion deposition modeling, jet printing and stereolithography, powder bed fusion, and DED. Specifically, attempts have been made to print nonmetallic SMMs, either individually or in combination, using various additive manufacturing processes such as micro-extrusion, stereolithography, and polyjet technology. The choice of suitable printing techniques primarily relies on the specific SMMs and the desired configuration. Herein, micro-extrusion printers utilize pneumatic or mechanical (piston or screw) dispensing systems to extrude continuous strands of material or cells in 4D bioprinting (Karvinen and Kellomäki 2023; Yang et al. 2020). Similarly, stereolithography has been used to fabricate stem cell research and biomedical scaffold preparation (Zhu et al. 2019). A shape memory scaffold made of polycaprolactone dimethacrylate manufactured with a UV-LED DLP printer has been reported (Greant et al. 2023). For metallic materials, DED, SLM, and EBSM techniques are most relevant in the case of 4D printing (DebRoy et al. 2018). Nonetheless, when printing smart materials, either single or multimaterials, the configuration of the printing setup varies; compared to single-material prints, the concept of multimaterials have a very limited number of printer types, thus reducing the material variety. Subsequently, the challenges associated with multimaterial prints are higher than with single material; in contrast, multimaterials have a divergence of application in various fields than the single material. In view of the current progress of research, with an increase in the choice of printers and the diverse materials used in many industries it is expected that in the future, hybrid approaches including combinations of multiple printing methods and multiple SMEs materials may be used to create complex 4D-printed structures.

4.6.2 Printing Variations

The printing parameters for various printing techniques can vary depending on the source of materials. Printing parameters include printing speed, extrusion rate, nozzle diameter, printing, and platform temperature. Such parameters can vary depending on the printer configurations, final constructs and material sources. Additionally, these parameters can be optimized based on the selected material characterization, such as thermal, crystallinity, particle size, rheology, and viscosity. For instance, in FDM-based 4D printing utilizing polymers, the printing speed plate build temperature and thickness of the filaments majorly influence the shape deformation process. Similarly, a lower build temperature can facilitate

Table 4.3 Printing techniques, processes, and their pros and cons.

Additive manufacturing techniques	Types/methodologies	Printing process/description	Advantages	Disadvantages
Vat photopolymerization	• Stereolithography (SLA)	SLA utilizes a photopolymer liquid as the material source within a vat. This liquid plastic is incrementally converted into a three-dimensional item by descending the construction platform into the vat and solidifying it with a UV laser.	1) Large parts fabrication 2) High accuracy 3) Good surface finish	1) Not stable over time and do not have well-defined mechanical properties (e.g., photopolymer)
	• Digital light processing (DLP)	DLP technology closely resembles SLA but employs a distinct light source and utilizes a liquid crystal display panel.	1) Higher print speed than SLA 2) High accuracy 3) Low-cost printers	1) Insecurity of consumable material 2) High-cost materials
	• Continuous direct light processing (CDLP)	CDLP operates similarly to DLP, except it depends on the constant upward movement of the printing bed in the z-direction. Accelerated construction times are achievable since the printer does not need to halt and detach the object from the printing bed once each layer is completed.	1) High print speed 2) High accuracy 3) Low-cost printers	1) Insecurity of consumable material 2) High-cost materials
Material extrusion	• Fused deposition modeling (FDM) • Fused filament fabrication (FFF) • Fused layer modeling/ manufacturing (FLM)	A plastic filament is liquefied and forced through a nozzle. Objects are constructed incrementally, layer-by-layer addition.	1) Functional parts fabrication	1) Printed parts have an anisotropy in the z-direction (vertical direction) and a step-structure on the surface.

(Continued)

Table 4.3 (Continued)

Additive manufacturing techniques	Types/methodologies	Printing process/description	Advantages	Disadvantages
	• Direct ink writing (DIW) • Robocasting (RC) • Direct-write assembly (DWA) • Microrobotic deposition (μRD) • Bioplotting • Low-temperature deposition manufacturing (LDM) • Freeform 3D printing • Extrusion freeform fabrication (EFF)	Substance in a semi-liquid or paste state can be extruded through a nozzle to print the cross-sections of a sliced 3D object.	1) Highest resolution for an extrusion system 2) Ideal for research environments	1) Limited geometry 2) High cost 3) Small build volume
Binder jetting (BJ)	• Binder jetting (BJ)	Inkjet printing heads dispense a liquid bonding agent onto the surface of the powder. The product is constructed by layer by layer adhering the particles together.	1) Fast and cheap 2) Wide material types 3) Parts in full color are possible	1) Limited mechanical properties
Material jetting (MJ)	• Multijet modeling • Drop-on-demand (DOD) • Thermojet • Inkjet printing	The inkjet printing head ejects molten wax onto the printing substrate. Upon cooling and solidification, the material permits the fabrication of layers a top one another.	1) High accuracy 2) Good surface finishes	1) Limited with wax-like materials

	• Polyjet modeling • Multijet modeling • Polyjetting • Multijetting • Jetted photopolymer	Analogous to multijet technology, it employs print head jets to deposit liquid photopolymers onto a printing substrate. The substance is promptly cured and solidified by UV radiation, enabling the construction of layers atop one another.	1) Multimaterial and multicolor objects	1) Only possible with UV-active photopolymers, which are not durable over time
Powder bed fusion (PBF)	• Laser sintering (LS) • Selective laser sintering (SLS)	SLS exhibits certain similarities with SLA. A laser selectively melts a thin coating of plastic powder. The components are constructed layer by layer within the powder bed.	1) Good mechanical properties 2) Wide variety of materials availability	1) Do not have the same properties as their injection-molded counterparts
	• Selective laser melting (SLM) • Direct metal laser sintering (DMLS) • Lasercusing	The LS procedure closely resembles the LM process. A laser selectively melts a thin layer of metal powder. The components are constructed incrementally in the powder bed.	1) High-density metal printing	1) Slow and expensive 2) Surface finishes are limited
	• Electron beam melting (EBM)	An electron beam selectively melts a small layer of metal powder. The components are constructed incrementally within the powder bed.	1) High-density metal printing	1) Limited material 2) Slow and expensive
	• Multijet fusion (MJF)	MJF essentially integrates SLS and MJ technologies. A carriage equipped with inkjet nozzles applies a fusing agent onto a small coating of plastic powder, which is then selectively melted using a high-power infrared energy source.	1) Fast process	1) Limited material

(Continued)

Table 4.3 (Continued)

Additive manufacturing techniques	Types/methodologies	Printing process/description	Advantages	Disadvantages
Direct energy deposition (DED)	• Laser-engineered net shaping (LENS)	Utilizes a high-powered laser to fuse metal powder that is applied to the substrate. Metal is applied to the laser's focal point, where it fuses together. An inert gas is employed to protect the metal from ambient gases. It employs a stratified methodology for component fabrication.	1) Repair parts and fabricate new ones 2) Good granular structure 3) Materials adaptability	1) Postprocessing necessitates 2) Remove some parts from the built substrate 3) Poor surface finish 4) Require machining or polishing 5) Low dimensional accuracy
	• Electron beam additive manufacture (EBAM)	Utilizes an electron beam as the thermal source to weld and fabricate metal components employing wire or metal powder. The method resembles LENS; however, electron beams exhibit greater efficiency than lasers.	1) Material availability 2) Less safety and procurement issues compared with LENS 3) Less energy consumption compared with the powder-feed method	1) Limited to single-material printing
Sheet lamination	• Laminated object manufacturing (LOM)	Layers of paper, plastic, or metal laminates are adhered and fused using heat and pressure, subsequently shaped with a computer-controlled laser or knife.	1) Larger-scale models' fabrication 2) Inexpensive 3) Fast and accurate 4) Good handling strength	1) Laborious task 2) Fire hazard 3) Not as good as materials used with other rapid prototyping methods

Source: Adapted from Rafiee et al. (2020).

attaining higher angle deformations than a higher build temperature (Kačergis et al. 2019; Rastogi and Kandasubramanian 2019). Similarly, other parameters also influence 4D shape shifting, either positively or negatively (Gao et al. 2024; Ren et al. 2023). Current industry requirement is focused on improving production speeds and therefore demand research in this direction. On the one hand, multimaterials are finding booming applications in the field of 4D printing in diverse industry fields, and on the other hand, they possess different characteristics, and such materials are more complex to be handled during printing processes. For instance, they may need adaptive or vibrative nozzles, high-end printers, and technical skills for better process control. However, currently, there are only a limited number of printing methods, and material selection is a challenge. Overall, compared to 3D printers, costs for printers used in 4D applications are higher as well, and most of the printers used in 4D, do not come with user-friendliness. This element highlights a key research need in the field of 4D printing technology.

4.7 Mathematical Modeling in 4D Printing

Modeling serves as a valuable tool that enables users to define the parameters and complexities of real-world applications within a virtual environment, facilitating the optimization of designs. In 4D printing, modeling outlines the trajectory of shape alterations while minimizing repetition, thus conserving materials and reducing costs. Many challenges can be effectively tackled using the inverse approach (predefined applications), while the forward strategy is prevalent in research areas where final applications are still uncertain (Momeni et al. 2017). Various tools are available for 4D printing design modeling, including Abaqus Finite Element Analysis, COMSOL Multiphysics, SolidWorks, Autodesk, and ANSYS. Although these tools effectively optimize printing variables, they often face credibility issues due to their inability to replicate real-world environmental conditions accurately. Discrepancies in the degradation of material properties during the modeling phase can lead to significant differences in the predicted number of transformation cycles before failure, as opposed to actual performance. Inherent variations in materials or printers, along with challenges in simulating larger dimensions, asymmetrical designs, and complex geometries, contribute to reduced reliability and extend the simulation process (Rastogi and Kandasubramanian 2019).

Mathematical modeling is also essential for the design of material distribution within structures. Specifically, 4D printing utilizes the arrangement of active and passive materials to achieve desired behaviors. In 4D-printed structures, there exist two or three stable states, and the entire structure can transition between these states when exposed to specific stimuli. The primary purposes of mathematical modeling in 4D printing are to predict time-dependent shape shifting, to prevent collisions between components during self-assembly and to minimize the need for extensive experimentation. Additionally, interaction mechanisms may be employed to induce specific movements in the printed structure. Also, the swelling effect can be integrated into SMPs through appropriate stimuli interactions through modeling (Fang et al. 2023). A fundamental flow chart for computational designs for multimaterial 4D printing applications has been reported by Athinarayanarao et al. (2023).

In 4D printing and related multimaterial structures, the physical voxel can be defined comparably. Digital material is characterized as a collection of diverse physical voxels. The physical voxel is the element that facilitates the transition from analog to digital materials. The fundamental and physical unit that occupies 3D space is defined as a bit. The actual voxel may possess any dimensions and configuration. The spatial configuration of voxels significantly influences the characteristics of a 4D-printed structure. In digital materials, each voxel comprises solely one material (active). Contiguous voxels may consist of distinct materials (active and passive). Each voxel possesses distinct properties, and the aggregation of various voxels culminates in a multimaterial structure.

Three fundamental types of modeling structures were utilized for 4D printing: uniform distribution, gradient distribution, and unique patterns. A key category of specialized patterns is the fiber and matrix structure. Each structure exists within a single layer but can be combined to create bi-layer or multilayer designs, utilizing more than two materials. An example of a gradient distribution multimaterial structure is shown in Figure 4.12.

Recently, researchers have proposed that the identified discrepancies may be mitigated through the utilization of advanced software techniques such as Abaqus or through specialized software developed by various universities globally, such as project Cyborg, collaboratively created by Stratasys, the Self Assembly Lab at MIT, and Autodesk (Athinarayanarao et al. 2023; Zafar and Zhao 2020). Nonetheless, augmenting the

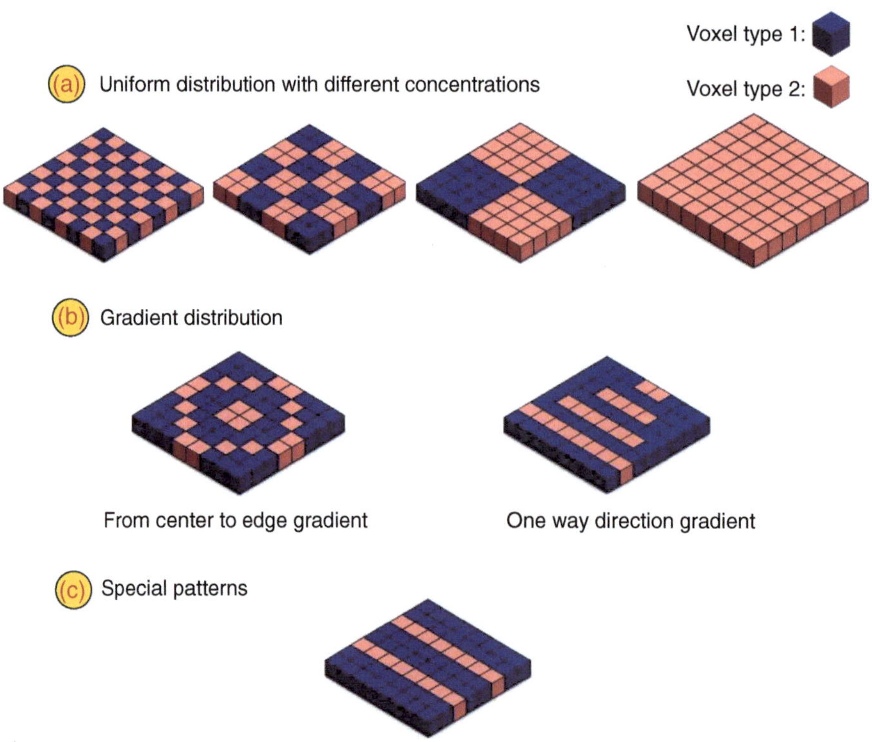

Figure 4.12 Digital material structures in 4D printing. *Source:* Momeni et al. (2017) / with permission of Elsevier.

algorithms with supplementary constraints related to print feed characteristics and comprehensive environmental conditions or condensing bigger algorithms into more concise versions may potentially alleviate the observable disparities. Another simulation software for the 4D printing process, called the 4D Modeler, has been developed by the Gwangju Institute of Science and Technology in South Korea (Chung et al. 2017; Teng et al. 2021). Research in the development of 4D printing software is limited, as this technology has not yet reached a sufficient degree of maturity. Significant research potentiality are required to recreate the SMEs in SMMs to assist researchers in the 4D printing field. Furthermore, the researcher suggested the implementation of six software solutions to encompass all phases of 4D printing. Each program will address a certain printing phase, encompassing simulation, modeling, slicing, host/firmware, monitoring, and printing management software to encompass all necessary processes in 4D printing (Chung et al. 2017). Recently, machine learning (ML) coupled algorithms have been used in 4D printing studies to predict the printability of materials (Lai et al. 2024).

4.8 Current State of the Art

4D printing technology is presently in the research phase in many industry segments and is a rapidly expanding topic of study globally. In the context of 4D printing, emphasis must be placed on the interdisciplinary issues associated with multimaterial printability at the voxel level and the optimization of actuation performance in composite structures under diverse stimuli. Moreover, the primary considerations for 4D printing materials are mechanical strength, potential unfavorable effects, and design methodologies tailored to specific quantitative data constraints. Nonetheless, extensive study is necessary in this domain concerning interdisciplinary cooperation, creative methodologies, and the investigation of engineering, design, and artificial intelligence. The horizontal extension of the vertical bars indicates a paradigmatic expansion. Effective 4D printing solutions require a combination of scientific hybridization and epistemic flexibility. This position entails complex design responsibilities, encompassing the utilization of experience, comprehensive research, and heuristic methods. This domain is essential for the success of 4D commercialization and necessitates concentrated attention.

Concerning printers, composite printing with fillers can obstruct the nozzle heads and hinder the free extrusion, hence impacting the quality and duration of printing. In contrast to 3D printing, 4D printing has achieved a 70–90% reduction in printing time and printing materials for the production of 3D objects (Ding et al. 2018). The advantages of 4D printing and the challenges related to the commercialization of printed entities have created opportunities for advanced research, as well as risks and warnings associated with the new technology and have been explained using strengths, weaknesses, opportunities, and threats (SWOT) analyses. Such SWOT analysis shows the many facets of a technology or system's potential. Strengths include its ability to adapt over time, preserve energy, materials, and time and demonstrate novel functionalities such as self-healing and sensing. It is cost effective for small-scale production, portable for self-deployable structures, feasible in dangerous conditions, and supports uncrewed operations. However, weaknesses include the necessity for complicated modeling and form prediction, reversibility concerns, short overall lifespan, material supply issues, and cost impediments to industrial production. Social issues, such as employment displacement resulting from automation and technical limits, including printer size and quality constraints, also exist. Conversely, there are considerable opportunities for this technology, especially in enhancing biocompatibility and environmental sustainability through rigorous research. These possess the potential in real-time applications and may significantly contribute to national security and sophisticated weaponry. Nonetheless, it encounters threats, including cyber vulnerabilities, potential weaponization, technological sabotage, and safety issues, such as self-dismantling structures leading to accidents, and these aspects must be focused for improvements. Such aspects jointly provide insights into its potential while emphasizing issues requiring careful evaluation and mitigation (Rastogi and Kandasubramanian 2019).

4.9 Conclusion

4D printing has evolved significantly since its inception and has made inroads into various industrial sectors. It combines 3D printing technology with smart materials to enable shape transformation over time. This approach offers enhanced flexibility and versatility, allowing a single structure to serve multiple functions. The future of 4D-printed materials is bright and full of potential. By addressing the inherent weaknesses of mechanical systems, 4D printing can revolutionize design. Smart materials can replace traditional components such as motors and gears, paving the way for innovative systems with greater freedom of movement. Additionally, advanced methods for integrating electronics into smart buildings could lead to the development of unique smart devices. The emergence of biocompatible smart materials has the power to transform the medical field and improve healing outcomes. Certainly, it is difficult to fully envision the profound impact that 4D printing will have on the industrial landscape in the future.

References

Agarwal, K., Srinivasan, V., Lather, V., Pandita, D., & Vasanthan, K. S. (2023). Insights of 3D bioprinting and focusing the paradigm shift towards 4D printing for biomedical applications. *Journal of Materials Research*, *38*(1), 112–141. https://doi.org/10.1557/s43578-022-00524-2

Ahmed, A., Arya, S., Gupta, V., Furukawa, H., & Khosla, A. (2021). 4D printing: Fundamentals, materials, applications and challenges. *Polymer*, *228*, 123926. https://doi.org/10.1016/j.polymer.2021.123926

Akbar, I., El Hadrouz, M., El Mansori, M., & Lagoudas, D. (2022). Toward enabling manufacturing paradigm of 4D printing of shape memory materials: Open literature review. *European Polymer Journal*, *168*, 111106. https://doi.org/10.1016/j.eurpolymj.2022.111106

Ali, M. H., Abilgaziyev, A., & Adair, D. (2019). 4D printing: A critical review of current developments, and future prospects. *The International Journal of Advanced Manufacturing Technology*, *105*(1–4), 701–717. https://doi.org/10.1007/s00170-019-04258-0

Alshebly, Y. S., Nafea, M., Mohamed Ali, M. S., & Almurib, H. A. F. (2021). Review on recent advances in 4D printing of shape memory polymers. *European Polymer Journal*, *159*, 110708. https://doi.org/10.1016/j.eurpolymj.2021.110708

Arsuffi, B., Magrini, T., Champeau, M., Siqueira, G., & Titotto, S. (2025). 4D printing of natural materials: A review. *Sustainable Materials and Technologies*, *44*, e01346. https://doi.org/10.1016/j.susmat.2025.e01346

Athinarayanarao, D., Prod'hon, R., Chamoret, D., Qi, H. J., Bodaghi, M., André, J.-C., & Demoly, F. (2023). Computational design for 4D printing of topology optimized multi-material active composites. *NPJ Computational Materials*, *9*(1), 1. https://doi.org/10.1038/s41524-022-00962-w

Ayushi, Kumar Vates, U., Mishra, S., & Jee Kanu, N. (2021). Biomimetic 4D printed materials: A state-of-the-art review on concepts, opportunities, and challenges. *Materials Today Proceedings*, *47*, 3313–3319. https://doi.org/10.1016/j.matpr.2021.07.148

Azher, K., Nazir, A., Farooq, M. U., Ali, Z., Dalaq, A. S., Abubakar, A. A., Hussain, S., Syed, M. N., Ullah, A., Laghari, R. A., & Khan, S. (2025). Revolutionizing the future of smart materials: a review of 4d printing, design, optimization, and machine learning integration. *Advanced Materials Technologies*, *10*(12), 2401369. https://doi.org/10.1002/admt.202401369

Biswas, M. C., Chakraborty, S., Bhattacharjee, A., & Mohammed, Z. (2021). 4D printing of shape memory materials for textiles: Mechanism, mathematical modeling, and challenges. *Advanced Functional Materials*, *31*(19). https://doi.org/10.1002/adfm.202100257

Champeau, M., Heinze, D. A., Viana, T. N., de Souza, E. R., Chinellato, A. C., & Titotto, S. (2020). 4D printing of hydrogels: A review. *Advanced Functional Materials*, *30*(31), 1910606. https://doi.org/10.1002/adfm.201910606

Chen, Q., Kalpoe, T., & Jovanova, J. (2024). Design of mechanically intelligent structures: Review of modelling stimuli-responsive materials for adaptive structures. *Heliyon*, *10*(14), e34026. https://doi.org/10.1016/j.heliyon.2024.e34026

Chu, H., Yang, W., Sun, L., Cai, S., Yang, R., Liang, W., Yu, H., & Liu, L. (2020). 4D printing: A review on recent progresses. *Micromachines*, *11*(9), 796. https://doi.org/10.3390/mi11090796

Chung, S., Song, S. E., & Cho, Y. T. (2017). Effective software solutions for 4D printing: A review and proposal. *International Journal of Precision Engineering and Manufacturing-Green Technology*, *4*(3), 359–371. https://doi.org/10.1007/s40684-017-0041-y

DebRoy, T., Wei, H. L., Zuback, J. S., Mukherjee, T., Elmer, J. W., Milewski, J. O., Beese, A. M., Wilson-Heid, A., De, A., & Zhang, W. (2018). Additive manufacturing of metallic components – process, structure and properties. *Progress in Materials Science*, *92*, 112–224. https://doi.org/10.1016/j.pmatsci.2017.10.001

Ding, A., Tang, F., & Alsberg, E. (2025). 4D printing: A comprehensive review of technologies, materials, stimuli, design, and emerging applications. *Chemical Reviews*, *125*(7), 3663–3771. https://doi.org/10.1021/acs.chemrev.4c00070

Ding, Z., Weeger, O., Qi, H. J., & Dunn, M. L. (2018). 4D rods: 3D structures via programmable 1D composite rods. *Materials & Design*, *137*, 256–265. https://doi.org/10.1016/j.matdes.2017.10.004

Fang, L., Yan, W., Chen, S., Duan, Q., Herath, M., Epaarachchi, J., Liu, Y., & Lu, C. (2023). Light and shape-memory polymers: Characterization, preparation, stimulation, and application. *Macromolecular Materials and Engineering*, *308*(12). https://doi.org/10.1002/mame.202300158

Gao, Y.-C., Duan, D.-X., Zeng, S.-Y., Zheng, H., Wang, L.-P., & Tan, J.-R. (2024). Programming time-dependent behavior in 4D printing by geometric and printing parameters. *Advances in Manufacturing*. https://doi.org/10.1007/s40436-024-00489-x

Ge, Q., Sakhaei, A. H., Lee, H., Dunn, C. K., Fang, N. X., & Dunn, M. L. (2016). Multimaterial 4D printing with tailorable shape memory polymers. *Scientific Reports*, *6*(1), 31110. https://doi.org/10.1038/srep31110

González-Henríquez, C. M., Sarabia-Vallejos, M. A., & Rodriguez-Hernandez, J. (2019). Polymers for additive manufacturing and 4D-printing: Materials, methodologies, and biomedical applications. *Progress in Polymer Science*, *94*, 57–116. https://doi.org/10.1016/j.progpolymsci.2019.03.001

Greant, C., Van Durme, B., Van Damme, L., Brancart, J., Van Hoorick, J., & Van Vlierberghe, S. (2023). Digital light processing of poly(ε-caprolactone)-based resins into porous shape memory scaffolds. *European Polymer Journal*, *195*, 112225. https://doi.org/10.1016/j.eurpolymj.2023.112225

Hu, J., Meng, H., Li, G., & Ibekwe, S. I. (2012). A review of stimuli-responsive polymers for smart textile applications. *Smart Materials and Structures*, *21*(5), 053001. https://doi.org/10.1088/0964-1726/21/5/053001

Jamróz, W., Szafraniec, J., Kurek, M., & Jachowicz, R. (2018). 3D printing in pharmaceutical and medical applications – recent achievements and challenges. *Pharmaceutical Research*, *35*(9), 176. https://doi.org/10.1007/s11095-018-2454-x

Joshi, S., Rawat, K., Karunakaran, C., Rajamohan, V., Mathew, A. T., Koziol, K., Kumar Thakur, V., & Balan, A. S. S. (2020). 4D printing of materials for the future: Opportunities and challenges. *Applied Materials Today*, *18*, 100490. https://doi.org/10.1016/j.apmt.2019.100490

Kačergis, L., Mitkus, R., & Sinapius, M. (2019). Influence of fused deposition modeling process parameters on the transformation of 4D printed morphing structures. *Smart Materials and Structures*, *28*(10), 105042. https://doi.org/10.1088/1361-665X/ab3d18

Karvinen, J., & Kellomäki, M. (2023). Design aspects and characterization of hydrogel-based bioinks for extrusion-based bioprinting. *Bioprinting*, *32*, e00274. https://doi.org/ 10.1016/j.bprint.2023.e00274

Khalid, M. Y., Arif, Z. U., Noroozi, R., Zolfagharian, A., & Bodaghi, M. (2022). 4D printing of shape memory polymer composites: A review on fabrication techniques, applications, and future perspectives. *Journal of Manufacturing Processes*, *81*, 759–797. https://doi.org/10.1016/j.jmapro.2022.07.035

Kim, Y., Yuk, H., Zhao, R., Chester, S. A., & Zhao, X. (2018). Printing ferromagnetic domains for untethered fast-transforming soft materials. *Nature*, *558*(7709), 274–279. https://doi.org/10.1038/s41586-018-0185-0

Kuang, X., Liu, G., Dong, X., & Wang, D. (2016). Triple-shape memory epoxy based on Diels–Alder adduct molecular switch. *Polymer*, *84*, 1–9. https://doi.org/10.1016/j.polymer.2015.12.033

Kuang, X., Roach, D. J., Wu, J., Hamel, C. M., Ding, Z., Wang, T., Dunn, M. L., & Qi, H. J. (2019). Advances in 4D printing: Materials and applications. *Advanced Functional Materials*, *29*(2). https://doi.org/10.1002/adfm.201805290

Lai, J., Liu, Y., Lu, G., Yung, P., Wang, X., Tuan, R. S., & Li, Z. A. (2024). 4D bioprinting of programmed dynamic tissues. *Bioactive Materials*, *37*, 348–377. https://doi.org/10.1016/j.bioactmat.2024.03.033

Li, X., Shang, J., & Wang, Z. (2017). Intelligent materials: A review of applications in 4D printing. *Assembly Automation*, *37*(2), 170–185. https://doi.org/10.1108/AA-11-2015-093

Magalhães, M. I., & Almeida, A. P. C. (2023). Nature-inspired cellulose-based active materials: From 2D to 4D. *Applied Biosciences*, *2*(1), 94–114. https://doi.org/10.3390/applbiosci2010009

Manshor, M. R., Alli, Y. A., Anuar, H., Ejeromedoghene, O., Omotola, E. O., & Suhr, J. (2023). 4D printing: Historical evolution, computational insights and emerging applications. *Materials Science and Engineering B*, *295*, 116567. https://doi.org/10.1016/j.mseb.2023.116567

Mathur, V., Agarwal, P., Srinivasan, V., Panwar, A., & Vasanthan, K. S. (2023). Facet of 4D printing in biomedicine. *Journal of Materials Research*, *38*(1), 2–18. https://doi.org/10.1557/s43578-022-00779-9

McLellan, K., Sun, Y.-C., & Naguib, H. E. (2022). A review of 4D printing: Materials, structures, and designs towards the printing of biomedical wearable devices. *Bioprinting*, *27*, e00217. https://doi.org/10.1016/j.bprint.2022.e00217

Momeni, F.S. M., Mehdi Hassani.N., Liu, X., & Ni, J. (2017). A review of 4D printing. *Materials & Design*, *122*, 42–79. https://doi.org/10.1016/j.matdes.2017.02.068

Momeni, F., & Ni, J. (2020). Laws of 4D printing. *Engineering*, *6*(9), 1035–1055. https://doi.org/10.1016/j.eng.2020.01.015

Nam, S., & Pei, E. (2019). A taxonomy of shape-changing behavior for 4D printed parts using shape-memory polymers. *Progress in Additive Manufacturing*, *4*(2), 167–184. https://doi.org/10.1007/s40964-019-00079-5

Osouli-Bostanabad, K., Masalehdan, T., Kapsa, R. M. I., Quigley, A., Lalatsa, A., Bruggeman, K. F., Franks, S. J., Williams, R. J., & Nisbet, D. R. (2022). Traction of 3D and 4D printing in the healthcare industry: From drug delivery and analysis to regenerative medicine. *ACS Biomaterials Science & Engineering*, *8*(7), 2764–2797. https://doi.org/10.1021/acsbiomaterials.2c00094

Pei, E., & Loh, G. H. (2018). Technological considerations for 4D printing: an overview. *Progress in Additive Manufacturing*, *3*(1–2), 95–107. https://doi.org/10.1007/s40964-018-0047-1

Pisani, S., Genta, I., Modena, T., Dorati, R., Benazzo, M., & Conti, B.(2022). Shape-memory polymers hallmarks and their biomedical applications in the form of nanofibers. *International Journal of Molecular Sciences*, *23*(3), 1290. https://doi.org/10.3390/ijms23031290

Rafiee, M., Farahani, R. D., & Therriault, D. (2020). Multi-material 3D and 4D printing: A survey. *Advanced Science*, *7*(12). https://doi.org/10.1002/advs.201902307

Rastogi, P., & Kandasubramanian, B. (2019). Breakthrough in the printing tactics for stimuli-responsive materials: 4D printing. *Chemical Engineering Journal*, *366*, 264–304. https://doi.org/10.1016/j.cej.2019.02.085

Ren, L., Wang, Z., Ren, L., Xu, C., Li, B., Shi, Y., & Liu, Q. (2023). Understanding the role of process parameters in 4D printing: A review. *Composites Part B: Engineering*, *265*, 110938. https://doi.org/10.1016/j.compositesb.2023.110938

Roy, D., Cambre, J. N., & Sumerlin, B. S. (2010). Future perspectives and recent advances in stimuli-responsive materials. *Progress in Polymer Science*, *35*(1–2), 278–301. https://doi.org/10.1016/j.progpolymsci.2009.10.008

Sadraei, A., & Naghib, S. M. (2025). 4D printing of physical stimuli-responsive hydrogels for localized drug delivery and tissue engineering. *Polymer Reviews*, *65*(1), 104–168. https://doi.org/10.1080/15583724.2024.2427184

Shanthamma, S., Preethi, R., Moses, J. A., & Anandharamakrishnan, C. (2021). 4D printing of sago starch with turmeric blends: A study on pH-triggered spontaneous color transformation. *ACS Food Science & Technology*, *1*(4), 669–679. https://doi.org/10.1021/acsfoodscitech.0c00151

Shraim, A. M., Ahmed, T. A., Rahman, M. M., & Hijji, Y. M. (2021). Determination of total flavonoid content by aluminum chloride assay: A critical evaluation. *LWT*, *150*, 111932. https://doi.org/10.1016/j.lwt.2021.111932

Slavkovic, V., Palic, N., Milenkovic, S., Zivic, F., & Grujovic, N. (2023). Thermo-mechanical characterization of 4D-Printed biodegradable shape-memory scaffolds using four-axis 3D-printing system. *Materials*, *16*(14), 5186. https://doi.org/10.3390/ma16145186

Subeshan, B., Baddam, Y., & Asmatulu, E. (2021). Current progress of 4D-printing technology. *Progress in Additive Manufacturing*, *6*(3), 495–516. https://doi.org/10.1007/s40964-021-00182-6

Sun, L., Huang, W. M., Ding, Z., Zhao, Y., Wang, C. C., Purnawali, H., & Tang, C. (2012). Stimulus-responsive shape memory materials: A review. *Materials & Design*, *33*, 577–640. https://doi.org/10.1016/j.matdes.2011.04.065

Sydney Gladman, A., Matsumoto, E. A., Nuzzo, R. G., Mahadevan, L., & Lewis, J. A. (2016). Biomimetic 4D printing. *Nature Materials*, *15*(4), 413–418. https://doi.org/10.1038/nmat4544

Tee, Y. L., & Tran, P. (2021). On bioinspired 4D printing: Materials, design and potential applications. *Australian Journal of Mechanical Engineering*, *19*(5), 642–652. https://doi.org/10.1080/14484846.2021.1988434

Teng, X., Zhang, M., & Mujumdar, A. S. (2021). 4D printing: Recent advances and proposals in the food sector. *Trends in Food Science & Technology*, *110*, 349–363. https://doi.org/10.1016/j.tifs.2021.01.076

Teoh, J., An, J., Feng, X., Zhao, Y., Chua, C., & Liu, Y. (2018). Design and 4D printing of cross-folded origami structures: A preliminary investigation. *Materials*, *11*(3), 376. https://doi.org/10.3390/ma11030376

Tibbits, S. (2013). The emergence of "4D printing." *TED Conference*.

Tibbits, S. (2014). 4D printing: Multi-material shape change. *Architectural Design*, *84*(1), 116–121. https://doi.org/10.1002/ad.1710

Tsujimoto, T., Takayama, T., & Uyama, H. (2015). Biodegradable shape memory polymeric material from epoxidized soybean oil and polycaprolactone. *Polymers*, *7*(10), 2165–2174. https://doi.org/10.3390/polym7101506

van Manen, T., Janbaz, S., & Zadpoor, A. A. (2017). Programming 2D/3D shape-shifting with hobbyist 3D printers. *Materials Horizons*, *4*(6), 1064–1069. https://doi.org/10.1039/C7MH00269F

Wan, X., He, Y., Liu, Y., & Leng, J. (2022). 4D printing of multiple shape memory polymer and nanocomposites with biocompatible, programmable and selectively actuated properties. *Additive Manufacturing*, *53*, 102689. https://doi.org/10.1016/j.addma.2022.102689

Wu, J., Yuan, C., Ding, Z., Isakov, M., Mao, Y., Wang, T., Dunn, M. L., & Qi, H. J. (2016). Multi-shape active composites by 3D printing of digital shape memory polymers. *Scientific Reports*, *6*(1), 24224. https://doi.org/10.1038/srep24224

Yang, Q., Gao, B., & Xu, F. (2020). Recent advances in 4D bioprinting. *Biotechnology Journal*, *15*(1). https://doi.org/10.1002/biot.201900086

Yu, K., Dunn, M. L., & Qi, H. J. (2015). Digital manufacture of shape changing components. *Extreme Mechanics Letters*, *4*, 9–17. https://doi.org/10.1016/j.eml.2015.07.005

Zafar, M. Q., & Zhao, H. (2020). 4D printing: Future insight in additive manufacturing. *Metals and Materials International*, *26*(5), 564–585. https://doi.org/10.1007/s12540-019-00441-w

Zhang, Q., Zhang, K., & Hu, G. (2016). Smart three-dimensional lightweight structure triggered from a thin composite sheet via 3D printing technique. *Scientific Reports*, *6*(1), 22431. https://doi.org/10.1038/srep22431

Zhang, W., Wang, H., Wang, H., Chan, J. Y. E., Liu, H., Zhang, B., Zhang, Y.-F., Agarwal, K., Yang, X., Ranganath, A. S., Low, H. Y., Ge, Q., & Yang, J. K. W. (2021). Structural multi-colour invisible inks with submicron 4D printing of shape memory polymers. *Nature Communications*, *12*(1), 112. https://doi.org/10.1038/s41467-020-20300-2

Zhu, W., Webster, T. J., & Zhang, L. G. (2019). 4D printing smart biosystems for nanomedicine. *Nanomedicine*, *14*(13), 1643–1645. https://doi.org/10.2217/nnm-2019-0134

5

Expanding Avenues in 4D Printing Technology

CHAPTER MENU

5.1 Introduction

Research and commercial interest in 4D printing technology is on the rise, and the field has already attracted significant funding and initiatives across various research and technology sectors. Importantly, materials can remarkably adapt to changing environmental conditions, such as humidity, temperature, and pressure, while also catering to user needs. Additionally, 4D-printed products may be designed to occupy less space, making shipping and storage more efficient. That is, printed structures facilitate transportation at a decreased cost and with diminished human resources requirements. Once used, these items can be reconfigured to meet new demands, and this is supported by their unique self-repair capabilities (Haleem et al., 2021; Wang et al., 2019). In the future, self-actuators produced

4D Printing of Foods: Fundamentals, Applications and Prospects, First Edition.
C. Anandharamakrishnan, Jeyan A. Moses, and P. Santhoshkumar.

using additive manufacturing may enable the development of smart devices devoid of batteries and external energy sources.

Accordingly, there have been numerous applications of 4D printing technology that have been reported in order to achieve unique capabilities in the manufacturing sector. It can facilitate designing and developing tool for practical purposes, for instance, to support essential services such as medical treatments and healthcare. In the current competitive landscape of manufacturing and its subdomains, emerging 4D printing technologies stand out as a promising candidate, given their remarkable features such as accuracy, precision, and quality responses to complex industrial problems. 4D printing technology and its services have diverse and innovative applications in a variety of industrial sectors, such as meta-materials, origami structures, smart textiles, adaptive scaffolds, tissue engineering, drug delivery systems, self-healing hydrogel, bio-origami, optical devices, pressure sensors, flexible devices, wearable electronics, edible electronics, microfluidic devices, food processing, coating protection, soft robotics, and smart grippers (Khalid et al., 2022; Santhoshkumar et al., 2024). Further, owing to its unique benefits and intricate design flexibility, the technology is set to find newer avenues in the future across different industry segments. Accordingly, in the upcoming sections of this chapter, such applications, spanning different application sectors, have been presented, highlighting fundamental concepts and emphasizing the wide capabilities of the 4D printing approach, presenting the scope for new technological applications.

5.2 Sector-Specific Applications

4D printing, an innovative progression of 3D printing, incorporates the dimension of time, allowing printed materials or objects to alter their shape, properties, or function in response to external stimuli, including heat, light, humidity, or pressure. This innovative technology has been implemented in various industries, providing distinctive benefits in adaptability, efficiency, and design. The following is a comprehensive description of sector-specific applications, highlighting the state of the art.

5.2.1 Aerospace and Aeronautics

Producing exceptionally precise components is definitely important in the aerospace industry for flight and also for space missions. The parts must be durable and also cost effective. Therefore, 4D printing has discovered many potential applications within space programs, all possessing great precision and importance. 4D-printed structures have great capabilities for space missions, as they can be designed to withstand harsh ecological conditions (Kang et al., 2024). The self-assembly capability in the space of many 4D-printed parts for satellites and aircraft makes the entire assembly process simpler. Furthermore, zero issues will arise from the deformations of those structures in space. These traits render them quite flexible and efficient for applications in space (Zhang et al., 2021). For example, the International Space Station (ISS) has successfully printed many 3D parts from plastic in microgravity conditions. This shows a large potential for these materials in all future space missions (Santhoshkumar et al., 2023). Components can be quickly printed directly in spacecraft, reducing the need for earth stations to supply key parts. Self-assembling 4D-printed structures could transform mission capabilities through the use of ecological conditions, an important aspect of space missions when self-sustaining materials are used. Certain materials usually work well to create self-sustaining parts, along with tools for satellites as well as space exploration. The mass of components can be reduced by up to 80% due to the lightweight nature of 4D-printed structures (Ahmed et al., 2021). Solar cells and antennas for satellites and spacecraft could also be created via 4D printing, allowing self-assembly in space and lower labor costs. Additionally, these materials can adapt to the varying conditions of the space environment, enhancing functionality and efficiency (Ellery and Elaskri 2025). In the future, 4D printing may facilitate the creation of self-deployable components for space systems, such as hinges, reflector antennas, flexible lunar homes, and truss booms. Moreover, in harsh space settings, spacesuits must maintain a pressurized atmosphere with adequate oxygen and effective carbon dioxide elimination. Supplementary needs encompass temperature regulation, micrometeoroid protection, radiation shielding, visibility enhancement, and facilitation of movement for essential mission activities. Contemporary spacesuit designs, although theoretically sufficient, frequently hinder astronaut mobility. 4D printing technology has the capability to produce prototype spacesuits and garments that improve performance and comfort.

Likewise, the aviation sector encounters stringent safety restrictions concerning materials utilized in aircraft interiors. These materials must possess flame-retardant, wear-resistant, stain-resistant, nontoxic properties and be as lightweight as feasible. Researchers are developing self-illuminating fabrics that integrate metal filaments, enabling them to generate light when electrified. Utilizing these fabrics for cabin ceiling upholstery could diminish the total weight of conventional lighting fixtures (Biswas et al., 2021). Moreover, 4D-printed composite springs in automobiles have demonstrated performance on par with conventional springs (Ahmed et al., 2021). In the past decade, ultrasonic additive manufacturing, with concepts integrating metals and advanced materials to fabricate intelligent structures, facilitating the development of smart airplanes, has evolved (Bournias-Varotsis et al., 2018). Substantial attempts are underway to incorporate active materials into aviation technology.

5.2.2 Fashion and Textiles

In the textile sector, 4D printing can generate fabrics that adjust to fluctuating ambient conditions, improving comfort and ventilation. These intelligent textiles may modify their color and texture in response to the surrounding environment by employing shape-memory materials (Biswas et al., 2021). In the film and fashion industries, 3D-printed clothing is becoming increasingly popular. Similarly, 4D printing has the potential to enhance the responsiveness of products to fluctuations in temperature and humidity. For example, 4D-printed footwear can respond to environmental variables (Manaia et al., 2023).

Despite numerous 4D-printed fashion and clothing goods on commercial platforms, scholarly publications regarding these products are relatively few due to commercial interests. Overall, 4D printing technology can produce a wide variety of products, including

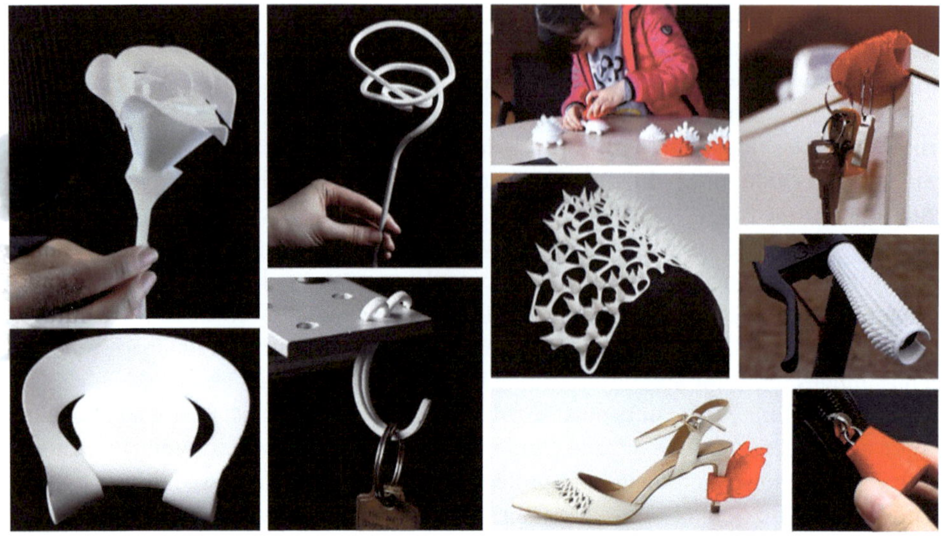

Figure 5.1 Photographic images of some 4D-printed household devices. *Source:* Joharji et al. (2022) / with permission of Elsevier.

footwear, apparel, jewelry, and accessories (Aldawood, 2023). Adidas is exploring the scope of using 4D printing to create lightweight, tailored sporting footwear (Mahmood et al., 2023). The mechanics, controllability, geometry, and memory attributes required to fabricate diverse shapes (Sun et al., 2020), including flowers (An et al., 2018) and chairs, for instance, 4D printed products are presented in Figure 5.1.

5.2.3 Defense

The army and other first responders, including firefighters, frequently function in very adverse situations, affected by climatic conditions and possible chemical or biological hazards. The US military is investigating the application of 4D printing for textiles to improve military functions. Recent works include the development of 4D-printed coatings for military vehicles with capabilities to modify shape according to different weather conditions. Moreover, soldier uniforms fabricated with 4D-printed materials can alter upon contact with sharp objects to improve safety (Ahmed et al., 2021). Similarly, materials are being engineered to respond to environmental conditions by changing color for camouflage, bending light to obscure soldiers, and modifying permeability or hardness for different defense applications. These innovations not only protect soldiers from severe conditions but also render them less noticeable to enemies (Ramezani & Mohd Ripin, 2023).

Furthermore, cutting-edge stimuli-responsive materials have the potential to transform the defense sector, revolutionizing uniforms, weaponry, vehicles, and even submarines. Future warriors may be outfitted with lightweight, intelligent weaponry that incorporates defensive shields, which engage solely when required (Zafar & Zhao, 2020). This technology may also be utilized for deployable and adaptive constructions such as self-deploying tents, robotic parachutes, and assistive gear, all of which are essential for military applications (Biswas et al., 2021; Li et al., 2025).

5.2.4 Ceramics and Construction

4D printing can be employed to rectify construction defects by allowing structures to self-assemble in reaction to particular stimuli. The self-repairing properties of 4D-printed materials may facilitate the creation of self-healing structures, hence improving their durability and longevity. Moreover, 4D printing can facilitate the development of buildings that emphasize sustainability, productivity, energy efficiency, and environmental friendliness with reduced lead time. Figure 5.2 shows an overview of 4D-printed living structures. A novel application is the intelligent water pipeline system, which modifies its diameter according to water pressure, facilitating expansion or contraction as required. Investigating the capabilities of 4D printing in engineered living materials may result in transformative advancements in military tactics and infrastructure development in isolated regions, employing self-repairing and self-adaptive materials capable of enduring severe environments (Niazy et al., 2024; Subeshan et al., 2021; Sun et al., 2021). Recently, the benefits and potential applications of 4D printing in art, architecture, and construction were exclusively documented by Niazy et al. (2024). Overall, there is a huge possibility to achieve sustainable development goals through 4D printing approaches in this construction field. Particularly, the application of 4D printing in the concrete sector remains relatively underexplored; however, it holds considerable potential for future studies and development.

5.2.5 Renewable Energy

4D printing can significantly contribute to the fabrication of renewable energy field components, such as wind turbine blades, eliminating the necessity for traditional electromechanical systems (actuators and sensors), hence enhancing control and efficiency in energy generation (Han et al., 2020; Momeni et al., 2019). For instance, an advanced photovoltaic system has been created that adapts its shape according to sunlight availability, hence maximizing solar energy production (Momeni & Ni, 2018). In several applications, 4D printing possesses the capability to mitigate the issues of low efficiency and elevated costs in renewable energy (Aldawood, 2023). At present, its applications in this domain, including the manufacturing of wind turbine blades and the production of solar cells, remain relatively underdeveloped in comparison to other sectors of 4D printing.

5.3 Applications in Life Sciences

Unlike several application sectors, life sciences and integration with biological aspects of living beings present new challenges and opportunities. 4D printing is emerging as a revolutionary advancement in manufacturing, especially within the life sciences, where flexible, responsive structures can greatly improve medicinal and biological applications. For instance, 4D printing can utilize materials capable of altering their characteristics, shape, or usefulness over time in reaction to environmental stimuli such as heat, moisture, pH, or

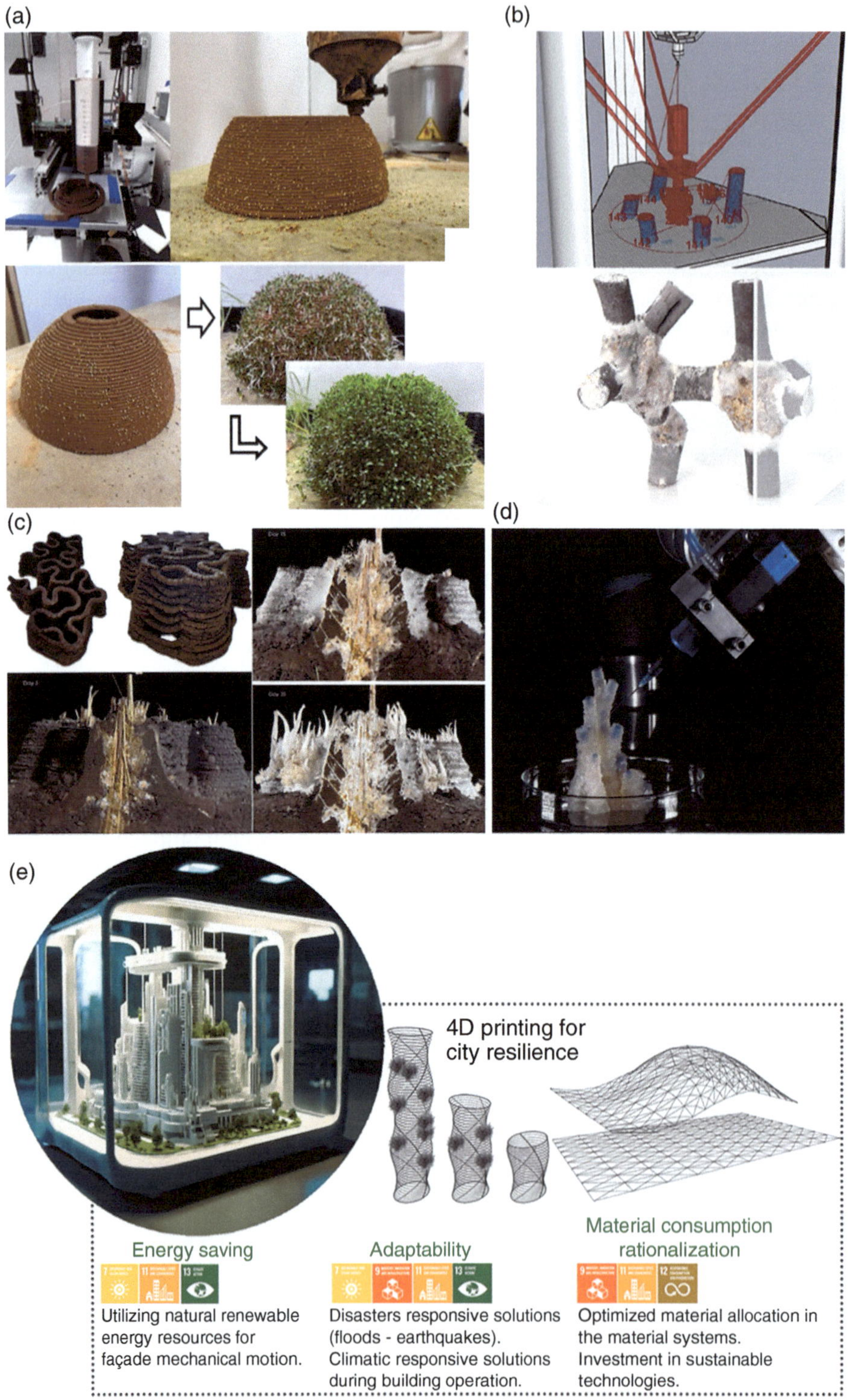

Figure 5.2 4D-printed constructions. (a) Earthen walls, (b) living joint with cells, (c) growth of walls for soil remedy, (d) illuminating wall, and (e) construction field alignment to UN-SDGs. *Source:* Niazy et al. (2024) / Elsevier / CC BY 4.0.

light. Accordingly, this section exclusively focuses on different applications reported in the field of life sciences, emphasizing recent advances while highlighting challenges and current research needs.

5.3.1 Drug Delivery

Targeted drug delivery refers to delivering drugs to a specific location inside the body with localized and effective treatment and diminished side effects. 3D printing has been uniquely applied to delivering pharmaceutical drugs (Muhindo et al., 2023) and nutraceuticals (Kavimughil et al., 2023) inside the human body. Some other studies have also described the adoption of multiple printing avenues, particularly in drug delivery. Through 4D printing, devices are produced by incorporating pharmaceutical entities and releasing the same on specific environmental inputs to targeted areas within the human body. A classic example is 4D-printed containers designed for drug storage and controlled release. Shape memory polymers and hydrogels have been utilized for drug encapsulation and subsequent release in biomedical applications (Joshi et al., 2020b; Tran et al., 2022; Zu et al., 2022), as depicted in Figure 5.3. The shape memory and recovery mechanisms of these drug delivery systems are engineered specifically to enhance the efficacy of drug release.

The principles of shape evolution and self-assembly are utilized in drug delivery systems across nano-, micro-, and macro-scales, including human applications. The self-assembly of molecules with diverse chemical properties such as lipids, peptides, and polymers – yields supramolecular structures such as micelles, liposomes, and nanoparticles. These structures are recognized for their capacity to encapsulate active compounds, illustrating shape alterations at the nanoscale (Jing et al., 2022). Typically, the basic requirements to develop 4D-printed drugs are as follows: the prompt achievement of the retentive configuration after dosing, maintenance over the desired time also despite the mechanical stresses associated with smooth muscle contraction undergone by the device during its residence in the organ, release of the drug in a rate controlled way, and spontaneous removal from the retaining organ after exhaustion (Gazzaniga et al., 2023). Moreover, some initial 4D-printed constructs have been created as drug-infused scaffolds or implantable devices. These products utilize *in vivo* alterations in shape and size to enable little surgical intervention for body insertion. These systems exist at the convergence of medical devices, scaffolds, and drug delivery systems, arising from the increasing inclination toward minimally invasive surgery without focusing on a particular administration location or therapeutic objective.

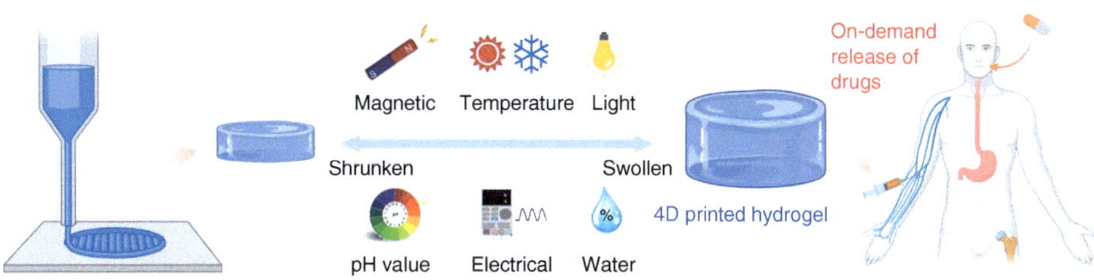

Figure 5.3 4D printing in drug delivery: stimulus responsiveness toward 4D-printed hydrogel. *Source:* Tran et al. (2022) / MDPI / CC BY 4.0.

Subsequently, to alleviate the discomfort associated with medication and improve patient adherence to complex therapies, drug delivery systems that allow for less frequent dosing are crucial. In this context, 4D printing has been utilized to create various organ-retentive devices designed to be positioned within hollow organs, releasing drugs in a controlled manner. For instance, theragrippers are gastrointestinal-resident mini- or microdevices made up of multilayer structures equipped with sharp microtips that can latch onto mucosal tissue in response to specific stimuli (Gazzaniga et al., 2023; Ghosh et al., 2020).

Recently, 4D printing has been applied to develop innovative drug delivery systems for different administration routes designed so as to retain their shape within target organs while providing controlled drug release behaviors. Examples of capabilities for varying delivery sites such as gastroretentive, esophageal, and intravesical delivery are presented in Figure 5.4 and 5.5. The potential of 4D printing for such demanding applications has been recently reviewed by Mahmoud & Schulz-Siegmund (2023). These approaches have recently booming in this field, and until now, there have not been major reports demonstrating human trials assessing the biological performance of such 4D printing drugs.

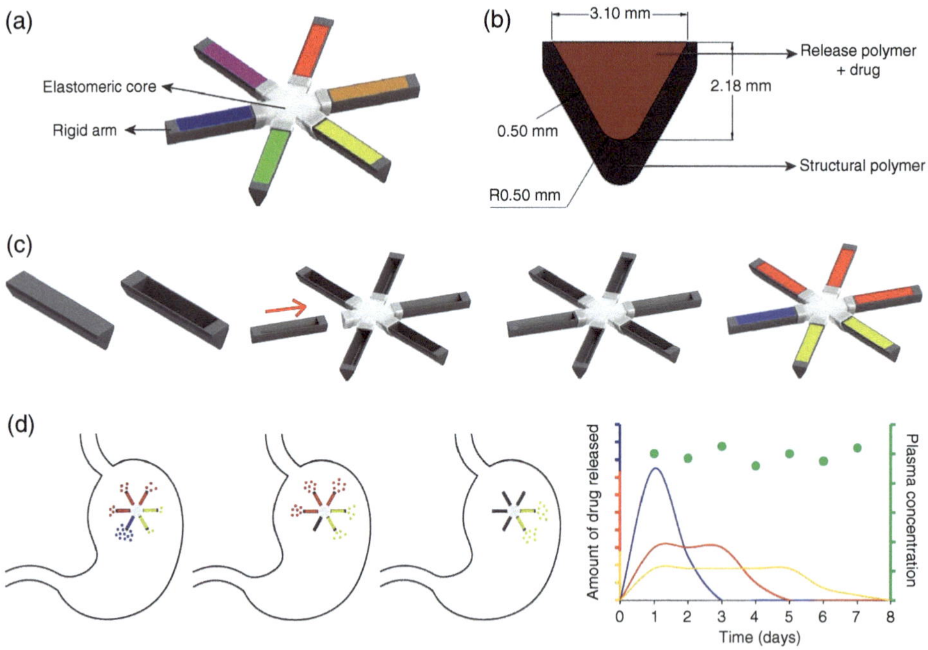

Figure 5.4 Oral long-acting antiretrovirals – shape-shifting mechanism. (a) Design of the gastric resident dosage forms (dosage form consists of an elastomeric core (gray) and six drug-loaded arms (multicolored)), (b) cross section of the arm (outer sleeve of the arms is made of a rigid structural polymer, which provides the arm its mechanical strength, and this sleeve is then filled with a drug-polymer matrix that releases the drug at a desired rate), (c) manufacturing scheme of the dosage form, (d) expected performance of the dosage form *in vivo*. (The dosage form is loaded with three different polymers (blue, red, and yellow), which release the drug at different rates. Selection of appropriate polymers may result in almost constant and sustained plasma drug concentrations. It should be noted that (d) is a schematic representing an ideal system and is not experimentally obtained data). *Source:* Kirtane et al. (2018) / Springer Nature / CC BY 4.0.

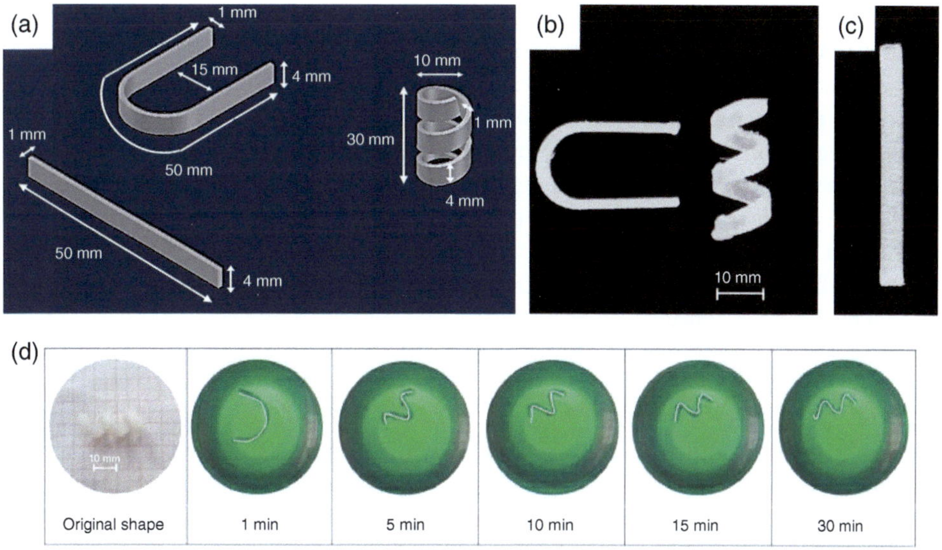

Figure 5.5 Intravesical drug delivery system – shape-shifting configurations. (a) Virtual model, (b) prototype, (c) programmed temporary shape, and (d) shape recovery observation at 37°C (shapes: helix (original), I-shape (temporary)). *Source:* Melocchi et al. (2019a) / with permission of Elsevier.

The urinary bladder and stomach are particularly appealing areas for 4D-printed devices because of the various obstacles that remain in drug administration to these anatomical sites (Palugan et al., 2021; Uboldi et al., 2022). Accordingly, there has been a rising focus on developing newer solutions for futuristic intravesical drug delivery applications. In a different line, gastroretentive systems can be developed as an advancement over existing expandable devices, typically utilizing mechanically driven mechanisms. Such designs intend to facilitate retention in the stomach during the inner digestive phase, featuring dimensions of at least 15 mm, a size commonly associated with an open pylorus. Additionally, the temporary shape of the system can be designed for compatibility with swallowing, making it suitable for insertion into commercially available hard-gelatin capsules. Once in the stomach, such devices can be expected to revert to their larger configuration upon contact with gastric fluids at body temperature after the capsule gets dissolved (Gazzaniga et al., 2023).

As of date, the most popularly used materials for drug delivery applications are polymers and hydrogels, especially for 4D printing approaches. Herein, the expansion in size can be obtained through swelling or transformation (e.g., hydrogel) of the temporary structures of expandable systems in response to particular stimuli in the stomach. In this instance, systems can be limited to a collapsed state, for example, folding during manufacture (Awasthi & Kulkarni, 2016). 4D printing methodologies, especially in drug delivery applications, represent a significant advancement in the field, offering innovative processes that prompt further development. Given this, in the last few years, numerous studies employing these novel approaches have been reported by researchers globally (Mahmoud & Schulz-Siegmund, 2023; Melocchi et al., 2019b; Osouli-Bostanabad et al., 2022; Pingale et al., 2023; Shi et al., 2019; Tran et al., 2022; Zhang et al., 2019).

4D bioprinting is one of the newer developments in this area, and it has made a tremendous impact; synergistic benefits of both methods can result in breakthrough technologies. New drug studies have utilized 4D bioprinting technology, possibly to report considerable synergistic effects enhancing drug delivery application. In this regard, biomimetic 3D/4D printing regulations for drug delivery purposes have been reported solely by Dowaidar (2024). The use of smart materials as biomaterials opens up previously unlimited possibilities for the creation of active pharmaceuticals (Agarwal et al., 2023a,b; Kim et al., 2024; Lukin et al., 2019; Moheb Afzali et al., 2024; Osouli-Bostanabad et al., 2022). However, several studies are further needed, including those focusing on human trials of these newly created medications or drugs.

5.3.2 Tissue Engineering

With its many uses in the medical field, especially in tissue engineering, 4D printing is becoming a fascinating technique. For example, this method makes it possible to implant living tissues into people, encouraging functional reactions through the effects of stimuli. Intelligent materials and 3D printing have led to the development of 4D printing, which makes it possible to create scaffolds that almost exactly mimic the dynamic characteristics of real tissues. These bioinspired materials, which include hydrogels and shape memory polymers, are widely used to create these active scaffolds. According to Hasanzadeh et al. (2023) and Zeenat et al. (2023), their dynamic properties enable a variety of applications, including tissue regeneration, self-assembly, and self-recovery. In addition, recent research has underscored the promising potential of shape memory scaffolds for minimally invasive administration of functional tissues (Chopra et al., 2024; Kalogeropoulou et al., 2024; Soleymani & Naghib, 2023; Zhou et al., 2023). As shown in Figure 5.6, these scaffolds can be integrated with bioelectronics and biodegradable components to enable precise, wireless control of the distribution process. One significant advancement in tissue regeneration is the direct implantation of stem cells into biological scaffolds using 4D printing and biotechnology or 4D bioprinting.

Avoiding certain stimuli, such as high temperatures and extremely high pH levels, is essential when working with living cells in tissue engineering (Chu et al., 2020). To advance in this field, 5D printing was introduced. Unlike 3D/4D printing, 5D printing acts as intelligent technology providing capabilities to communicate with the surroundings and cause positive changes (Lai & Wang, 2023). The use of 4D printing in tissue engineering is presently in the proof-of-concept phase, and considerable advancements are required before it can be routinely utilized in clinical environments.

5.3.3 Bioprinting

The swift rise in the global population and the issues associated with the aging population have led to a growing demand for medical developments and organ availability. In this context, 4D bioprinting facilitates the precise deposition of biomaterials and physiologically active elements that can undergo morphological transformations over time. These temporal transitions, for instance, enable the folding and unfolding of encapsulated entities, along with the regulated encapsulation and release of pharmaceuticals or cells in a

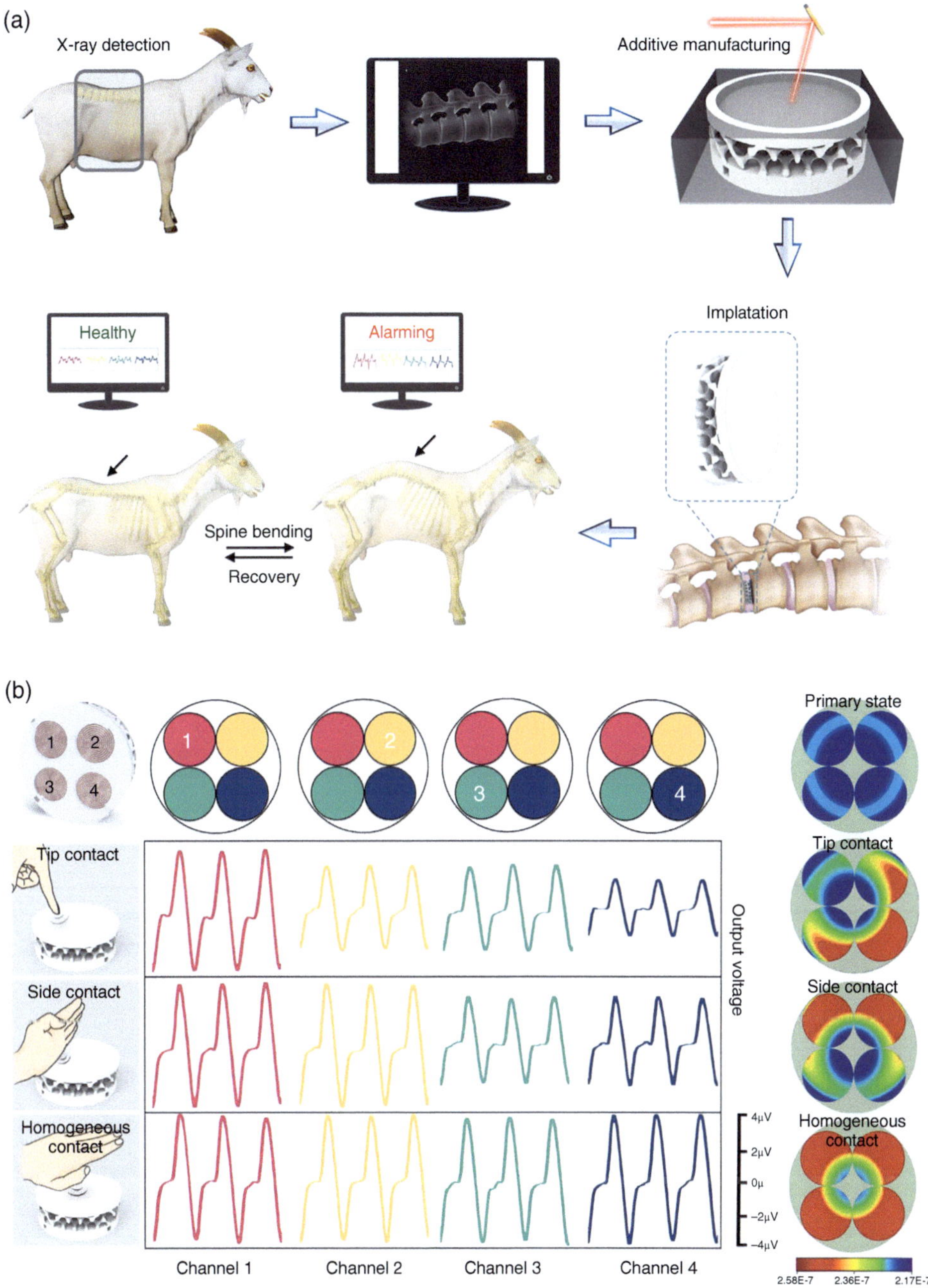

Figure 5.6 (a) Schematic illustrations of a 4D-printed scaffold implanted in a goat's spine to monitor, (b) a simulation model for different compressive forces on a scaffold. *Source:* Shi et al. (2022) / with permission of Elsevier.

programmable manner. This technique facilitates the development of biologically pertinent 3D structures that mimic the dynamic alterations occurring in the human body or regulate the prolonged release of encapsulated pharmaceuticals (Taylor et al., 2024).

4D bioprinting must support dynamic processes at every stage of the printing process to produce complex, dynamic tissues and organs both *in vitro* and *in situ* (Lai et al., 2024; Murphy et al., 2019). To respond to environmental stimuli such as light, voltage, temperature, magnetic fields, and pH, 4D bioprinting uses intelligent materials, such as shape memory polymers. To achieve the necessary architecture and functionality in 4D-printed objects, stimuli-responsive materials are highly sought after. Natural and synthetic biomaterials are designed to interact with living tissues and must be safe and nontoxic for recipients. As potential 4D printing possibilities, several smart biomaterials have been investigated; each exhibits unique properties and sensitivity to different stimuli (Kim et al., 2024).

A schematic summary of the 4D bioprinting process and the associated challenges are illustrated in Figure 5.7. A bioprinting approach utilizing a bilayer architecture facilitates

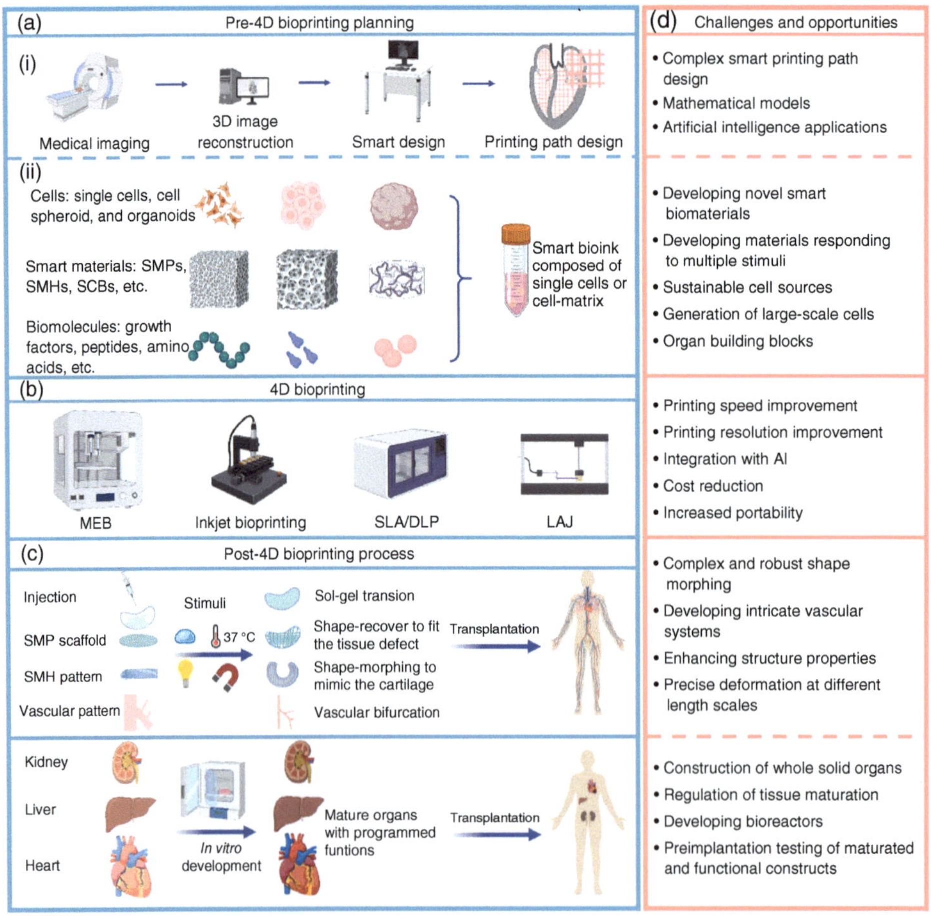

Figure 5.7 4D bioprinting of dynamic tissues and organs. *Source:* Lai et al. (2024) / with permission of Elsevier / CC BY-NC-ND 4.0.

shape morphing in three-dimensional constructions comprising an actuation layer for driving transformation and a biodegradable, photocurable microgel layer that supports cellular growth. The microgel layer facilitates the printing of a cell-exclusive bioink that generates a condensate, ultimately resulting in the development of mature tissue as it degrades (Zeenat et al., 2023). In the context of the fabrication of blood vessels using 3D bioprinting, obstacles concerning material properties, dimensions, and structural integrity of the resultant conduits have been encountered. Interestingly, 4D bioprinting mitigates these challenges by facilitating the fabrication of flat constructs that metamorphose into conduits upon external stimulation. An exemplary case is the utilization of cell traction forces to deform sheets into tubular configurations, demonstrating the evolution of shape and functioning in constructs (Lai et al., 2024). These scaffold-free methods of cell condensation eliminate the necessity for synchronization between hydrogel degradation and tissue development, facilitating programmable modification of the construct via shape transformation and tissue maturation (Ding et al., 2022).

Moreover, research has demonstrated the scope of encapsulating brain cells within convoluted renal proximal tubules on perfusable chips (Homan et al., 2016). The issues caused by conventional implantable devices, especially stents, can be overcome by bioprinting approaches. For instance, stents, small tubular devices used for the treatment of narrowed vessels, are frequently constructed from shape-memory alloys such as nitinol. Nonetheless, these alloys exhibit disadvantages, such as an elevated incidence of restenosis and complications related to cellular regeneration and thrombus development. Shape-memory polymers offer a superior option owing to their biocompatibility, biodegradability, and capacity for shape modulation (Petrini & Migliavacca, 2011).

Although 4D bioprinting offers the potential for manipulating the morphology and dimensions of constructs over time, it is still an evolving technique that encounters numerous obstacles requiring resolution to improve its utility in tissue engineering and the biomedical fields. A significant problem is the optimization of stimuli-responsive biomaterials for bioprinting (Rahimnejad et al., 2024). Furthermore, modern shape transformations such as folding and assembly might not adequately meet the complex requirements of biological systems. These materials need to retain their properties over time for long-term applications, but very few structures can return to their former shape under specific circumstances. Moreover, the mechanical characteristics of existing biomaterials often fail to withstand the high pressures associated with ongoing stimuli-responsive changes (Ionov, 2018). The field of biomedical engineering field has adopted several such innovative concepts, and details of advances have been exclusively documented by many researchers (Agarwal et al., 2023; Noroozi et al., 2023; Pourmasoumi et al., 2023; Zhou et al., 2023).

5.3.4 Biomedical and Healthcare

Medical sciences hold significant promise for the application of 4D-printed structures. For instance, 4D-printed stents can be inserted into the body through a small incision, and, once they reach the target site, they can change shape in response to external stimuli, achieving the desired therapeutic effect (He et al., 2023). Additionally, studies have demonstrated that nano 4D-printed devices can be introduced into the body with the capabilities to self-assemble upon reaching the intended location, enabling specific medical interventions to be performed effectively (Bajpai et al., 2020; Zhou et al., 2021). In summary, their fabrication entails many intricate steps because stents have complex, patient-specific geometries. The development of 4D printing facilitates the quick production of customized stents. These stents' shape memory properties make implantation less invasive from a surgical standpoint. For ease of insertion and accurate placement, stents can be programmed to a smaller diameter after being initially manufactured at their final diameter. The stent returns to its initial diameter at body temperature after implantation (Hatami et al., 2024).

Geometry is crucial in designing biomedical devices, such as splints, where proper fit is essential for functionality. However, conventional fabrication methods often find it challenging to handle large subject growth over time. 4D printing presents an innovative solution, allowing the adaptation of devices to account for changes in size as growth occurs (Javaid & Haleem, 2020). Similarly, nerve guidance conduits, tubular structures designed to facilitate nerve regeneration, must satisfy various physical, chemical, and biological criteria to promote effective tissue formation. Furthermore, the shape memory characteristics of the shape memory materials provide the axial tension necessary to support nerve regrowth (Cui et al., 2024). In addition, comparable approaches have been used to create scaffolds for smart-implantation tissue engineering applications. Such scaffolds have shown outstanding shape-changing capabilities and superior human cell adhesion and proliferation (Qu et al., 2023; Zhang et al., 2023). In addition to these uses, 4D printing has been employed in biomedical applications such as medical devices, orthopedics, dentistry, smart implants and prostheses, rehabilitation devices, diagnostic tools, responsive surgical equipment, and regenerative medicine. Several researchers have documented several applications of 4D printing, with an emphasis on the design and manufacture of medical devices and healthcare packages (Afzali Naniz et al., 2022; Agarwal et al., 2021; Javaid & Haleem, 2019; Sajjad et al., 2024).

5.4 Interdisciplinary Approaches

4D printing, which integrates 3D printing with temporal shape or property modifications, also gains strength through its versatility to handle interdisciplinary concepts. These approaches may integrate multiple disciplines to address this cutting-edge technology's scientific, technological, and application-oriented challenges. Considering current trends, this section overviews the multidisciplinary approaches driving 4D printing.

5.4.1 Soft Robotics

An intriguing area for 4D printing applications is soft robotics. It is possible to create programmable transformational shape-memory materials with actuation capabilities. These materials have many uses, from programmable robotics designed for difficult situations to medical-delivery devices. They are especially beneficial in severe weather conditions such as floods or snow and uninhabitable environments such as high altitudes and outer space. Concepts of soft robotics have frequently been influenced by biological systems that have evolved to adapt to their surroundings. These materials can mimic the functions of living things by taking on structures that resemble artificial entities that function similarly to biological ones because of their adaptability and reactivity (Mahmood et al., 2023). Technological improvements are facilitated by the substantial attention given to soft robotics and actuators. In robotics, 4D-printed structures can replace expensive parts such as motors, sensors, and intricate assembling materials. Additionally, these materials can be used to make robots smaller (Soleimanzadeh et al., 2023).

Conventional robotics, which are made of rigid materials, struggle to perform delicate and flexible jobs such as mimicking the intricate movements of an octopus tentacle or a human hand's grasp. These limitations have encouraged the development of soft robotics, which uses particular soft materials, primarily elastomers, to interface robots and their environment. Compared to traditional robots, these soft substances exhibit improved resilience to damaging forces and allow for delicate interactions with delicate objects. Soft actuators or grippers, for instance, are inexpensive, have simple production methods, and are environmentally friendly in this field. For example, it was stated that a 13 g artificial gripper could lift 1 kg of weight (Miriyev et al., 2017).

Similarly, several other applications have been reported in diverse fields. For example, 4D printing soft robotics, along with machine learning studies, were reported (Zolfagharian et al., 2021), and nanoparticles have been used for 4D-printed hybrid robots in liquid metal applications (Huang et al., 2024). In addition, soft robots hold benefits in the medical field (Hann et al., 2020), and environmental applications to monitor the climate (Cecchini et al., 2023) were reported, as earlier mentioned. Nonetheless, their form renders them less appropriate for traditional robotic applications that emphasize speed and accuracy. Conversely, shape-memory materials, which can be activated by diverse inputs, may prove beneficial as actuators in specific applications. Figure 5.8 represents different varieties of 4D-printed soft robots.

In the future, soft robotics can potentially substitute rigid robotic frameworks with intelligent, highly responsive elements. Their exceptional adaptability makes them suitable for a wide array of jobs in challenging environments. Current advancements in modeling and printing technologies seek to augment the significance of soft robotics in the contemporary industrial sector. Anticipated advancements will derive from the incorporation of artificial intelligence (AI) and machine learning (ML), resulting in the development of highly versatile, self-learning robotic systems.

5.4.2 Sensors and Electronics

4D printing is promising for creating advanced sensors, as printed structures may react to diverse stimuli, including humidity, temperature, stress, strain, and pH levels (Mahmud et al., 2021). This responsiveness enables them to operate efficiently as sensors for various environmental alterations. They are economically efficient, yielding lightweight, sensitive, precise, and highly responsive structures, enhancing sensors' sophistication and futurism (Mahmood et al., 2023). Additionally, the concept can be used to manufacture circuits (Joshi et al., 2020a), flexible electronics (Xin et al., 2020), wearable electronics (McLellan et al., 2022), implantable electronics (Mahmud et al., 2021), and edible electronics (Santhoshkumar et al., 2024) fabrications. These devices facilitate electronics integration into daily life and are essential in fields such as implantable bioelectronics, artificial skin, and health monitoring. In this regard, in recent years, researchers have reviewed and

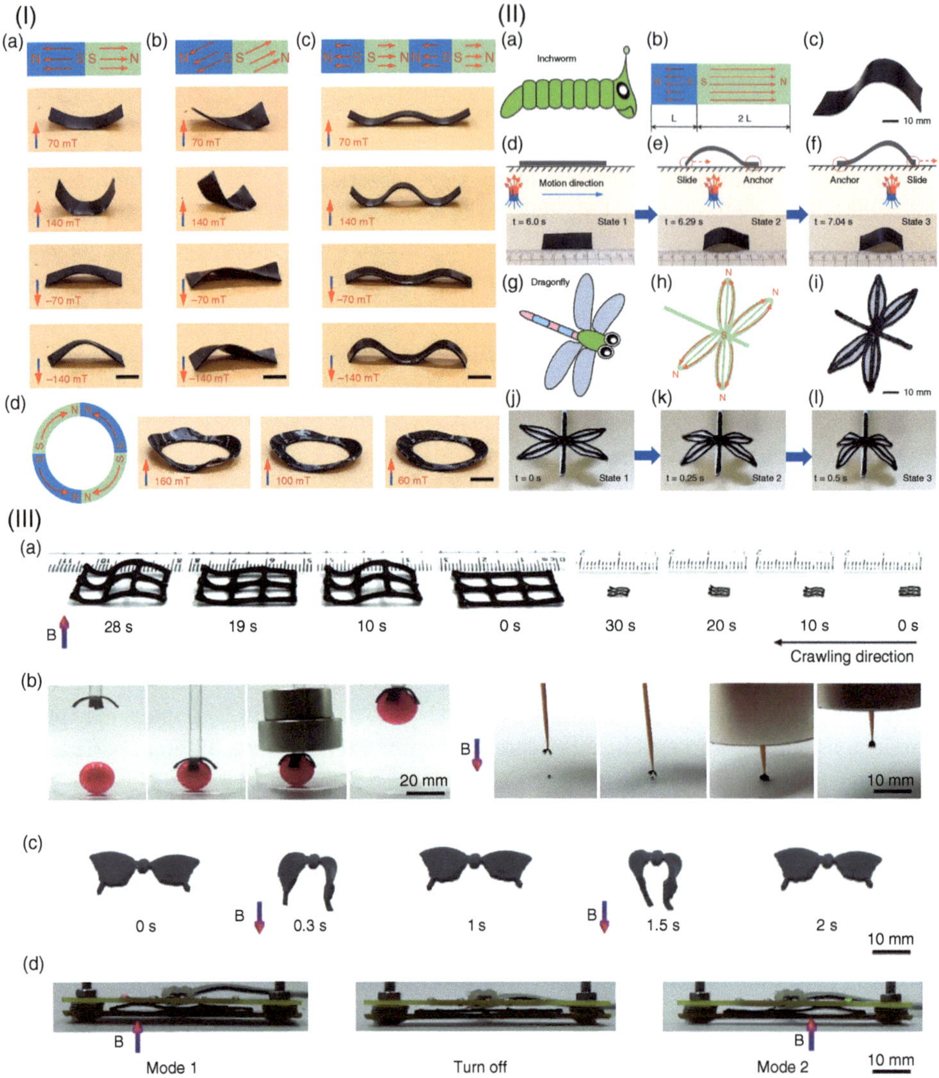

Figure 5.8 (I) Magneto response – soft actuators (a to d); (II) bionic soft robots: inchworm robot (a–to f), dragonfly robot (g– to j) *Source:* Wang et al., 2022 / with permission of Elsevier; (III) smart material-based robot: soft crawling robots (a), flexible gripper (b), bionic butterfly (c), magnetic switch with multistate (d) *Source:* Zhu et al., 2022 / with permission of Elsevier.

documented the concepts and emerging applications of soft and sustainable electronics fabrication through 4D printing (Han & Yoon, 2021; Qiao et al., 2023; Tan et al., 2023; Wang et al., 2021).

Similarly, copious works have been reported in sensor fabrications, creating higher accuracy objects, and reporting higher efficient results of the printed sensors. These include biosensors, biodegradable, self-powered, and flexible sensor fabrications. A notable advancement propelled by wearable technology is the development of customizable apparel. Self-healing circuits for illumination, audio, and sensing can be integrated into textiles, markedly prolonging the durability of wearable electronics by rapidly addressing damage during high-stress activities (McLellan et al., 2022). Current research is focused on sustainable energy production using solar cells, supercapacitors, lithium-ion batteries, and self-healing grids (Wang et al., 2013; Wang et al., 2017). Emerging advances not only streamline the design but also diminish the expense of soft electronic gadgets. However, to date, there is limited knowledge on adapting these fabrication methods for edible electronics. While there has been some research into the use of edible materials, fully edible electronic devices remain largely unexplored. Additionally, the selection of materials and circuit design poses challenges that must be addressed to enable practical applications. Further research is needed to investigate new materials and advancements in edible electronics.

5.4.3 Microfluidics

Microfluidics refers to systems that control extremely small fluid volumes through extremely small channels, which range in size from tens to hundreds of micrometers. The idea has several applications in areas such as diagnosis, biomolecular and cell analysis, high-throughput screening, and therapy (Bhusal et al., 2024). For microfluidic devices to operate and be managed as efficiently as possible, low power consumption, cost-effectiveness, and unrestricted shrinking are essential. These objectives can be accomplished with the aid of the creative micro and smart components (Chu et al., 2020). Microfluidic systems that use soft and stimulus-responsive hydrogels are promising for future advancements. Such hydrogels can accurately control the flow of fluids by modifying their shape in reaction to outside forces (D'Eramo et al., 2018). Recently, a smart hydrogel actuator was created by researchers for this use, with capabilities to control microfluidic systems using pH variations (Beck et al., 2023). Figure 5.9 illustrates examples of microfluidic devices for 4D printing applications. At the microscale level, a recent study reported employing 4D printing to accomplish unidirectional liquid movement at a greater angle (0°–180°). Bioinspired structures can be created on both soft and stiff substrates by producing these ultra-wide-angle liquid transports in three dimensions. This characteristic facilitates the creation of microfluidic devices, such as scalable oil-water separation, large-area evaporation, and self-propelled liquid transport and mixing (Liu et al., 2023). Coaxial-based microfluidic printing strategies have been used in tissue applications to develop 4D-printed thermo responsive hydrogel-based actuators (Podstawczyk et al., 2024). Nevertheless, 4D printing's use in microfluidics is in its infancy. There is great promise for numerous industries in the realm of research that integrates 4D printing with microfluidics (e.g., lab-on-chip).

5.4.4 Material Handling and Transportation

Compared to raw materials, 4D printing reduces bulk and facilitates the effective movement of constructions, which significantly benefits the transportation industry (Haleem et al., 2021). Building activation on-site might become feasible in the future. Because it requires less heavy machinery and equipment for production, this technology may represent a new paradigm in the Industrial Revolution 4.0. Flat-pack distribution reduces transit

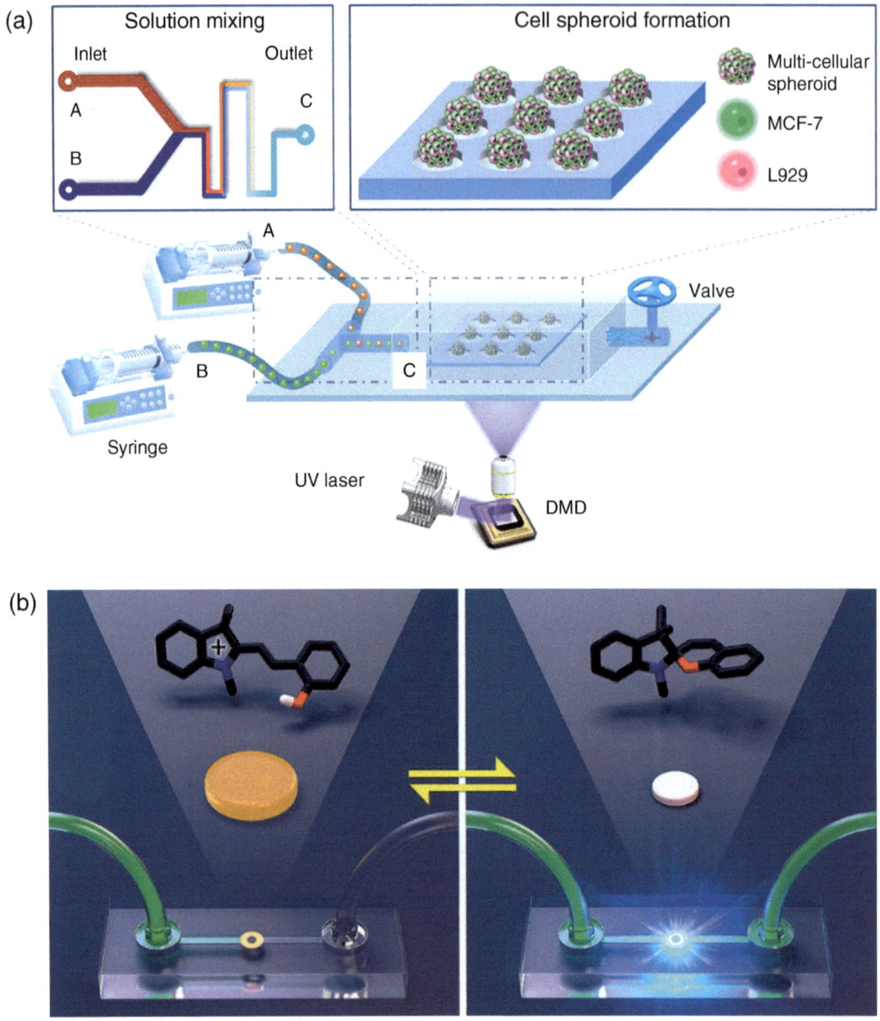

Figure 5.9 4D printing in microfluidics. (a) Microfluidic chip for solution mixing, transportation, and providing conditions for long-term culture *Source:* Yang et al., 2019 / with permission of Elsevier / CC BY-NC-ND 4.0; (b) microwell fabricating system includes the following components, molecular design – light-responsive hydrogels *Source:* ter Schiphorst et al. (2015) / with permission of American Chemical Society.

volume dramatically, allowing things to grow into their 3D shapes and functionalities when they arrive. This transition could result in lower inventory, fewer replacement parts needed, a faster time to market, and, ultimately, better business efficiency. One of the primary advantages is the capacity to manufacture on-demand, which reduces the need for storage. Additionally, 4D printing's inherent self-assembly capabilities can significantly reduce labor costs, delivery times, and manufacturing duration (Ramesh et al., 2018; Zafar & Zhao, 2020).

5.5 Challenges and Research Needs

As a nascent field, 4D printing faces several limitations and challenges that must be addressed. The main hurdles are technological, material, and design challenges.

- *Technological barrier*: Currently, 4D printing technologies are not extensively accessible, with only a limited number of research organizations equipped with the requisite printers. A majority of 4D printing utilizes PolyJet and selective laser melting technology. These methods restrict the range of smart materials that can be utilized, as they must meet particular criteria to be compatible with the technology. Progress in printing technology is crucial for broadening the spectrum of smart materials applicable in 4D printing.
- *Material barrier*: Certain smart materials may disintegrate under deformation, while others, such as hydrogels, might degrade with successive wetting and drying cycles. Furthermore, specific materials necessitate merge with additional components to achieve printability. Therefore, investigation is required to identify novel smart materials that exhibit durability, minimum degradation, and effective interaction with 3D printing methodologies.
- *Design barrier*: Designing adaptive smart structures that respond to environmental changes is critical for attaining desired functions. There is limited technical know-how on printing techniques to fabricate prestrained membranes directly, underscoring the necessity for enhanced design tactics in 4D printing.

Moreover, the ecological implications of 4D printing remain inadequately examined and necessitate scrutiny. Upon resolving these restrictions, 4D printing certainly has the capabilities to transform product manufacture and design. It may decrease capital expenditures, expedite time to market, improve transportability, and diminish product dimensions, facilitating the development of more efficient business models. 4D printing can greatly streamline the creation of 3D structures, enabling user-responsive goods that adapt to particular needs and environmental conditions. In this heterogeneous field, increased collaboration across the diverse disciplines engaged in 4D printing will improve our mastery of shape transformation and functioning. In the future, 4D printing may significantly influence everyday life, altering the methods of product creation and manufacturing.

5.6 Conclusion

4D printing, as an extension of current 3D printing technologies, integrates the dimension of time, enabling structures to alter in reaction to diverse stimuli. Its applications are not limited to medical sciences, soft robotics, self-healing materials, and active origami. Particularly in tissue engineering, 4D bioprinting has advanced considerably in organ and tissue development for transplants, potentially emulating biological growth patterns and making organ printing more feasible. Also, 4D printing is being investigated for precise drug delivery systems. Notwithstanding its potential, 4D printing encounters constraints concerning printing methodologies, material compatibility, and design strategies. Nonetheless, substantial development potential exists as innovative smart materials and enhanced printing methods develop. Presently, 4D printing predominantly emphasizes structural transformations; nonetheless, it possesses the potential to advance into multifunctional applications. 4D printing, being a multidisciplinary area, is poised to have numerous applications in various domains such as engineering, materials science, medical science, and nanoscience. In the future, 4D printing will extend beyond mere shape-changing functionalities, diversifying into several applications and solidifying its status as an advanced domain of controlled smart structure printing.

References

Afzali Naniz, M., Askari, M., Zolfagharian, A., Afzali Naniz, M., & Bodaghi, M. (2022). 4D printing: A cutting-edge platform for biomedical applications. *Biomedical Materials*, *17*(6), 062001. doi:10.1088/1748-605X/ac8e42

Agarwal, K., Srinivasan, V., Lather, V., Pandita, D., & Vasanthan, K. S. (2023a). Insights of 3D bioprinting and focusing the paradigm shift towards 4D printing for biomedical applications. *Journal of Materials Research*, *38*(1), 112–141. doi:10.1557/s43578-022-00524-2

Agarwal, T., Chiesa, I., Costantini, M., Lopamarda, A., Tirelli, M. C., Borra, O. P., Varshapally, S. V. S., Kumar, Y. A. V., Koteswara Reddy, G., De Maria, C., Zhang, L. G., & Maiti, T. K. (2023b). Chitosan and its derivatives in 3D/4D (bio) printing for tissue engineering and drug delivery applications. *International Journal of Biological Macromolecules*, *246*, 125669. doi:10.1016/j.ijbiomac.2023.125669

Agarwal, T., Hann, S. Y., Chiesa, I., Cui, H., Celikkin, N., Micalizzi, S., Barbetta, A., Costantini, M., Esworthy, T., Zhang, L. G., De Maria, C., & Maiti, T. K. (2021). 4D printing in biomedical applications: Emerging trends and technologies. *Journal of Materials Chemistry B*, *9*(37), 7608–7632. doi:10.1039/D1TB01335A

Ahmed, A., Arya, S., Gupta, V., Furukawa, H., & Khosla, A. (2021). 4D printing: Fundamentals, materials, applications and challenges. *Polymer*, *228*, 123926. doi:10.1016/j.polymer.2021.123926

Aldawood, F. K. (2023). A comprehensive review of 4D printing: State of the arts, opportunities, and challenges. *Actuators*, *12*(3), 101. doi:10.3390/act12030101

An, B., Tao, Y., Gu, J., Cheng, T., Chen, X. A., Zhang, X., Zhao, W., Do, Y., Takahashi, S., Wu, H.-Y., Zhang, T., & Yao, L. (2018). *Thermorph: democratizing 4D printing of self-folding materials and interfaces*. Proceedings of the 2018 CHI Conference on Human Factors in Computing Systems, 1–12. https://doi.org/10.1145/3173574.3173834

Awasthi, R., & Kulkarni, G. T. (2016). Decades of research in drug targeting to the upper gastrointestinal tract using gastroretention technologies: Where do we stand? *Drug Delivery*, *23*(2), 378–394. doi:10.3109/10717544.2014.936535

Bajpai, A., Baigent, A., Raghav, S., Brádaigh, C. Ó., Koutsos, V., & Radacsi, N. (2020). 4D printing: Materials, technologies, and future applications in the biomedical field. *Sustainability*, *12*(24), 10628. doi:10.3390/su122410628

Beck, A., Obst, F., Gruner, D., Voigt, A., Mehner, P. J., Gruenzner, S., Koerbitz, R., Shahadha, M. H., Kutscher, A., Paschew, G., Marschner, U., & Richter, A. (2023). Fundamentals of hydrogel-based valves and chemofluidic transistors for lab-on-a-chip technology: A tutorial review. *Advanced Materials Technologies*, *8*(3). doi:10.1002/admt.202200417

Bhusal, A., Yogeshwaran, S., Goodarzi Hosseinabadi, H., Cecen, B., & Miri, A. K. (2024). Microfluidics for high throughput screening of biological agents and therapeutics. *Biomedical Materials & Devices*, *3*(1). doi:10.1007/s44174-024-00169-1

Biswas, M. C., Chakraborty, S., Bhattacharjee, A., & Mohammed, Z. (2021). 4D printing of shape memory materials for textiles: Mechanism, mathematical modeling, and challenges. *Advanced Functional Materials*, *31*(19). doi:10.1002/adfm.202100257

Bournias-Varotsis, A., Wang, S., Hutt, D., & Engstrøm, D. S. (2018). The effect of ultrasonic additive manufacturing on integrated printed electronic conductors. *Electronic Materials Letters*, *14*(4), 413–425. doi:10.1007/s13391-018-0053-y

Cecchini, L., Mariani, S., Ronzan, M., Mondini, A., Pugno, N. M., & Mazzolai, B. (2023). 4D printing of humidity-driven seed inspired soft robots. *Advanced Science*, *10*(9). doi:10.1002/advs.202205146

Chopra, V., Fuentes-Velasco, V., Nacif-Lopez, S. R., Melendez-Malpicca, J., Mendez-Hernandez, A. S., Ramos-Mendez-Iris, L. F., Arroyo-Jimenez, D. A., Reyes-Segura, D. G., Gonzalez-Y-Mendoza, P., Sanchez-Hernandez, K. A., Spinola-Corona, E., Vazquez-del-Mercado-Pardiño, J. A., & Chauhan, G. (2024). Advancements in 3D-4D printing of hydroxyapatite composites for bone tissue engineering. *Ceramics International*, *50*(20), 38819–38840. doi:10.1016/j.ceramint.2024.07.266

Chu, H., Yang, W., Sun, L., Cai, S., Yang, R., Liang, W., Yu, H., & Liu, L. (2020). 4D printing: A review on recent progresses. *Micromachines*, *11*(9), 796. doi:10.3390/mi11090796

Cui, H., Zhu, W., Miao, S., Sarkar, K., & Zhang, L. G. (2024). 4D printed nerve conduit with in situ neurogenic guidance for nerve regeneration. *Tissue Engineering Part A*, *30*(11–12), 293–303. doi:10.1089/ten.tea.2023.0194

D'Eramo, L., Chollet, B., Leman, M., Martwong, E., Li, M., Geisler, H., Dupire, J., Kerdraon, M., Vergne, C., Monti, F., Tran, Y., & Tabeling, P. (2018). Microfluidic actuators based on temperature-responsive hydrogels. *Microsystems & Nanoengineering*, *4*(1), 17069. doi:10.1038/micronano.2017.69

Ding, A., Lee, S. J., Tang, R., Gasvoda, K. L., He, F., & Alsberg, E. (2022). 4D cell-condensate bioprinting. *Small*, *18*(36). doi:10.1002/smll.202202196

Dowaidar, M. (2024). Guidelines for biomimetic 3D/4D printing in drug delivery. *Materials Chemistry and Physics*, *325*, 129692. doi:10.1016/j.matchemphys.2024.129692

Ellery, A., & Elaskri, A. (2025). 3D printing modules for self-assembling space systems. *Progress in Aerospace Sciences*, *101128*. doi:10.1016/j.paerosci.2025.101128

Gazzaniga, A., Foppoli, A., Cerea, M., Palugan, L., Cirilli, M., Moutaharrik, S., Melocchi, A., & Maroni, A. (2023). Towards 4D printing in pharmaceutics. *International Journal of Pharmaceutics: X*, *5*, 100171. doi:10.1016/j.ijpx.2023.100171

Ghosh, A., Li, L., Xu, L., Dash, R. P., Gupta, N., Lam, J., Jin, Q., Akshintala, V., Pahapale, G., Liu, W., Sarkar, A., Rais, R., Gracias, D. H., & Selaru, F. M. (2020). Gastrointestinal-resident, shape-changing microdevices extend drug release in vivo. *Science Advances*, *6*(44). doi:10.1126/sciadv.abb4133

Haleem, A., Javaid, M., Singh, R. P., & Suman, R. (2021). Significant roles of 4D printing using smart materials in the field of manufacturing. *Advanced Industrial and Engineering Polymer Research*, *4*(4), 301–311. doi:10.1016/j.aiepr.2021.05.001

Han, M., Yang, Y., & Li, L. (2020). Energy consumption modeling of 4D printing thermal-responsive polymers with integrated compositional design for material. *Additive Manufacturing*, *34*, 101223. doi:10.1016/j.addma.2020.101223

Han, M. J., & Yoon, D. K. (2021). Advances in soft materials for sustainable electronics. *Engineering*, *7*(5), 564–580. doi:10.1016/j.eng.2021.02.010

Hann, S. Y., Cui, H., Nowicki, M., & Zhang, L. G. (2020). 4D printing soft robotics for biomedical applications. *Additive Manufacturing*, *36*, 101567. doi:10.1016/j.addma.2020.101567

Hasanzadeh, R., Mihankhah, P., Azdast, T., Rasouli, A., Shamkhali, M., & Park, C. B. (2023). Biocompatible tissue-engineered scaffold polymers for 3D printing and its application for 4D printing. *Chemical Engineering Journal*, *476*, 146616. doi:10.1016/j.cej.2023.146616

Hatami, H., Almahmeed, W., Kesharwani, P., & Sahebkar, A. (2024). Exploring the potential of 3D and 4D printing in advancing stent manufacturing for cardiovascular diseases. *European Polymer Journal*, *212*, 113035. doi:10.1016/j.eurpolymj.2024.113035

He, W., Zhou, D., Gu, H., Qu, R., Cui, C., Zhou, Y., Wang, Y., Zhang, X., Wang, Q., Wang, T., & Zhang, Y. (2023). A biocompatible 4D printing shape memory polymer as emerging strategy for fabrication of deployable medical devices. *Macromolecular Rapid Communications*, *44*(2). doi:10.1002/marc.202200553

Homan, K. A., Kolesky, D. B., Skylar-Scott, M. A., Herrmann, J., Obuobi, H., Moisan, A., & Lewis, J. A. (2016). Bioprinting of 3D convoluted renal proximal tubules on perfusable chips. *Scientific Reports*, *6*(1), 34845. doi:10.1038/srep34845

Huang, X., Zhang, L., Hang, J., Quinn, T., Nasar, N. K. A., Lin, Y., Hu, C., Pang, X., Chen, X., Davis, T. P., & Qiao, R. (2024). 4D printing hybrid soft robots enabled by shape-transformable liquid metal nanoparticles. *Advanced Materials*, *36*(46). doi:10.1002/adma.202409789

Ionov, L. (2018). 4D biofabrication: Materials, methods, and applications. *Advanced Healthcare Materials*, *7*(17). doi:10.1002/adhm.201800412

Javaid, M., & Haleem, A. (2019). 4D printing applications in medical field: A brief review. *Clinical Epidemiology and Global Health*, *7*(3), 317–321. doi:10.1016/j.cegh.2018.09.007

Javaid, M., & Haleem, A. (2020). Significant advancements of 4D printing in the field of orthopaedics. *Journal of Clinical Orthopaedics and Trauma*, *11*, S485–S490. doi:10.1016/j.jcot.2020.04.021

Jing, Z., Ma, Y., & Zhu, J. (2022). Application of a novel electrostatic dry powder coating technology on capsules for enteric release. *Journal of Drug Delivery Science and Technology*, *68*, 103058. doi:10.1016/j.jddst.2021.103058

Joharji, L., Mishra, R. B., Alam, F., Tytov, S., Al-Modaf, F., & El-Atab, N. (2022). 4D printing: A detailed review of materials, techniques, and applications. *Microelectronic Engineering*, *265*, 111874. doi:10.1016/j.mee.2022.111874

Joshi, A., Goh, J. K., & Goh, K. E. J. (2020a). Polymer-based conductive composites for 3D and 4D printing of electrical circuits. In K. K. Sadasivuni, K. Deshmukh, & M. A. Almaadeed (Eds.), *3D and 4D printing of polymer nanocomposite materials* (pp. 45–83). Elsevier. doi:10.1016/B978-0-12-816805-9.00003-X

Joshi, S., Rawat, K., Karunakaran, C., Rajamohan, V., Mathew, A. T., Koziol, K., Kumar Thakur, V., & Balan, A. S. S. (2020b). 4D printing of materials for the future: Opportunities and challenges. *Applied Materials Today*, *18*, 100490. doi:10.1016/j.apmt.2019.100490

Kalogeropoulou, M., Díaz-Payno, P. J., Mirzaali, M. J., van Osch, G. J. V. M., Fratila-Apachitei, L. E., & Zadpoor, A. A. (2024). 4D printed shape-shifting biomaterials for tissue engineering and regenerative medicine applications. *Biofabrication*, *16*(2), 022002. doi:10.1088/1758-5090/ad1e6f

Kang, J., Li, R., Zheng, D., Wang, M., Niu, P., & Yuan, T. (2024). 4D printed NiTi variable-geometry inlet for aero engines. *Virtual and Physical Prototyping*, *19*(1). doi:10.1080/17452759.2024.2382166

Kavimughil, M., Mahalakshmi, L., Moses, J. A., & Anandharamakrishnan, C. (2023). Delivery of nutraceutical ingredients through three-dimensional food printing. In *Industrial application of functional foods, ingredients and nutraceuticals* (pp. 253–278). Elsevier. https://doi.org/10.1016/B978-0-12-824312-1.00008-X

Khalid, M. Y., Arif, Z. U., Noroozi, R., Zolfagharian, A., & Bodaghi, M. (2022). 4D printing of shape memory polymer composites: A review on fabrication techniques, applications, and future perspectives. *Journal of Manufacturing Processes, 81*, 759–797. doi:10.1016/j.jmapro.2022.07.035

Kim, J., DA, G., Debnath, P., & Saha, P. (2024). Smart multi-responsive biomaterials and their applications for 4D bioprinting. *Biomimetics, 9*(8), 484. doi:10.3390/biomimetics9080484

Kirtane, A. R., Abouzid, O., Minahan, D., Bensel, T., Hill, A. L., Selinger, C., Bershteyn, A., Craig, M., Mo, S. S., Mazdiyasni, H., Cleveland, C., Rogner, J., Lee, Y.-A. L., Booth, L., Javid, F., Wu, S. J., Grant, T., Bellinger, A. M., Nikolic, B., ... Traverso, G. (2018). Development of an oral once-weekly drug delivery system for HIV antiretroviral therapy. *Nature Communications, 9*(1), 2. doi:10.1038/s41467-017-02294-6

Lai, J., Liu, Y., Lu, G., Yung, P., Wang, X., Tuan, R. S., & Li, Z. A. (2024). 4D bioprinting of programmed dynamic tissues. *Bioactive Materials, 37*, 348–377. doi:10.1016/j.bioactmat.2024.03.033

Lai, J., & Wang, M. (2023). Developments of additive manufacturing and 5D printing in tissue engineering. *Journal of Materials Research, 38*(21), 4692–4725. doi:10.1557/s43578-023-01193-5

Li, J., Liang, Z., Li, T., Wan, Q., Huang, X., & Kan, Q. (2025). A comprehensive review on 4D printing and applications of thermo-induced shape memory polymers. *International Journal of Smart and Nano Materials*, 1–60. doi:10.1080/19475411.2025.2458833

Liu, X., Li, B., Gu, Z., & Zhou, K. (2023). 4D printing of butterfly scale–inspired structures for wide-angle directional liquid transport. *Small, 19*(34). doi:10.1002/smll.202207640

Lukin, I., Musquiz, S., Erezuma, I., Al-Tel, T. H., Golafshan, N., Dolatshahi-Pirouz, A., & Orive, G. (2019). Can 4D bioprinting revolutionize drug development? *Expert Opinion on Drug Discovery, 14*(10), 953–956. doi:10.1080/17460441.2019.1636781

Mahmood, A., Akram, T., Shenggui, C., & Chen, H. (2023). Revolutionizing manufacturing: A review of 4D printing materials, stimuli, and cutting-edge applications. *Composites Part B: Engineering, 266*, 110952. doi:10.1016/j.compositesb.2023.110952

Mahmoud, D. B., & Schulz-Siegmund, M. (2023). Utilizing 4D printing to design smart gastroretentive, esophageal, and intravesical drug delivery systems. *Advanced Healthcare Materials, 12*(10). doi:10.1002/adhm.202202631

Mahmud, M. A. P., Tat, T., Xiao, X., Adhikary, P., & Chen, J. (2021). Advances in 4D-printed physiological monitoring sensors. *Exploration, 1*(3). doi:10.1002/EXP.20210033

Manaia, J. P., Cerejo, F., & Duarte, J. (2023). Revolutionising textile manufacturing: A comprehensive review on 3D and 4D printing technologies. *Fashion and Textiles, 10*(1), 20. doi:10.1186/s40691-023-00339-7

McLellan, K., Sun, Y.-C., & Naguib, H. E. (2022). A review of 4D printing: Materials, structures, and designs towards the printing of biomedical wearable devices. *Bioprinting, 27*, e00217. doi:10.1016/j.bprint.2022.e00217

Melocchi, A., Inverardi, N., Uboldi, M., Baldi, F., Maroni, A., Pandini, S., Briatico-Vangosa, F., Zema, L., & Gazzaniga, A. (2019a). Retentive device for intravesical drug delivery based on water-induced shape memory response of poly(vinyl alcohol): Design concept and 4D printing feasibility. *International Journal of Pharmaceutics*, *559*, 299–311. doi:10.1016/j.ijpharm.2019.01.045

Melocchi, A., Uboldi, M., Inverardi, N., Briatico-Vangosa, F., Baldi, F., Pandini, S., Scalet, G., Auricchio, F., Cerea, M., Foppoli, A., Maroni, A., Zema, L., & Gazzaniga, A. (2019b). Expandable drug delivery system for gastric retention based on shape memory polymers: Development via 4D printing and extrusion. *International Journal of Pharmaceutics*, *571*, 118700. doi:10.1016/j.ijpharm.2019.118700

Miriyev, A., Stack, K., & Lipson, H. (2017). Soft material for soft actuators. *Nature Communications*, *8*(1), 596. doi:10.1038/s41467-017-00685-3

Moheb Afzali, A., Kheradmand, M. A., & Naghib, S. M. (2024). Bioreactor design-assisted bioprinting of stimuli-responsive materials for tissue engineering and drug delivery applications. *Bioprinting*, *37*, e00325. doi:10.1016/j.bprint.2023.e00325

Momeni, F., & Ni, J. (2018). Nature-inspired smart solar concentrators by 4D printing. *Renewable Energy*, *122*, 35–44. doi:10.1016/j.renene.2018.01.062

Momeni, F., Sabzpoushan, S., Valizadeh, R., Morad, M. R., Liu, X., & Ni, J. (2019). Plant leaf-mimetic smart wind turbine blades by 4D printing. *Renewable Energy*, *130*, 329–351. doi:10.1016/j.renene.2018.05.095

Muhindo, D., Elkanayati, R., Srinivasan, P., Repka, M. A., & Ashour, E. A. (2023). Recent advances in the applications of additive manufacturing (3D printing) in drug delivery: A comprehensive review. *AAPS PharmSciTech*, *24*(2), 57. doi:10.1208/s12249-023-02524-9

Murphy, S. V., De Coppi, P., & Atala, A. (2019). Opportunities and challenges of translational 3D bioprinting. *Nature Biomedical Engineering*, *4*(4), 370–380. doi:10.1038/s41551-019-0471-7

Niazy, D., Ashraf, M., Bodaghi, M., & Zolfagharian, A. (2024). Resilient city perspective: 4D printing in art, architecture and construction. *Materials Today Sustainability*, *26*, 100708. doi:10.1016/j.mtsust.2024.100708

Noroozi, R., Arif, Z. U., Taghvaei, H., Khalid, M. Y., Sahbafar, H., Hadi, A., Sadeghianmaryan, A., & Chen, X. (2023). 3D and 4D bioprinting technologies: A game changer for the biomedical sector? *Annals of Biomedical Engineering*, *51*(8), 1683–1712. doi:10.1007/s10439-023-03243-9

Osouli-Bostanabad, K., Masalehdan, T., Kapsa, R. M. I., Quigley, A., Lalatsa, A., Bruggeman, K. F., Franks, S. J., Williams, R. J., & Nisbet, D. R. (2022). Traction of 3D and 4D printing in the healthcare industry: From drug delivery and analysis to regenerative medicine. *ACS Biomaterials Science & Engineering*, *8*(7), 2764–2797. doi:10.1021/acsbiomaterials.2c00094

Palugan, L., Cerea, M., Cirilli, M., Moutaharrik, S., Maroni, A., Zema, L., Melocchi, A., Uboldi, M., Filippin, I., Foppoli, A., & Gazzaniga, A. (2021). Intravesical drug delivery approaches for improved therapy of urinary bladder diseases. *International Journal of Pharmaceutics: X*, *3*, 100100. doi:10.1016/j.ijpx.2021.100100

Petrini, L., & Migliavacca, F. (2011). Biomedical applications of shape memory alloys. *Journal of Metallurgy*, *2011*, 1–15. doi:10.1155/2011/501483

Pingale, P., Dawre, S., Dhapte-Pawar, V., Dhas, N., & Rajput, A. (2023). Advances in 4D printing: From stimulation to simulation. *Drug Delivery and Translational Research*, *13*(1), 164–188. doi:10.1007/s13346-022-01200-y

Podstawczyk, D., Nizioł, M., Śledzik, P., Simińska-Stanny, J., Dawiec-Liśniewska, A., & Shavandi, A. (2024). Coaxial 4D printing of vein-inspired thermoresponsive channel hydrogel actuators. *Advanced Functional Materials*, *34*(10). doi:10.1002/adfm.202310514

Pourmasoumi, P., Moghaddam, A., Nemati Mahand, S., Heidari, F., Salehi Moghaddam, Z., Arjmand, M., Kühnert, I., Kruppke, B., Wiesmann, H.-P., & Khonakdar, H. A. (2023). A review on the recent progress, opportunities, and challenges of 4D printing and bioprinting in regenerative medicine. *Journal of Biomaterials Science, Polymer Edition*, *34*(1), 108–146. doi:10.1080/09205063.2022.2110480

Qiao, Y., Luo, J., Cui, T., Liu, H., Tang, H., Zeng, Y., Liu, C., Li, Y., Jian, J., Wu, J., Tian, H., Yang, Y., Ren, T.-L., & Zhou, J. (2023). Soft electronics for health monitoring assisted by machine learning. *Nano-Micro Letters*, *15*(1), 66. doi:10.1007/s40820-023-01029-1

Qu, G., Huang, J., Gu, G., Li, Z., Wu, X., & Ren, J. (2023). Smart implants: 4D-printed shape-morphing scaffolds for medical implantation. *International Journal of Bioprinting*, *9*(5), 764. doi:10.18063/ijb.764

Rahimnejad, M., Jahangiri, S., Zirak Hassan Kiadeh, S., Rezvaninejad, S., Ahmadi, Z., Ahmadi, S., Safarkhani, M., & Rabiee, N. (2024). Stimuli-responsive biomaterials: Smart avenue toward 4D bioprinting. *Critical Reviews in Biotechnology*, *44*(5), 860–891. doi:10.1080/07388551.2023.2213398

Ramesh, S., Kiran reddy, S., Usha, C., Naulakha, N. K., Adithyakumar, C., & Lohith Kumar Reddy, M. (2018). Advancements in the research of 4D printing-A review. *IOP Conference Series: Materials Science and Engineering*, *376*, 012123. doi:10.1088/1757-899X/376/1/012123

Ramezani, M., & Mohd Ripin, Z. (2023). 4D printing in biomedical engineering: Advancements, challenges, and future directions. *Journal of Functional Biomaterials*, *14*(7), 347. doi:10.3390/jfb14070347

Sajjad, R., Chauhdary, S. T., Anwar, M. T., Zahid, A., Khosa, A. A., Imran, M., & Sajjad, M. H. (2024). A review of 4D printing – technologies, shape shifting, smart polymer based materials, and biomedical applications. *Advanced Industrial and Engineering Polymer Research*, *7*(1), 20–36. doi:10.1016/j.aiepr.2023.08.002

Santhoshkumar, P., Negi, A., & Moses, J. A. (2023). 3D printing for space food applications: Advancements, challenges, and prospects. *Life Sciences in Space Research*, *40*. doi:10.1016/j.lssr.2023.08.002

Santhoshkumar, P., Ramu, D., Mahalakshmi, L., & Moses, J. A. (2024). 3D printed edible electronics: Components, fabrication approaches and applications. *Biosensors and Bioelectronics*, *117059*. doi:10.1016/j.bios.2024.117059

Shi, Q., Liu, H., Tang, D., Li, Y., Li, X., & Xu, F. (2019). Bioactuators based on stimulus-responsive hydrogels and their emerging biomedical applications. *NPG Asia Materials*, *11*(1), 64. doi:10.1038/s41427-019-0165-3

Shi, Y., Wang, K., Feng, X., Li, S., Liang, H., Li, G., Li, W., Ke, W., Chen, A., Yang, X., Yang, L., Yan, C., Su, B., & Yang, C. (2022). Additive manufactured self-powered mechanoelectric sensor as the artificial nucleus pulposus for monitoring tissue rehabilitation after discectomy. *Nano Energy*, *96*, 107113. doi:10.1016/j.nanoen.2022.107113

Soleimanzadeh, H., Rolfe, B., Bodaghi, M., Jamalabadi, M., Zhang, X., & Zolfagharian, A. (2023). Sustainable robots 4D printing. *Advanced Sustainable Systems*, *7*(12). doi:10.1002/adsu.202300289

Soleymani, S., & Naghib, S. M. (2023). 3D and 4D printing hydroxyapatite-based scaffolds for bone tissue engineering and regeneration. *Heliyon*, *9*(9), e19363. doi:10.1016/j.heliyon.2023.e19363

Subeshan, B., Baddam, Y., & Asmatulu, E. (2021). Current progress of 4D-printing technology. *Progress in Additive Manufacturing*, *6*(3), 495–516. doi:10.1007/s40964-021-00182-6

Sun, L., Li, J., Chen, Y., Yang, Y., Tao, Y., Wang, G., & Yao, L. (2020). *4DTexture: a shape-changing fabrication method for 3D surfaces with texture*. Extended abstracts of the 2020 CHI conference on human factors in computing systems, 1–7. https://doi.org/10.1145/3334480.3383053

Sun, L., Yang, Y., Chen, Y., Li, J., Luo, D., Liu, H., Yao, L., Tao, Y., & Wang, G. (2021). *ShrinCage: 4D printing accessories that self-adapt*. Proceedings of the 2021 CHI conference on human factors in computing systems, 1–12. https://doi.org/10.1145/3411764.3445220

Tan, J. M. R., Farraj, Y., Kamyshny, A., & Magdassi, S. (2023). Fabrication approaches of soft electronics. *ACS Applied Electronic Materials*, *5*(3), 1376–1393. doi:10.1021/acsaelm.2c01728

Taylor, S., Mueller, E., Jones, L. R., Makela, A. V., & Ashammakhi, N. (2024). Translational aspects of 3D and 4D printing and bioprinting. *Advanced Healthcare Materials*, *13*(27). doi:10.1002/adhm.202400463

ter Schiphorst, J., Coleman, S., Stumpel, J. E., Ben Azouz, A., Diamond, D., & Schenning, A. P. H. J. (2015). Molecular design of light-responsive hydrogels, for in situ generation of fast and reversible valves for microfluidic applications. *Chemistry of Materials*, *27*(17), 5925–5931. doi:10.1021/acs.chemmater.5b01860

Tran, T. S., Balu, R., Mettu, S., Roy Choudhury, N., & Dutta, N. K. (2022). 4D printing of hydrogels: Innovation in material design and emerging smart systems for drug delivery. *Pharmaceuticals*, *15*(10), 1282. doi:10.3390/ph15101282

Uboldi, M., Melocchi, A., Moutaharrik, S., Palugan, L., Cerea, M., Foppoli, A., Maroni, A., Gazzaniga, A., & Zema, L. (2022). Administration strategies and smart devices for drug release in specific sites of the upper GI tract. *Journal of Controlled Release*, *348*, 537–552. doi:10.1016/j.jconrel.2022.06.005

Wang, C., Wu, H., Chen, Z., McDowell, M. T., Cui, Y., & Bao, Z. (2013). Self-healing chemistry enables the stable operation of silicon microparticle anodes for high-energy lithium-ion batteries. *Nature Chemistry*, *5*(12), 1042–1048. doi:10.1038/nchem.1802

Wang, G., Tao, Y., Capunaman, O. B., Yang, H., & Yao, L. (2019). *A-line: 4D printing morphing linear composite structures*. Proceedings of the 2019 CHI conference on human factors in computing systems, 1–12. https://doi.org/10.1145/3290605.3300656

Wang, J., Liu, F., Tao, F., & Pan, Q. (2017). Rationally designed self-healing hydrogel electrolyte toward a smart and sustainable supercapacitor. *ACS Applied Materials & Interfaces*, *9*(33), 27745–27753. doi:10.1021/acsami.7b07836

Wang, Y., Yokota, T., & Someya, T. (2021). Electrospun nanofiber-based soft electronics. *NPG Asia Materials*, *13*(1), 22. doi:10.1038/s41427-020-00267-8

Wang, Z., Wu, Y., Wu, D., Sun, D., & Lin, L. (2022). Soft magnetic composites for highly deformable actuators by four-dimensional electrohydrodynamic printing. *Composites Part B: Engineering*, *231*, 109596. doi:10.1016/j.compositesb.2021.109596

Xin, X., Liu, L., Liu, Y., & Leng, J. (2020). 4D printing auxetic metamaterials with tunable, programmable, and reconfigurable mechanical properties. *Advanced Functional Materials*, *30*(43). doi:10.1002/adfm.202004226

Yang, W., Cai, S., Yuan, Z., Lai, Y., Yu, H., Wang, Y., & Liu, L. (2019). Mask-free generation of multicellular 3D heterospheroids array for high-throughput combinatorial anti-cancer drug screening. *Materials & Design*, *183*, 108182. doi:10.1016/j.matdes.2019.108182

Zafar, M. Q., & Zhao, H. (2020). 4D printing: Future insight in additive manufacturing. *Metals and Materials International*, *26*(5), 564–585. doi:10.1007/s12540-019-00441-w

Zeenat, L., Zolfagharian, A., Sriya, Y., Sasikumar, S., Bodaghi, M., & Pati, F. (2023). 4D printing for vascular tissue engineering: Progress and challenges. *Advanced Materials Technologies*, *8*(23). doi:10.1002/admt.202300200

Zhang, B., Li, H., Cheng, J., Ye, H., Sakhaei, A. H., Yuan, C., Rao, P., Zhang, Y., Chen, Z., Wang, R., He, X., Liu, J., Xiao, R., Qu, S., & Ge, Q. (2021). Mechanically robust and UV-curable shape-memory polymers for digital light processing based 4D printing. *Advanced Materials*, *33*(27). doi:10.1002/adma.202101298

Zhang, X., Yang, Y., Yang, Z., Ma, R., Aimaijiang, M., Xu, J., Zhang, Y., & Zhou, Y. (2023). Four-dimensional printing and shape memory materials in bone tissue engineering. *International Journal of Molecular Sciences*, *24*(1), 814. doi:10.3390/ijms24010814

Zhang, Z., Demir, K. G., & Gu, G. X. (2019). Developments in 4D-printing: A review on current smart materials, technologies, and applications. *International Journal of Smart and Nano Materials*, *10*(3), 205–224. doi:10.1080/19475411.2019.1591541

Zhou, Y., Zhou, D., Cao, P., Zhang, X., Wang, Q., Wang, T., Li, Z., He, W., Ju, J., & Zhang, Y. (2021). 4D printing of shape memory vascular stent based on β CD-g-Polycaprolactone. *Macromolecular Rapid Communications*, *42*(14). doi:10.1002/marc.202100176

Zhou, Z., Tang, W., Yang, J., & Fan, C. (2023). Application of 4D printing and bioprinting in cardiovascular tissue engineering. *Biomaterials Science*, *11*(19), 6403–6420. doi:10.1039/D3BM00312D

Zhu, H., He, Y., Wang, Y., Zhao, Y., & Jiang, C. (2022). Mechanically-guided 4D printing of magnetoresponsive soft materials across different length scale. *Advanced Intelligent Systems*, *4*(3). doi:10.1002/aisy.202100137

Zolfagharian, A., Durran, L., Gharaie, S., Rolfe, B., Kaynak, A., & Bodaghi, M. (2021). 4D printing soft robots guided by machine learning and finite element models. *Sensors and Actuators A: Physical*, *328*, 112774. doi:10.1016/j.sna.2021.112774

Zu, S., Wang, Z., Zhang, S., Guo, Y., Chen, C., Zhang, Q., Wang, Z., Liu, T., Liu, Q., & Zhang, Z. (2022). A bioinspired 4D printed hydrogel capsule for smart controlled drug release. *Materials Today Chemistry*, *24*, 100789. doi:10.1016/j.mtchem.2022.100789

6

Printing Foods in the Fourth Dimension

CHAPTER MENU

4D Printing of Foods: Fundamentals, Applications and Prospects, First Edition.
C. Anandharamakrishnan, Jeyan A. Moses, and P. Santhoshkumar.

6.1 Introduction

Typically, 3D printing technology utilizes computer-aided design and numerical control systems to construct complex three-dimensional foods incrementally in a layer-by-layer manner (Nachal et al. 2019). Accordingly, the term "additive manufacturing" comes into play, but foods can also be 3D printed using other approaches. This innovative technique has extensive applications across multiple sectors, including healthcare, aerospace, and materials engineering (Santhoshkumar et al. 2023). Advancements in material science have led to the emergence of 3D printing in the food industry, enabling customized food derived from digital designs (Derossi et al. 2024). This facilitates accurate shape, texture alteration, and personalization of food products to accommodate varied consumer tastes (Waseem et al. 2024). Nonetheless, conventional 3D food printing predominantly produces static food structures and encounters difficulties with dynamic alterations over time (Wang et al. 2024). Four-dimensional (4D) food printing, as an extension, produces food structures that experience time-dependent alterations in response to internal or external stimuli (Anandharamakrishnan et al. 2022b). This allows for modifying a printed food's shape, texture, flavor, color, or other attributes after printing. Such transformations are activated by external factors, including heat, moisture, light, pH, or magnetism, and they produce an interactive and dynamic culinary experience with significant potential for a variety of applications (Srivastava et al. 2023). It is important to note that given the properties of foods and the biological functions it has, some of these stimuli are not currently viable for food applications and may be so in the future as well.

To implement 4D food printing at a successful level, it must meet three critical considerations. Initially, the use of smart materials is essential, as they enable self-transformation in response to specific stimuli, thereby determining the nature of changes in the food's properties and functionality. Subsequently, the selected stimulus must effectively initiate the desired transformation in the innovative material. Lastly, time is a critical factor, as the metamorphosis in the printed food may require a specific duration to fully manifest. Overall, the technology can produce functional and personalized foods specifically tailored to each individual's sensory experiences, preferences, and dietary needs. Its prospective applications are particularly promising in specialized contexts that require efficient, compact, and nutrient-rich food solutions (Shabir et al. 2024).

Despite its widespread use in a variety of fields, the complexity of this method is significantly greater than that of the 3D printing counterpart. Worldwide, researchers are increasingly emphasizing the development of 4D food printing as a solution to the obstacles associated with printing. These structures are meticulously designed and constructed to seamlessly transition from one configuration to another in the process of 4D food printing. Instances of designed 4D behaviors encompass biological development, including vitamin conversion; nutritional enhancement; color, shape, and texture modulation (Chen et al. 2022a, 2023a; Oral et al. 2021). This chapter covers the basics of 4D food printing and its related terminologies, applications, and associated challenges.

6.2 From 3D to 4D Printing

Researchers from MIT are recognized as the pioneers in the implementation of 4D printing for food applications, creating a 2D film made of starch, cellulose, and protein that transforms into a 3D structure when exposed to water. Since then, significant advances have happened, and the field has evolved phenomenally (Wang et al. 2017). Although 4D food printing has its roots in 3D food printing, with the potential to personalize food products and develop unique flavors, 4D food printing can also perform intelligent functions, for instance, nutrient improvements based on individual health conditions. The diverse combinations of food components in the printing ink, tailored to specific structures and formulae, can be used to implement modifications in 4D-printed food (Teng et al. 2021). In response to stimuli such as temperature and pH, these variations may include changes in color, texture, nutrition, shape, aroma, and structure (Guo et al. 2025; Niu et al. 2025; Oral et al. 2021; Xie et al. 2025). For instance, microwave heating was reported to result in the 4D transformation of 3D-printed buckwheat bread and lotus root powder gel (Chen et al. 2021a). Similarly, 3D-printed composites of beetroot, pumpkin, and soy protein isolate have been observed to undergo chromatic alterations induced by pH (Phuhongsung et al., 2020).

In contrast to traditional food shape morphing, 4D food printing employs a 3D printing technique that meticulously regulates the shape, structure, and composition of food, allowing for individualized customization or personalization. Figure 6.1 presents a transition of advancements from 3D food printing to 4D food printing. By incorporating a "space-time axis" into the 3D coordinate system (x, y, and z axes), 4D food printing allows meals/snacks/foods to modify their structure, functionality, and qualities in response to different stimuli (Wang et al. 2024). In contrast to 3D, 4D food printing generates dynamic structures for improved esthetics and enhances sensory experience and nutritional design. Moreover, it can diminish capital expenditures, minimize inventory requirements, conserve time, and enhance operational efficiency.

6.3 Defining 4D Food Printing

3D printing entails the sequential layering of materials to fabricate engineering and culinary structures according to specified designs. This layer-by-layer methodology yields static 3D structures, wherein the geometric characteristics and mechanical qualities are determined by the selected procedures, parameters, and material interactions (Santhoshkumar et al. 2024a). The use of "static structures" implies that the functions and characteristics of these additively created materials are invariant over time. Conversely, 4D printing surpasses 3D printing by incorporating time as an additional dimension. That is, while 3D printing functions within three spatial dimensions, 4D printing facilitates the fabrication of dynamic structures capable of temporal transformation. This dynamism is attained through intelligent design and the use of custom materials that react to diverse stimuli, including chemical, thermal, and biological conditions (Singh et al. 2024). The alterations in 4D-printed structures are deliberate and preordained, signifying a targeted modification in attributes.

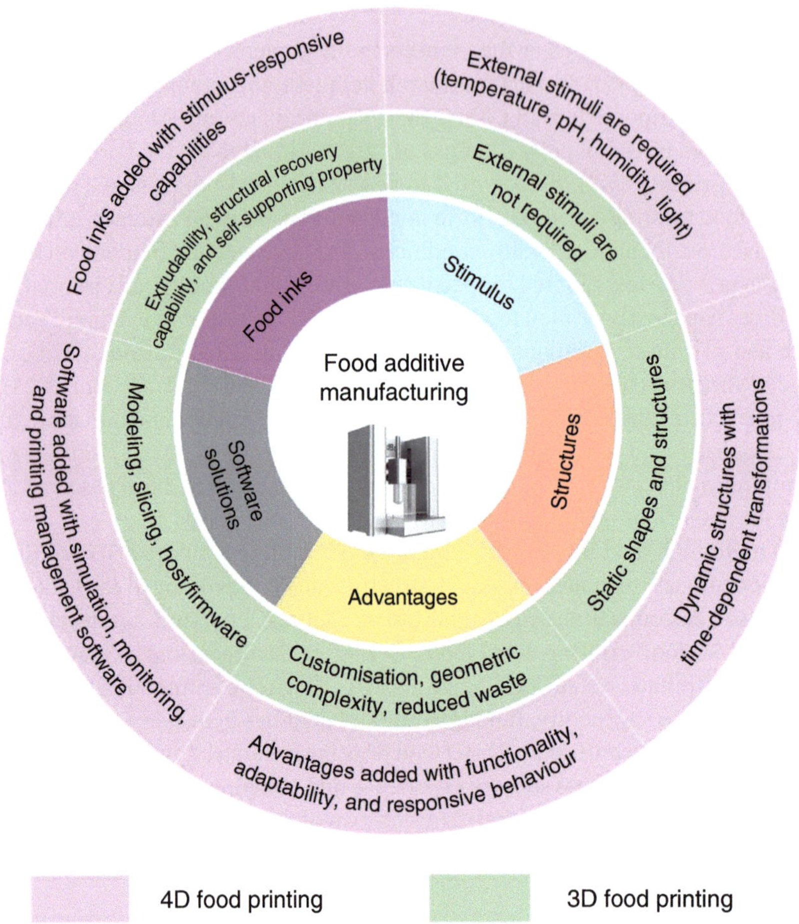

Figure 6.1 Illustration of fundamental differentiation between 3D and 4D printing. *Source*: Wang et al. (2024) / with permission of Elsevier.

Initially, 4D printing was conceived for engineering purposes before its application in food structuring, motivated by the objective of producing intelligent materials capable of adapting their functioning in response to stimuli over time. The advent of 4D food printing has led to the emergence of multiple definitions and expansions, such as 5D and 6D, underscoring a lack of consensus over the terminology (Nida et al. 2022). Although some may allude to further dimensions, the cosmos is fundamentally characterized by the space–time continuum, indicating that any pertinent transformations must be regarded within the four dimensions: three spatial and one temporal.

Nida et al. (2022) reported that 5D and 6D printing denote the incorporation of one or two rotational axes into the printer's capabilities. This is more accurately interpreted as an enhancement in degrees of freedom rather than genuine additional dimensions. Consequently, any incorporation of time-dependent effects should still be categorized as

4D food printing. For instance, a study demonstrated enhanced color alterations in food by probiotic development, designating it as 5D printing (Chen et al. 2023a). From one perspective, this may be viewed as a 4D process as both impacts are regulated by time; therefore, such clear distinctions and terminologies must be better delineated.

Moreover, conventional processing techniques and 3D food printing frequently necessitate heat stabilization to attain particular modifications in texture, flavor, and nutritional content. In recent years, researchers have also considered such changes, for instance, thermally induced effects, as 4D printing concepts. From a different perspective, 4D food printing involves the structuring of food items that contain one or more programmable and controlled time-dependent transformations, including sensory or nutritional properties, using 3D printing techniques. The alterations must arise from external or internal stimuli and are not characteristic of traditional processing or conventional 3D printing (Fahmy et al. 2023).

6.4 Stages in 4D Printing of Foods

Optimizing and refining critical elements in 4D food printing, such as printing technology, digital models, food inks, and postprocessing procedures, is vital for effective deployment and futuristic advancements. An integrated strategy that highlights the synergistic enhancement of these elements and the printing process will foster a more thorough and resilient research framework in the domain of 4D food printing. Accordingly, a 4D food printing management process comprises three components: preprinting, printing, and postprinting/processing. Typically, 3D and 4D printers are the same in terms of materials selection and preparation in preprinting, but the postprinting stage varies.

6.4.1 Preprinting

Before the actual 4D printing process, stages such as material selection and pretreatment, as in regular 3D printing processes, are considered. In addition to these, simulation and modeling are integral to predicting smart materials' functionality when exposed to a stimulus. For example, heat distribution in materials during material extrusion and the distribution of stresses on the material has been simulated using modeling approaches (Sinha et al. 2021). In 4D food printing, this aspect of research is lacking, demanding future work. Additionally, to control printers, advanced software has been introduced for shape-memory materials, not limited to food printing. Notably, control of a stimulus on a printed food significantly improves the dynamic/intelligent functions of the printed material. Based on these aspects, this section details three aspects: software requirements, simulation and modeling, and smart/intelligent food materials.

6.4.1.1 Software

The fundamental notion of 4D food printing entails the development of a 3D model that integrates many forms of information, including materials, structures, and stimulation signals. The model is subsequently simulated to examine its printing process and potential deformations, color alterations, or other characteristics, facilitating the optimization of the

3D model. Upon optimization, the model is segmented utilizing slicing software, which instructs the printer during the printing procedure. Subsequent to the object's printing, methodologies and instruments are employed to evaluate the modifiable alterations. The quality of 4D food printing depends on the interaction of printing software (host/firmware), simulation software, modeling software, slicing software, and monitoring software, and these aspects are together denoted as printing management software.

Currently, software for 4D printing predominantly includes forward design, reverse design, and inverse design. Forward design emphasizes anticipating ultimate modifications depending on material structures, properties, and stimuli. This methodology constitutes preliminary research intended to elucidate concepts, theories, and linkages. In other words, it is predictive and goal oriented, focusing on developing designs based on material properties and expected outcomes (Figure 6.2). Conversely, the reverse design prioritizes biomimicry, drawing inspiration from the traits of flora and fauna. For example, flowers can modify their mechanical asymmetry in reaction to environmental stimuli such as heat, water, and light. Accordingly, reverse design is focused on attaining particular functionalities or forms. It emphasizes learning from nature to create new designs, focusing on mimicking natural systems for functionality. Lastly, the inverse design (Figure 6.2) utilizes computational algorithms to explore various configurations and material compositions, allowing designers to simulate how different factors affect the food's behavior under environmental stimuli. In contrast to industries such as automotive, aviation, military, or fashion, 4D food printing relies more on forward design due to constraints in the availability of intelligent food materials.

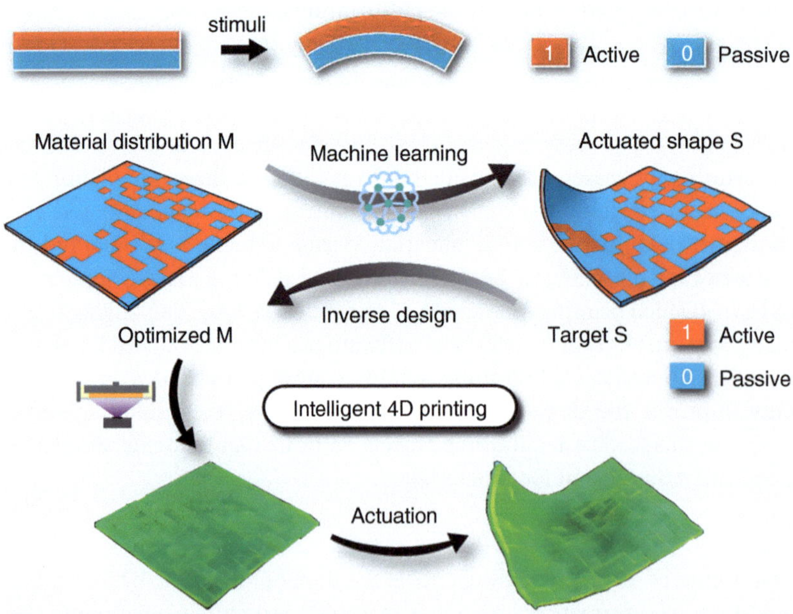

Figure 6.2 Fundamental 4D printing design models. *Source*: Sun et al. (2024) / Springer Nature / CC BY 4.0.

The printing commences with the development of a 3D model utilizing CAD software, subsequently transformed into a "Standard Tessellation Language" or "Standard Triangle Language" (STL) file. This STL file is segmented into 2D layers to strategize the printing trajectory for the product. Nonetheless, STL is restricted to a singular color and material, as it solely encompasses surface geometry devoid of color or texture data. OBJ (wavefront object) files can encapsulate color, material, and texture information, but Polygon (PLY) files are specifically intended for 3D-scanned objects. Moreover, novel file formats, including the 3D Manufacturing Format (3MF) and the Additive Manufacturing File Format (AMF), have been introduced. Both formats can encapsulate internal data regarding 3D models, encompassing the constructs' color, materials, and textures. In contrast to STL, AMF encodes geometric data to facilitate curved triangular tessellation, yielding more precise representations. 3MF enhances the functionalities of AMF by integrating all its technical attributes and utilizing a human-readable XML format for improved data management (Park et al. 2022).

The slicing software divides the 3D object into 2D segments and creates machine codes that communicate with the printer. This includes layering information, extrusion fan speed, printing speed, and deposit amounts of materials on the printing platform (Anukiruthika et al. 2020). The most familiar software used in 3D food printing is Crafware, Cura, KISSSlicer, Simplify 3D, and Slic3r (Anandharamakrishnan et al. 2022a). Cura and Crafware are intuitive, whereas Slic3r and KISSSlicer have faster slicing velocities. Guo et al. (2019) recognize KISSSlicer and Simplify 3D for their outstanding slicing quality. Currently, KISSSlicer and Simplify 3D are widely used slicing software for 4D food printing research.

6.4.1.2 Simulation and Modeling

One of the most critical and challenging functions in 4D food printing is determining the printing parameters. In addition to conventional techniques, such as evaluating the properties of materials and fine-tuning printing parameters, simulation and modeling software can offer more effective optimal solutions without requiring much trial and error. Software technologies that explain a model's deformation prior to printing, such as Virfac AM, Simufact Additive, and 3D Experience, can be used for this. For instance, a 4D-printed model created using simulation software employed a tilting mechanism induced by parallel stresses at the base of the petals. Additionally, an infrared thermal imager monitored temperature fluctuations during the deformation of the 3D-printed flowers (Figure 6.3) (Guo et al. 2022). These approaches can simulate the textural modifications in 4D-printed foods.

By simulating physical properties such as viscosity, pressure, and material homogeneity, the simulation software may also enhance the printing process and reduce the likelihood of printing errors (Srivastava et al. 2023). This software is primarily classified into two categories: those employing finite element analysis, including Abaqus, POLYFLOW, and COMSOL, and those created using programming languages such as JavaScript, C++, C#, and OpenFL (Teng et al. 2021). Recently, the "4D Printing Simulation Technology" project at the Gwangju Institute of Science and Technology in South Korea has created a software named 4D Modeler, which supports standardized evaluations based on external conditions and 3D printing materials to generate shareable data. These details were presented in previous chapters. Additionally, modeling software functions as the framework for printing, with tools such as 123D, Autodesk AutoCAD, CATIA, SolidWorks, Rhinoceros, and Blender employed for 4D food printing applications (Teng et al. 2021).

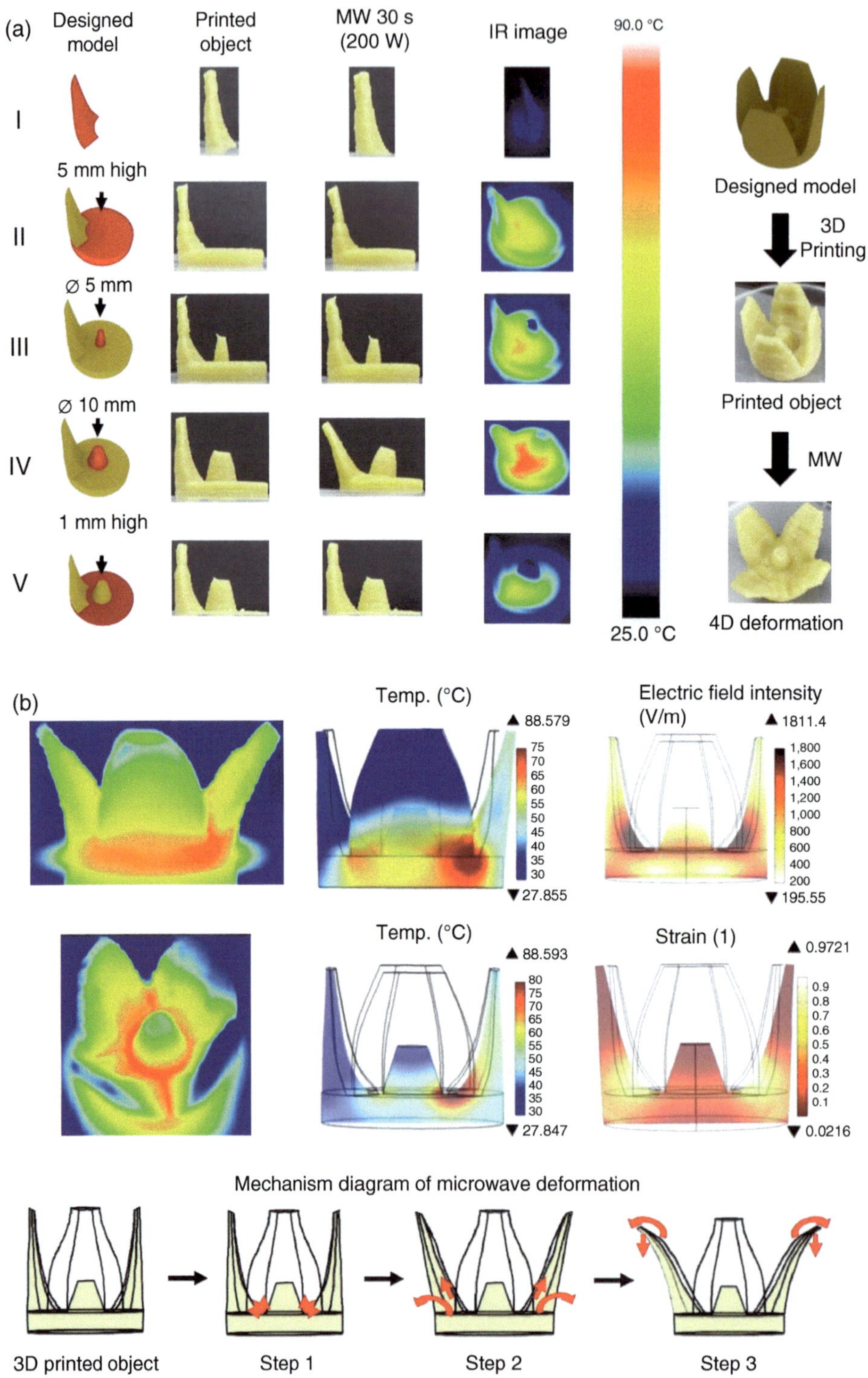

Figure 6.3 (a) Schematic illustration of 3D-designed and -printed models and 4D food printing shape deformation, (b) simulation model for different stimuli responses. *Source*: Guo et al. (2022) / with permission of Elsevier.

Within the past few years, various 3D scanners have been developed to facilitate the production of personalized things. Numerous digitization solutions employ laser scanners such as HandyScan and Faro, structured light technologies such as Vorum, Artec, and Sense 3D, or photogrammetry software utilizing conventional cameras, including PhotoModeler, 3DSOM, My 3D Scanner, PhotoScan, 123D Catch, Hypr3D, and RhinoPhoto. These tools produce a 3D point cloud of the item by optical triangulation to accurately represent its inherent geometry. However, the selection of the most suitable 3D scanner is contingent upon the particular application and the necessary degree of precision (Silva et al. 2024).

6.4.1.3 Smart/Intelligent Food Materials

The printability of materials is a crucial consideration for the progression of 4D food printing. Food materials can significantly modify their properties in reaction to external or internal stimuli. Early materials utilized in 4D food printing include soy protein isolate, starch gels, and hydrogel systems, including gelatin, sodium alginate, pectin, xanthan gum, carrageenan, konjac gum, oleogel, etc (Cheng et al. 2022). Essentially, the selected materials should possess the basic rheological characteristics for printing (e.g., shear-thinning for extrusion-based food printing applications) (Figure 6.4). Increase in dimensions can come with increased complexities in printing materials. These regulated modifications can be accomplished through the design of their components and structures, utilizing uniform distribution, gradient distribution, or distinct patterns in simulation software. To date, many materials have been reported in food printing, and these were detailed in previous chapters.

A crucial attribute for extruding food materials in 4D printing is their printability, which pertains to both the material's shear-induced behavior within the printer head and its performance and stability after printing. This consideration guarantees a seamless extrusion flow and adequate mechanical strength to preserve quality and texture postprinting (Kokane et al. 2024a). Optimal printing materials must possess three primary characteristics: (i) appropriate viscosity and shear-thinning behavior for enhanced flowability, (ii) adequate yield stress and viscoelastic modulus for self-supporting printed structures, (iii) exceptional thixotropy, facilitating the swift recovery of molecular structures in edible inks following shear force application. Many researchers have documented the food materials' formulations, and it is evident that their characteristics significantly impact printability (Derossi et al. 2024; Phuhongsung et al. 2022). Further, adding hydrocolloids can significantly improve the materials' printability and printing resolution (Santhoshkumar and Moses 2025). Therefore, optimizing printing formulations, including ingredient selection, levels of addition, and preprocessing, are essential in developing materials for 4D printing of foods.

The realization of 4D food printing necessitates a focus on manageable transformations and the intended attributes of the final product, in addition to fulfilling the basic printability requirements for 3D food printing. To achieve stimulus responsiveness, food ink's constituent ratios and types are crucial. For instance, changes in pH can cause a 3D-printed food's color to change (Shanthamma et al. 2021). During the 4D food printing process, careful attention must be paid to the moderate inclusion of certain components, as excessive amounts can significantly hinder the desired transformations of the food. For instance, high levels of proteins and chitin can result in a grainy texture, complicating uniformity and heat distribution when using microwave heating (Noree et al. 2023). 4D food printing-based studies have reported utilizing starch-based materials and protein hydrogels to execute the intelligent function in printed food materials (Table 6.1).

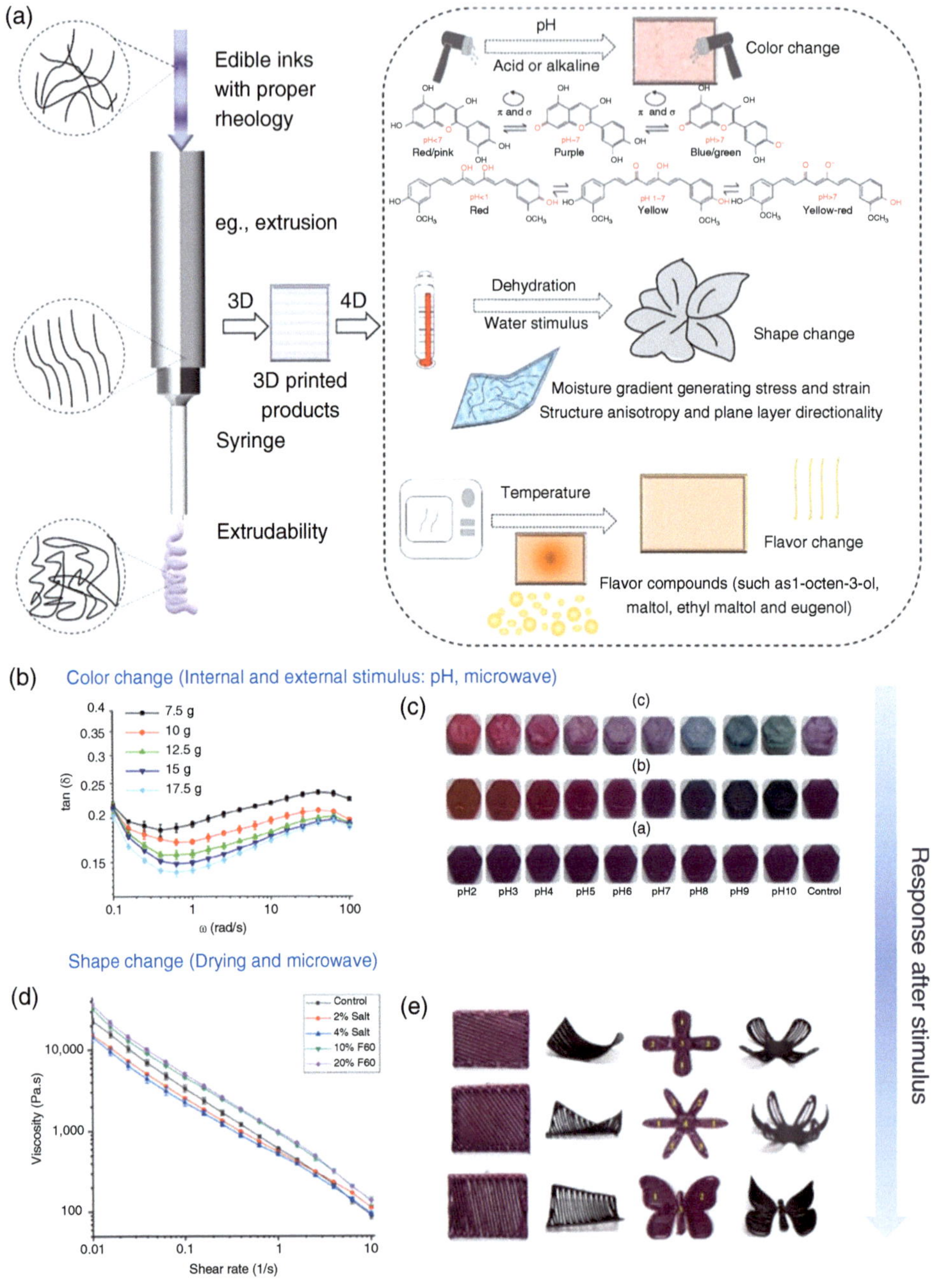

Figure 6.4 Rheology and intelligent changes in 4D food printing. (a) Schematic flow of 4D food printing process. *Source*: Cheng et al. (2022) / with permission of Elsevier, (b) loss tangent factor. *Source:* Ghazal et al. (2021) / with permission of Elsevier, (c) color changes due to internal stimulus. *Source*: Ghazal et al. (2021) / with permission of Elsevier, (d) shear ramp, (e) shape changes due to external stimulus. *Source*: He et al., (2020) / with permission of American Chemical Society.

Table 6.1 4D food printing materials and their applications.

Material formulations	Types of printers	Stimuli	Intelligent action	Application/description	References
Edible insect inks (cricket and silkworm), sodium alginate	Extrusion-based or FDM	Thermal (heating at 80 °C)	Shape	• Multinutrient structures	Noree et al. (2023)
Fruits and vegetable paste (yellow flesh peach-buckwheat past), calcium chloride	Extrusion-based or FDM	Thermal (microwave – 700 W)	Color, shape, aroma/flavor	• Interact with consumers in a short time • Highly attracted by childrens	Guo et al. (2022)
Purple sweet potato puree, sodium alginate, fructose	Extrusion-based or FDM	Thermal (microwave – 2 W/g)	Shape	• Enhance consumers' experience when consuming food	He et al., (2020)
Ethylcellulose, gelatin	Modified extrusion-based 3D printer	Hydro (water immersion)	Shape, color	• Complex structure fabrication	Pulatsu et al. (2022)
Mashed potatoes, lemon juice–starch gel, anthocyanin, baking soda, curcumin	Dual-nozzle extrusion-based printer	Thermal (microwave)	Color	• Colorful and attractive personalized foods	Zhu et al. (2023)
Red cabbage juice, vanillin powder, potato starch, lemon juice, orange juice, apple juice	Dual-extruder food printer	pH	Color, texture, flavor, taste	• Attractive sensory attributes	Ghazal et al. 2021
Edible gel (xanthan gum, potato flakes, potato starch, fructose syrup, oil, salt)	Dual-nozzle food printer	Thermal (microwave, infrared, hot air)	Shape changes	• Bending accumulation effect on food materials • Spontaneous shape changes	Liu et al. (2021)
Lotus root powder gel (lotus root powder, curcumin emulsion, sodium bicarbonate)	Extrusion-based food printer with bed temperature	Thermal (microwave)	Color, texture	• Two different new sensory characteristics increase the sensory acceptance of printed products	Chen et al. (2021a)

(Continued)

Table 6.1 (Continued)

Material formulations	Types of printers	Stimuli	Intelligent action	Application/description	References
Purple sweet potato paste, mashed potato, salt, butter	Extrusion-based 3D printer (double-nozzle or multinozzle printers)	Thermal (microwave)	Shape	• Shape shifting due to the addition of additives (salt or butter)	He et al. (2021)
Cricket powder, purple potato puree	Extrusion-based or FDM	pH	Color, texture	• 4D-printed insect-based protein-rich human food	Kang et al. (2023)
Ergosterol, purple sweet potato pastes	Extrusion-based or FDM	Nonthermal (UV-C irradiation)	Nutrition	• Conversion of ergosterol into vitamin D_2 due to stimuli	Chen et al. (2021d)
Buckwheat dough, yellow flesh peach, sodium alginate, gelatin, gum arabic, corn oil, pepper red pigment, cinnamaldehyde essential oil	Extrusion-based or FDM	Thermal (microwave)	Color, aroma	• Vitamin-based microcapsules	Guo et al. (2021)
Surimi, xylooligosaccharides, calcium chloride, carrageenan, xanthan gum, hydroxymethyl cellulose, anthocyanins	Dual-nozzle food printer	pH	Color	• Freshness monitoring	Zhan et al. (2023)
Beef paste, xanthan gum, salt	Dual-nozzle food printer	Thermal (oven drying, freeze drying)	Shape	• Meat based • No textural variations were observed between 3D and cast models	Dick et al. (2021)
Phytosterol nanoparticles, κ-carrageenan, soybean oil, cocoa powder, sugar, cocoa butter	Direct-ink-writing	Thermal (microwave)	Shape	• Highly customized • Chocolate printing	Jiang et al. (2022)

High internal phase emulsions (sunflower seed oil, β-cyclodextrin, sodium chloride, curcumin, sodium hydrochloride)	Extrusion-based food printer or FDM	pH	Color	• Emulsion-based 4D printing	Li et al., (2023)
β-cyclodextrin, sodium carboxymethyl cellulose	Extrusion-based food printer or FDM	Thermal (hot plate)	Color	• Personalized pickering emulsion food with bright colors	Guo et al. (2025)
Chestnut powder composite paste, sodium alginate	Extrusion-based food printer or FDM	Thermal (microwave, infrared heating)	Shape	• Increased flavor, texture, and swallowability • Personalized and rapidly shape-changing behavior of printed food	Lv et al. (2024)
Gelatin/β-cyclodextrin, curcumin, mushroom powder, soy protein isolate	Extrusion-based food printer or FDM	Thermal (microwave)	Color, flavor	• 4D-printed food with special color and target flavor based on microcapsule technology	Xiao et al. (2024)
Betanin powder, gelatin, nano-chitin, surimi	Dual-nozzle food printer	Thermal (superheated steam device, microwave device, steam heating device, steam-microwave device)	Color	• Surimi is potential transitional dysphagia food	Zhu et al. (2024)
Yam gel (soybean protein isolate-chitosan-maltodextrin, betacyanin, rose essence)	Extrusion-based food printer or FDM	Thermal (microwave)	Color, shape, flavor	• Microcapsules • Releasing internal pigments and essence and enhancing the redness and flavor of printed samples	Chen et al. (2024b)

(*Continued*)

Table 6.1 (Continued)

Material formulations	Types of printers	Stimuli	Intelligent action	Application/description	References
Pickering-type high internal phase emulsions (starch nanomaterials, including starch nanocrystals or starch nanoparticles)	Extrusion-based food printer or FDM	Thermal (freeze-drying)	Nutrients or bioactive	• Drug delivery • Smart releasing of bioactive	Shahbazi et al. (2024b)
Coix seeds, purple potato powders, xanthan gum, citric acid	Extrusion-based food printer or FDM	Thermal (microwave, infrared)	Shape	• Magical bidirectional deformation was observed	Niu et al. (2023)
Peanut protein, carrageenan or gellan gum, purple sweet potato flour, anthocyanins	Hot extrusion-based food printer or FDM (at 65°C)	pH	Color	• Plant protein-based 3D-printed food	Lin et al. (2023)
Apple, potato starch gel with red cabbage juice, whey protein isolate, gum arabic, anthocyanin	Extrusion-based food printer or FDM	Thermal (microwave)	Color	• Novel, colorful, and appealing 4D healthy food products that stimulate consumer appetite	Ghazal et al. (2023)
Yam gel (yam powder, gellan gum), pumpkin gel (pumpkin powder, gellan gum), spinach gel (spinach powder, gellan gum)	Extrusion-based food printer or FDM	Thermal (microwave)	Shape	• Vegetable-based gel using 4D-printed food	Chen et al. (2023b)
Soy protein isolate, oat flour, glycerol	Extrusion-based food printer or FDM	Thermal (microwave)	Shape	• Lower hardness resulted in a better sensory property (texture-modified product)	Guo et al. (2023)
Curcumin, whey protein isolate, potato starch, xanthan gum	Extrusion-based food printer or FDM	Thermal (catalytic infrared drying, microwave drying, hot air drying)	Color	• Higher nutritional retention • Stability of curcumin enhanced	Shen et al., (2023)

Mixed vegetable gels (Chinese cabbage puree, carrot powder), xanthan gum, white vinegar	Extrusion-based food printer or FDM	Thermal (hot air drying)	Shape, color	• Chlorophyll-rich vegetarian dishes show deformation and discoloration before consumption	Huang et al. (2022)
Purple potato puree, oleogel	Dual-nozzle extrusion-based food printer	Thermal (microwave)	Shape	• Oleogel anticipated vegetable-based materials	Shi et al. (2022)
Sago starch, turmeric	Extrusion-based food printer or FDM	pH	Color	• Customization of foods for enhanced consumer acceptability	Shanthamma et al. (2021)
Potato starch, anthocyanin, lemon juice gel	Extrusion-based food printer or FDM	pH	Color	• Healthy and attractive products through 4D printing approaches	Ghazal et al. (2019)
β-cyclodextrin, citrus pectin, sunflower seed oil, curcumin	Extrusion-based food printer or FDM	Thermal (microcomputer-controlled heating), pH	Color	• High flexibility produces novel food design and format	Cen et al. (2022)

6.4.2 Printers and Printing Types

The principal categories of printers utilized in 3D food printing are fusion deposition modeling (FDM), stereolithography, straight ink writing, inkjet printing, digital light processing, and selective laser melting. At present, FDM is the predominant technique for 4D food printing owing to its versatility, cost-effectiveness, and ease of usage, especially in extrusion-based 3D printing approaches (Anandharamakrishnan et al. 2022a). However, in 4D food printing, intelligent materials and design play a pivotal role, necessitating the consideration of mechanisms, anticipated behaviors, and specific parameters in relation to material qualities. These are accomplished by utilizing multimaterial printing approaches in other applications; in food systems, this aspect is crucial due to the challenges associated with the material selection in printing. Additionally, multimaterial printing can fabricate more intricate structures and create multifunctional duties in a single object (Hertafeld et al. 2019). An interesting feature of 4D food printing is the capability to produce multimaterial constructs, improving the adaptability and usefulness of the product. Multimaterial printers may fabricate hierarchical structures by incorporating two or three distinct materials in a single construct (Dick et al. 2019; Lee et al. 2023). 4D food printing through multimaterial still requires special specifications in the context of printers. For instance, specific biopolymers successfully react to external stimuli; nevertheless, they necessitate particular printing settings that may require modifications to the printer (Patrocinio et al. 2023).

Printers are integrated with supplementary devices such as cooling platforms, lasers, and microwaves. The application of infrared or laser technology in conjunction with the printing nozzle facilitates accurate localized heating, permitting distinct regions of the printed product to undergo variable degrees of thermal treatment while printing the food ingredients, thus finally delivering the ready-to-eat products (Blutinger et al. 2021, 2023; He, Zhang, & Fang, 2020; Hertafeld et al. 2019; Tong et al. 2024; Waseem et al. 2024). This capability can potentially enhance heat-stimuli changes, including shape, color, and nutrients. Changes in the properties of food components in printers (e.g., microwave, laser) can affect heat absorption rates and alter the final attributes of the printed product, as well as modifications in any stimuli such as microwave power or thermal addition, as well as in material formulation changes (Domżalska and Jakubczyk 2025). Furthermore, ongoing research in additive manufacturing (AM) of food involves the usage of multimaterial and multinozzles to print single food constructs with multiple materials, adaptive nozzles to change the layer thickness while printing and multiheads to enhance the production of printed constructs. Such advances are anticipated to take proper shape in the upcoming decade of food printing advancements, providing substantial benefits for 4D food printing and beyond.

6.4.3 Postprinting/Processing

The controlled responses of food substances to external stimuli in 4D food printing are significantly impeded by the intricate interactions between the components of food materials and these stimuli. To ensure that printed food structures can effectively demonstrate desired alterations, such as changes in shape, structure, and functionality, it is essential to assess the properties of the stimulus-responsive components meticulously. As 4D food printing necessitates state transitions triggered by external stimuli, the quality of the printed models and the integrity of the transformations that have occurred are both vulnerable to excessive postprocessing. This may result in increased food waste and reduced economic efficiency. Therefore, regulating postprocessing conditions is essential for preserving the longevity of the modified products (Theagarajan et al. 2021). Most 4D food printing is in the aspect of color and shape changes that are obtained by adapting thermal and ion concentration variations (pH) as a stimulus. Thermal and dehydration procedures, such as microwave, infrared, and hot air treatments, considerably influence sample deformation, underscoring the necessity of choosing suitable post-treatment methods (Liu et al. 2021). For instance, due to microwave treatment, heating, cooking, and drying can take place (Figure 6.5a); further, shape shifting was observed in different edible gels

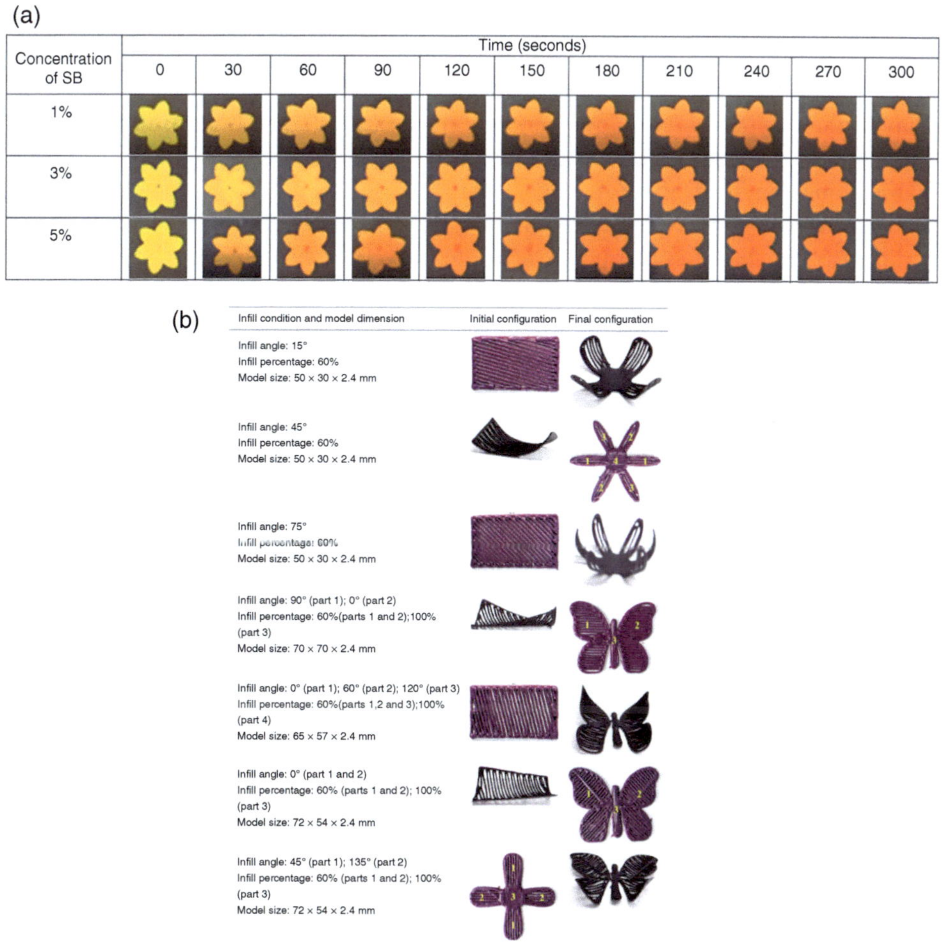

Figure 6.5 4D food printing. (a) Color changes due to pH stimuli. *Source*: Shanthamma et al. (2021), (b) shape changes due to thermal stimuli. *Source*: He et al., (2020) / with permission of American Chemical Society.

(Table 6.1). Similarly, color variations were observed in sago starch infused with curcumin with changes in the pH of the printed sample (Shanthamma et al. 2021) (Figure 6.5b).

Enhancing postprocessing consistency is essential for achieving more uniform 4D transformations. Proposed strategies include altering the microwave setup, regulating the electromagnetic field, and improving the microwave drying technique. Similarly, for infrared and other drying methods, different process parameters that affect the heating rate must be controlled. Furthermore, modifying the printed samples' structure/design can improve uniformity; for instance, modifications in the sample's center can dissipate excess heat while reducing the margins can result in consistent color variations. Apart from designs and geometries, ingredient composition, such as variations in sugar, salt, and moisture, impacts the dielectric characteristics of foods (He et al., 2020). For instance, enhancing ion conduction by altering material formulations can improve heat transmission and homogeneity in microwave processing. In addition, regulating the postprocessing treatment's duration is essential; prolonged periods may result in adverse deformations, including hardening, over-expansion, or significant shrinkage (Santhoshkumar et al. 2024b; Theagarajan et al. 2021). It is imperative to recognize that the quality and safety of food can be compromised by excessive pH levels when pH is used as an external stimulus. For example, the functionalities of food may be influenced by pH levels, as unsuitable levels may reduce the viscosity and rheological behavior of the printing ink, resulting in structural failure during printing (Chen et al. 2021b; He et al., 2020). Overall, it is crucial to thoroughly evaluate various postprocessing conditions, including their duration, intensity, and relationship with the internal structure of printed objects. Such a comprehensive approach is vital for the effective implementation of 4D food printing technology, ensuring quality consistency of printed foods and, ultimately, improvements in the efficiency of postprocessing.

6.4.3.1 Postprinting Analysis

Currently, 4D food printing reports color, shape, texture, flavor/aroma, and nutrition as a fourth dimension of 3D-printed foods. Various instrumental and image analysis approaches are used to assess these fourth dimensions of printed foods. For instance, colorimeters, spectrophotometers, chroma meters, and photomicrographs have been utilized to evaluate the color of printed constructs. Similarly, shape deformation has been assessed by considering parameters such as bending angle, shrinkage ratio, infill structure and dimensions, dielectric properties, aspect ratio, film thickness, swelling characteristics, height changes, and bending curvature. Further, the deformation of printed foods has also been quantified with image analysis software, while a transparent online protractor can assess variations in bending angle, porosity, and microstructure (Chen et al. 2024b).

In terms of flavor/aroma, alterations have been estimated through the use of electronic noses (E-nose), electronic tongues (E-tongue), biosensors, and gas chromatography (Chen et al. 2024b; Ghazal et al. 2021, 2024; Phuhongsung et al., 2020). Additionally, nutrition changes have been quantified by using advanced instruments such as high-performance liquid chromatography (HPLC) (Kokane et al. 2024b). The bending angle for shape transformation was assessed using image analysis methods; the captured image was imported to Rhino 7.0 software (McNeel Asia, Seattle, WA, USA) to measure the initial angle as (A1) and the bending angle (A2) after the treatment. The angle change rate (A) was calculated using Equation (6.1) (Guo et al. 2023). Then, the same researchers calculated mass variations after treatments. In addition, an equation to assess the extent of completion of the 4D transformation process, which includes alterations in color, flavor/aroma, and deformation, was used (Eq. 6.2; Guo et al. 2022).

$$A = \frac{A1 - A2}{A1} \times 100\% \tag{6.1}$$

$$Completion\ degree = \frac{C_t - C_0}{C_{\max} - C_0} \times 100\% \tag{6.2}$$

where C_t denotes the values of color, flavor, and bending angle immediately before the administration of stimuli at a time (t), C_0 signifies the value subsequent to the application of stimuli, and C_{max} represents the greatest value attained. Consequently, 4D transformation can be defined by the previously specified tools, techniques, equations, and characteristics of printed food. However, overall, there is a lack of in-depth studies on such aspects involving the assessment of transformations.

6.4.3.2 Stimulus

Stimuli are crucial for inducing alterations in the functions, characteristics, and geometries of 4D-printed materials. These stimuli may originate from several sources, encompassing physical, chemical, and biological components. Herein, physical stimuli include temperature, humidity, illumination, magnetism, and electrical energy; chemical stimuli encompass pH levels, ionic strength, and the presence of certain compounds; and biological stimuli are linked with chemicals such as glucose and enzymes. Based on their mode of action and activation mechanisms, they are also classified as internal and external stimuli. Popular stimuli reported in food printing applications are mentioned in Table 6.1. The mechanisms behind such stimuli and consequent changes in food materials are detailed in the upcoming chapter.

6.5 Role of Smart/Intelligent Materials in 4D Food Printing

Major food ingredients and additives utilized in 4D food printing are majorly classified into four categories based on their nutritional components: hydrocolloids, polysaccharides, proteins, and natural colors and taste components. Each such component represents an array of unique and diverse properties. Typically, formulations for 4D food printing are developed by integrating different ratios of these components. The remaining components are selected according to their printability and reactivity to stimuli. A primary goal in food printing research is to determine the most efficient printable formulation suited for certain applications.

6.5.1 Polysaccharides

Polysaccharides such as starch, pectin, and cellulose derivatives are essential in 4D food printing because of their gelling, thickening, and stabilizing characteristics. Polysaccharides are polymeric carbohydrates composed of at least 10 monosaccharides and possess many active groups that facilitate noncovalent interactions, including hydrogen bonding, electrostatic interactions, and van der Waals forces. Under specific conditions such as pH, temperature, and time, these polysaccharides can achieve complete hydration, yielding solutions that can function as the matrix for print inks (Cheng et al. 2022). The water-soluble polymers that form gels and modify viscosity are called hydrocolloids. Agar, carrageenan, gelatin, xanthan gum, guar gum, and sodium alginate are among the most popular hydrocolloids in food printing applications. The characteristics of various gels are contingent upon their ability to establish a bond with the food ingredients/inks.

The sources of hydrocolloids significantly influence the characteristics of printing inks. For instance, agar, derived from marine algae, is renowned for producing resilient gels that liquefy at elevated temperatures (Donn et al. 2022). Gelatin, derived from animal collagen, can form thermoreversible gels that solidify upon cooling and liquefy when heated, providing the flexibility necessary to develop dynamic food textures (Lueyot et al. 2021). Similarly, plant-origin hydrocolloids, including guar gum, konjac gum, gellan gum, pectin, and sodium alginate, are suitable for food printing applications and find applications in improving the printability of the material supply and intricate shape transformation (Chen et al. 2022b; Santhoshkumar and Moses 2025; Santhoshkumar et al., 2025). Furthermore, the starch, derived from various plants, undergoes gelatinization when heated, allowing the modification of the structural texture (Rashwan et al. 2024); when acid and sugar are combined with pectin extracted from fruits, polymers form, making it an ideal ingredient for creating responsive desserts or gummies (Cen et al. 2022; Said et al. 2023; Keerthy et al., 2025). Additionally, cellulose derivatives, such as ethyl cellulose, regulate viscosity and stabilize emulsions, an essential process for precisely printing intricate designs (Pulatsu et al. 2022). However, the precise applications of polysaccharides-based derivatives in non-food and 3D food printing applications are relatively under-explored in 4D food printing applications.

6.5.2 Protein

Proteins comprise amino acid residues that are connected by peptide bonds. These result from the guided hierarchical assembly of weak contacts, including hydrogen bonding and van der Waals forces. Random coils, α-helices, β-folds, and β-turns are structural components that impart unique functionalities to the print inks. Additionally, hydrophobic interactions, salt bridges, and van der Waals forces contribute to the structural integrity of 3D-printed foods after printing, thereby preserving the intended morphologies (Cheng et al. 2022). The gelling, emulsifying, and foaming properties of proteins, such as soy protein isolate, whey protein, and casein, are essential materials for 4D food printing. Soy protein isolate can generate heat-induced polymers that aid in shaping printed foods (Phuhongsung et al., 2020). In another study, the authors reported that whey protein gelling properties improve the texture and stability of printed foods (Shen et al. 2023a). The structural integrity of printed food is enhanced by the esteemed capacity of casein, the predominant protein in milk, to generate stable polymers and foams (Cheng et al. 2022). Recently, the synergistic effects of addition with other materials were detailed focusing on the printability aspect by Cheng et al. (2022).

6.5.3 Bioactive Components

Bioactive compounds sourced from various organic materials, including curcumin, betalains, and anthocyanins, are valued for chromatic diversity. Anthocyanins, found in various fruits and vegetables, change color in response to pH fluctuations, which allows for the evaluation of flavor or freshness modifications of the product. Most studies on 4D printing of foods reported the color changes of printed constructs with a specific compound (Li et al. 2023a; Navaf et al. 2022; Shanthamma et al. 2021). Additionally, the esthetic appeal of printed foods is further enhanced by the intense colors from betalains, which are present in beets and other flora and can undergo modifications in response to pH fluctuations (Zhu et al. 2024). Furthermore, other than these bioactive compounds, caffeine, phytosterols, quercetin, and resveratrol have been reported in 3D-printed food (Kavimughil et al. 2022; Tambe et al. 2023). Figure 6.6 presents some examples of pH-based stimuli changes in 3D printed foods induced with natural pigments.

The essential factor for inducing color change in 4D printing is the reaction of pigments to pH variations. Consequently, the incorporation of pigments in food substances is essential. In particular, structuring oils (oleogel, bigel, emulsion gel) provide considerable benefits owing to their distinctive encapsulating characteristics, which are compatible with a majority of oil-soluble and water-soluble pigments (Swathika et al., 2025). In such applications, encapsulation boosts stimulus-response properties and improves the stability of pigments, especially natural ones. Structuring oils have applications beyond the food processing sector; for example, they can be utilized to develop intelligent indication labels for food freshness. Decomposing food produces diverse acid-base gases, and 4D printing inks composed of structural oils can respond to these gases, facilitating visual assessment of food freshness (Cen and Meng 2024). Nonetheless, additional studies are required to actualize such concepts and bring them out for commercial-level applications.

6.5.4 Lipids

Lipids and fats are gaining increasing attention in the realm of 4D printing. In AM applications, these materials are commonly referred to as "structuring oils." Structuring oils are a solid or semisolid gel with plasticity. These gels primarily consist of natural and edible colloids or particles that encapsulate vegetable oils within a three-dimensional network. This structure is formed through interfacial and intermolecular interactions. These categories include oleogel, emulsion gel, and bigel (Cen and Meng 2024). For instance, emulsion gels, a type of structuring oil, demonstrate considerable potential in 4D printing (Cui et al. 2023). Nonetheless, investigations into the utilization of structural oils such as emulsion gels, oleogels, and bigels (Figure 6.7) as inks for 4D applications remain limited. Notwithstanding this, the remarkable rheological characteristics and distinctive architectures of these oils indicate their potential significance in the advancement of individualized 4D-printed products. Their exceptional rheological properties establish a robust basis for 3D printing, while their unique biphasic structure presents intriguing opportunities for developing 4D printing inks with stimulus-responsive functionalities.

Moreover, recent progress in 4D printing utilizing structuring oils underscore their thermal sensitivity and the contribution of biomacromolecules in facilitating deformation. Jiang et al. (2022) explained that the thermal treatment of palm kernel stearin compromises

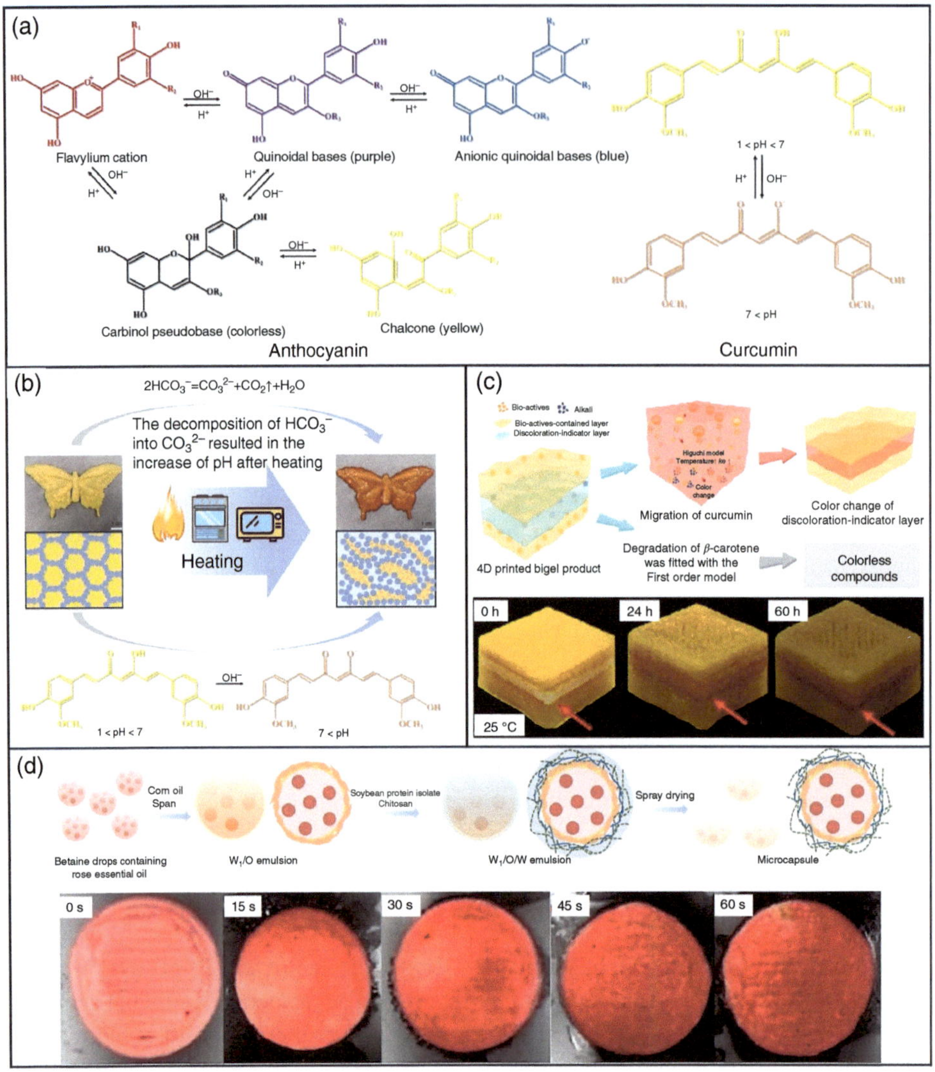

Figure 6.6 Examples of natural pigment-induced color changes due to pH change. (a) Anthocyanins and curcumin. *Source*: Cen and Meng (2024) / with permission of Elsevier, (b) curcumin-based high internal pickering emulsion. *Source*: Cen and Meng (2024), Li et al. (2023) / with permission of Elsevier, (c) curcumin based on bigels. *Source*: Cen and Meng (2024), Qiu et al. (2024) / with permission of Elsevier, (d) microcapsule and food gels. *Source*: Cen and Meng (2024), Chen et al. 2024a / with permission of Elsevier.

its fat crystal network, facilitating the morphological alteration of a pickering emulsion gel, with the model's tilt angle governing the deformation. Shahbazi et al. (2024a) created thermoresponsive poly-pickering high internal phase emulsions by leveraging the temperature-sensitive characteristics of methylcellulose and kappa-carrageenan to enable shape memory in printed scaffolds. Similarly, they used high pickering emulsion gel for drug delivery application to deliver bioactive compounds with a cold plasma treatment process for 4D printing applications (Shahbazi et al. 2024b). Both findings highlight that the primary deformation factors originate from the materials' intrinsic properties rather than the structuring oils utilized in derivative forms (Figure 6.7).

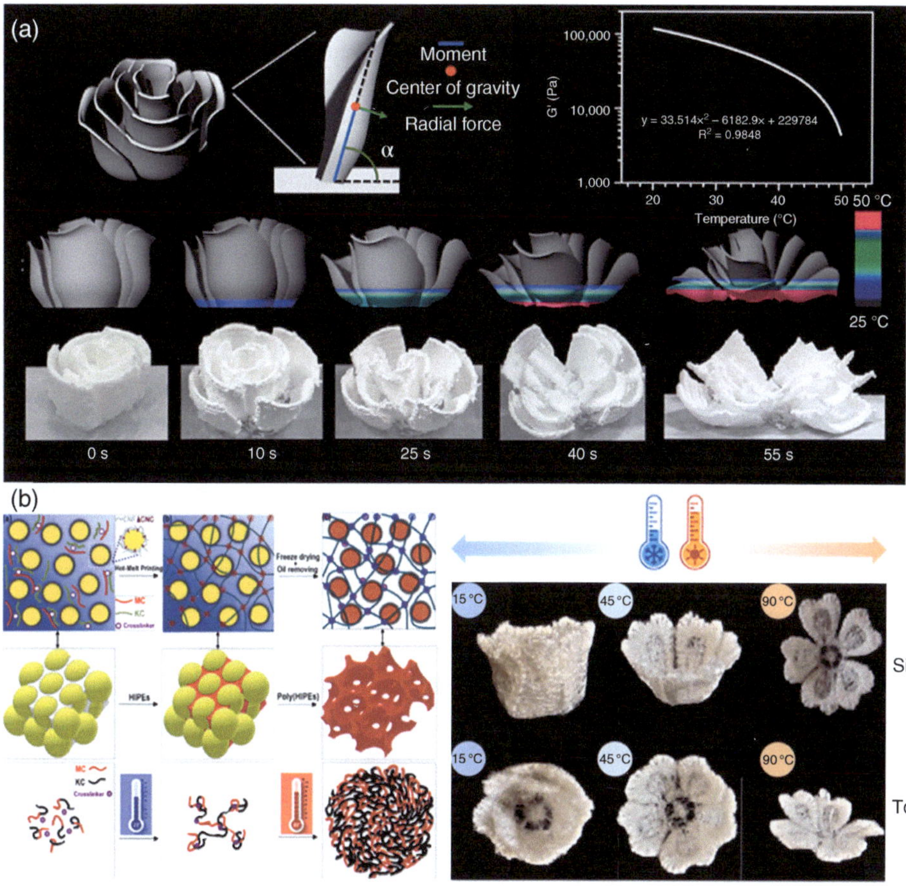

Figure 6.7 Structuring oils in 4D food printing. (a) Destruction of fat crystal structure. *Source*: Jiang et al. (2022) / with permission of Elsevier, (b) scaffold from emulsions. *Source*: Shahbazi et al. (2024a) / Elsevier / CC BY 4.0.

Chocolates with a high percentage of solid fat are more susceptible to the effects of temperature fluctuations than other types of chocolate. When fat crystals are heated, a change occurs in the network structure that comes into being, and the chocolate transforms from a solid to a liquid state (Qin et al., 2025). It appears that water migration has less of an effect on the properties of structuring oils when it comes to 4D printing, which suggests that these features are more closely tied to the fundamental properties of the materials. Overall, such intricacies illustrate that it is vital to carefully monitor process conditions, such as the temperature during the process of 4D printing, to prevent degradation effects and unpredictable results. The fact that this is the case highlights that more studies are required to successfully develop inks made from structural oils.

6.5.5 Essential Oils

The flavor characteristics of printed foods can be altered due to the implementation of microcapsules that contain flavor compounds and essential oils. These microcapsules are designed to release their contents under specific conditions. Such concepts perfectly extend to molecular gastronomy combined with food printing applications; there is ample scope in this field to merge these technologies into food sensory science (Subhasri et al. 2022). For instance, a study involving a gelatin-gum arabic-flavor oil complex demonstrated the temperature responsiveness of 3D-printed buckwheat dough with yellow-fleshed peach. The researchers found that when the product was treated with microwaves, there was a simultaneous change in both the color and the flavor of the product (Guo et al. 2021). Similarly, in work on yam gel-based microcapsules containing maize flavor oil, researchers observed multisensory effects, including color and shape, while applying microwave stimuli (Chen et al. 2024b). Further, in a study on a mushroom-based gel containing gelatin and β-cyclodextrin, researchers observed color change using essential oil (Xiao et al. 2024). While printed products are subjected to heat, the encapsulated compounds can be released, resulting in an explosion of flavor that can occur either during the preparation process or when the food is consumed (Tang et al. 2022). Overall, essential oils play a pivotal role, particularly in flavor release in 4D food printing applications, and several studies have demonstrated this idea. Similarly, other dynamic, intelligent functions and multisensory culinary experiences can be brought about with 4D food printing technology.

6.5.6 Cells

Bioprinting was an early advancement in 4D printing that used food-grade materials with different biological cells instead of conventional printing inks. A nutrient-rich substrate, under perfect conditions, including temperature, humidity, and light, is the ideal environment for the proliferation of edible cells. Rutzerveld (2014) reported that printed objects, composed of both plant and animal cells, can build tissue-like structures that match the nutritional profiles of actual food. In 2014, Rutzerveld introduced the concept of "edible growth food," which refers to snacks that are ready to consume and are made from seeds, fungal spores, yeast, and bacterial cells that are integrated inside a nutrient-dense matrix consisting of vegetables, fruits, nuts, and agar (Figure 6.8). This idea is in line with the growing interest in artificial meat and has the potential to benefit senior citizens who may have difficulties swallowing regular diets. For instance, a combination of carrot tissue and sodium alginate to generate inks and a chilled gelatin solution with adequate $CaCl_2$ served as the curing agent. In simulated conditions, at 37 °C, the cells proliferated, facilitating plant tissue regeneration that emulates the texture and nutritional value of authentic food (Park et al. 2020). The workflow for the material preparation is represented in Figure 6.9.

Earlier, using the 3D food printing technique, probiotics (*Lactiplantibacillus plantarum*) were encapsulated and subjected to different temperature-based postprocessing approaches (Yoha et al. 2021). Recently, using a 5D food printing approach, probiotic cells were reportedly cultured in starch-based yam-milk mixed gels along with the two probiotics bacteria (*Streptococcus thermophiles*, *Lactobacillus bulgaricus*), and their multidynamic changes were observed (Chen et al. 2024a). Additionally, animal cells can be printed and incubated to produce imitation meat at suitable temperatures (Benny et al. 2022; Rubio et al. 2020; Shabir et al. 2024). This innovative technique is gaining popularity, offering enhancements in flavor, texture, and nutritional quality while being quicker, more resource-efficient, and environmentally sustainable, potentially decreasing animal husbandry by 30% and greenhouse gas emissions by 15–24% (Teng et al. 2021). Overall, the usage of live cells in the printing medium remains an untapped field for applications in the field of 4D printing of foods.

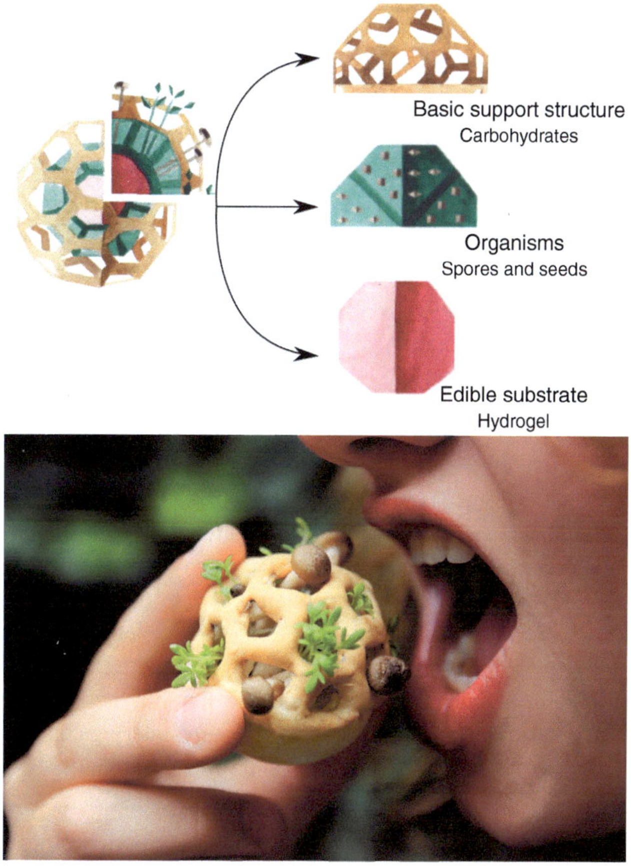

Figure 6.8 Illustration of edible food growth in a starch-supported structure. *Source*: Reproduced from Fahmy et al. (2023), Rutzerveld 2014 / MDPI / CC BY 4.0.

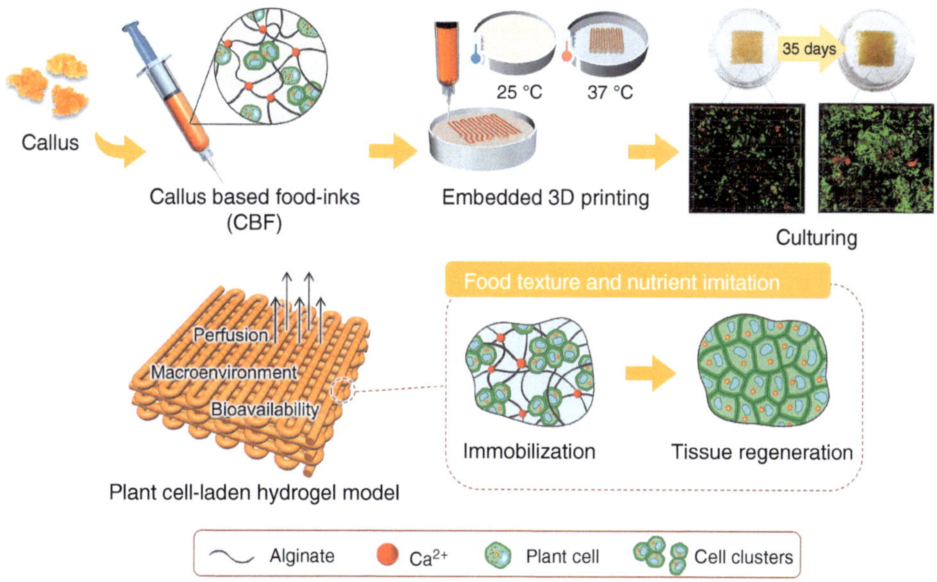

Figure 6.9 4D printing of living cells (4D bioprinting). *Source*: Park et al. (2020) / with permission of Elsevier.

6.6 Dynamic/Intelligent Structure Changes in 4D Food

Dynamic and intelligent structure changes in 4D food denote the capability of food to alter its shape, texture, or other characteristics in reaction to external stimuli, such as temperature, humidity, or pH levels. This concept integrates interdisciplinary principles of food science, materials science, and engineering to develop foods that can adapt or transform in innovative manners. Accordingly, this section details such changes, highlighting key findings from recent research and presenting avenues for futuristic development.

6.6.1 Shape

4D printing can refer to the programmable structural modifications of 3D-printed products postmanufacturing. Shape transformations may happen over many dimensions, including transitions from 1D to 1D, 1D to 2D, 2D to 2D, 2D to 3D, or higher levels (Momeni et al. 2017). Accomplishing such morphological alteration generally necessitates the formation of a bilayer configuration consisting of two materials exhibiting disparate expansion or contraction characteristics in reaction to stimuli. Three essential concepts dictate the properties of 4D-printed materials: (i) the differential expansion of active and passive materials results in diverse morphological alterations, including coiling, curling, twisting, and bending; (ii) there are four fundamental factors: mass diffusion, thermal expansion, molecular transformation, and organic growth that influence the shape-altering characteristics of these multimaterials, and (iii) the shape changes of 4D-printed multimaterials are governed by two temporal parameters: printing materials and stimuli (Ahmed et al. 2021; Momeni and Ni 2020). Foods prepared from purple sweet potato, potato, pumpkin, and cabbage were reported to undergo deformation with the inclusion of edible salt, sugar, and butter. The absorptive abilities of water and oil, together with the binding characteristics of the meals, enable shape transformations or deformations of the printed foods.

In food applications, morphological alterations can be induced by water absorption or desorption during cooking (hydration) or moisture removal (dehydration). For instance, hydrogels, characterized by their asymmetric swelling properties, are frequently employed in food printing as materials capable of altering their shape, which is suitable for 4D food printing applications. Hydration-induced morphological alteration enables programmed transformation, distinctive textures, enhanced shelf-life, and transportation efficiency (Navaf et al. 2022). These hydrogels possess a 3D network structure formed through chemical or physical interactions and comprise a hydrophilic material that facilitates morphological changes in response to various stimuli (Singh et al. 2024). Similarly, cooking can induce asymmetric swelling in composite cellulose, protein, or starch composite materials, altering 2D to 3D forms. Wang et al. (2017) conducted a study in which they developed pasta with ethyl cellulose to facilitate shape alterations during cooking in water.

Moreover, dehydration, a popular food processing method, can also cause alterations in the shape and texture of 3D-printed food substances. Water diffuses from the specimen during dehydration, resulting in shrinkage and distortion. The folding angle of a 3D-printed food positively correlates with moisture content and shrinkage ratio while demonstrating a negative correlation with the dielectric constant loss factor. Popular dehydration techniques reported in 4D food printing studies include microwave, infrared radiation, and air drying for different shape-shifting applications (Koirala et al. 2025; Liu et al. 2021). Tao et al. (2019) subsequently created an array of flour-based treats, including self-wrapping tacos, flavor-infused biscuits that fold autonomously, self-assembling noodles, and bespoke semolina pasta forms. This innovation utilizes dehydration to induce structural alterations in the printed food.

4D food printing has the potential to transform the manufacturing of textured proteins and meat substitutes. Conventional meat production is often explained to be resource-demanding and ecologically detrimental. In this context, 4D food printing can employ plant-derived proteins and alternative materials to fabricate meat analogs, presenting a sustainable solution (Miller et al. 2024; Reuben et al. 2025). Recently, different shape-morphing effects in food printing have been exclusively documented by Koirala et al. (2023). Figure 6.10 presents recent works in shape-shifting applications of 4D food printing.

6.6.2 Color

Color is an important factor in the sensory preference of food products and may positively or negatively influence consumers' thoughts of the printed food. The color alteration capabilities of 4D-printed foods through their intelligent function are key for the popularity of the approach in food applications. 4D-printed foods can be developed in a manner to attract the consumer by varying the color function. Dynamic pigment variations may also influence customers' expectations regarding flavor and taste. Especially, bioactive compounds such as curcumin and anthocyanin have been the primary focus of color change applications in most studies on 4D food printing related. Such nutraceutical compounds can transform from green to red in alkaline environments and exhibit red and yellow tints in acidic conditions (Shen et al. 2023a). By altering stimuli such as pH and temperature, this capability can produce dynamic color changes in 4D-printed foods.

With curcumin as a responsive material, a study was conducted on the microwave stimulation of 4D-printed lotus root powder gel. Influenced by the curcumin content and treatment duration, the microwave exposure transformed the printed food coloration from yellow to red. The alkalinity of the curcumin-containing medium was elevated due to the microwave-induced agitation of water molecules, which resulted in color changes. Enhanced release of curcumin into the surrounding medium and larger particle sizes can result from increased temperatures destabilizing emulsion structures (Chen et al. 2021a). Similarly, studies also reported microwaves as stimuli for 4D-printed buckwheat dough printability study (Guo et al. 2021).

Anthocyanin, a water-soluble pigment known for its pH-dependent color changes, was utilized in a purple sweet potato study (Ghazal et al. 2021; Su et al. 2025). A dual-extrusion printer was used to fabricate a composite of purple sweet potato puree and mashed potato, altering the pH of the mashed potato for the above-mentioned study. As the quantity of purple sweet potato puree increased, the color intensity increased, accompanied by an elevation in color ($L^*$and b^*) values. However, the a^* value diminished with greater concentrations of mashed potatoes in the printing ink. This phenomenon was attributed to the reduced diffusion of

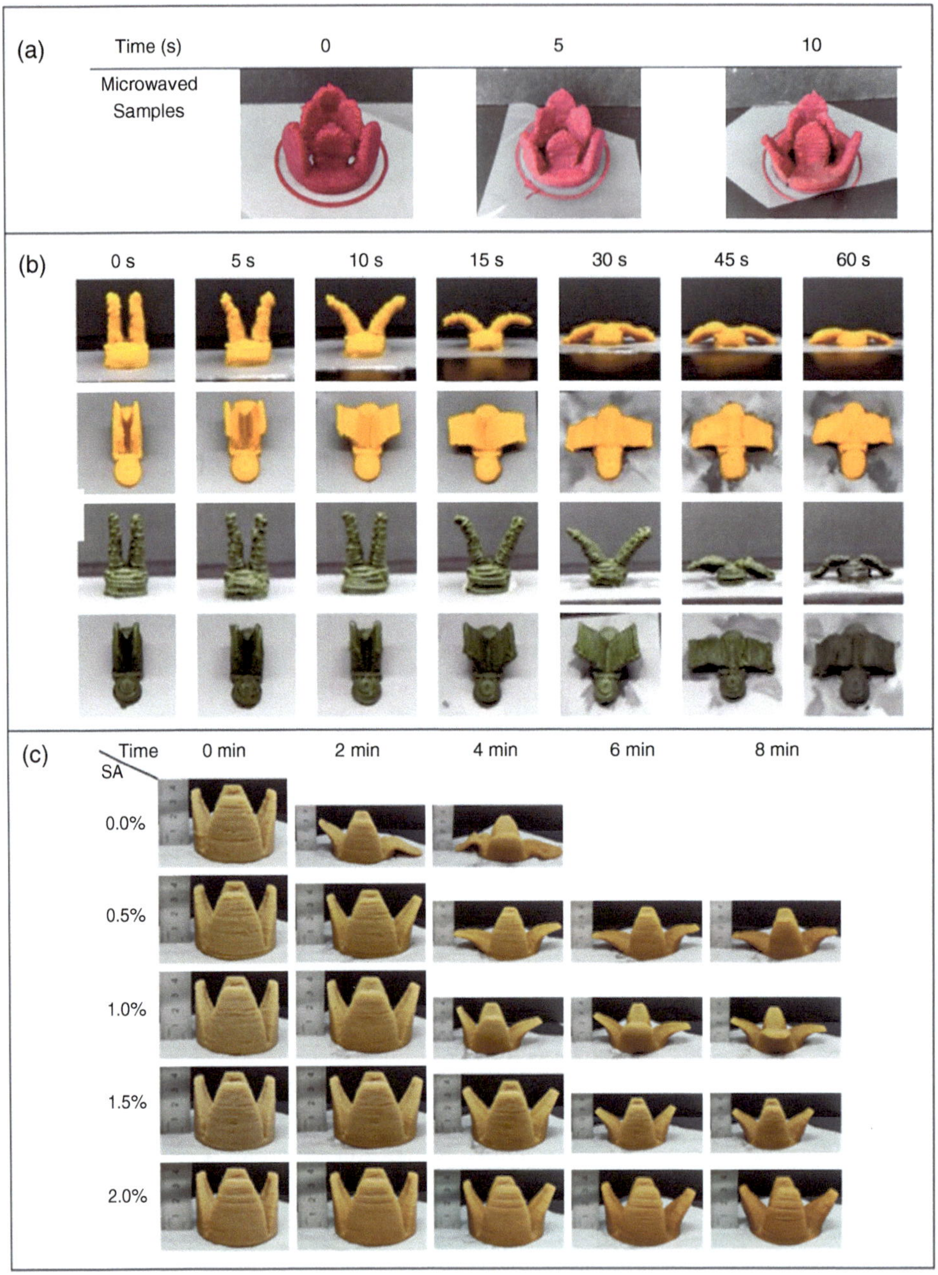

Figure 6.10 Shape transformation in 4D food printing. (a) Purple potato. *Source*: Niu et al. (2023) / with permission of Springer Nature, (b) pumpkin and spinach powder gel. *Source*: Chen et al. (2023b) / MDPI / CCBY 4.0, (c) sodium alginate. *Source*: Lv et al. (2024) / with permission of Elsevier.

anthocyanin between the two materials. As storage progressed, the mashed potato transformed from white to red, showing varying anthocyanin concentrations in various product zones (He et al., 2020). Recently, Niu et al. (2025) reported that anthocyanin microcapsules' color changes indicate that the growth of the probiotics provides 5D dynamic changes in food.

Similarly, multimaterial compounds consisting of gelatin-gum and arabic-oil complex coacervates can serve as stimulus-responsive materials. Microwave treatment boosted the redness of the dough while diminishing its lightness due to capsule destruction and oil release. The influence of surface pH on color variations in a 3D-printed mixture of soy protein isolate, pumpkin, and beetroot (rich in betalain) was also investigated. Researchers discovered that at pH values above 6, there were no significant color changes after one hour of storage; however, at lower pH levels, notable alterations occurred due to the oxidative degradation of pigments (Phuhongsung et al., 2020a, b). Significant color changes were also observed in a 3D-printed multismart material structure in response to internal and external pH stimuli. The application of various pH-controlled solutions on the printed structure changed the color across a spectrum from blue to red, purple, violet, blue, blue-green, and green-yellow (Figure 6.11). The color of a 3D-printed bilayer structure of starch and fruit juice gels varied over time due to internal pH stimuli from lower layers moving into the product, producing reversible structural changes in the anthocyanin molecules. The continuous deterioration of color pigments, particularly under acidic conditions, further exacerbated these alterations over time (Ghazal et al. 2021). Recently, utilizing the batanin complexes, a study on functionalized surimi reported 4D printing color changes, and Figure 6.12 illustrates the mechanisms and color variations thereof (Zhu et al. 2024). Similarly, in food packaging, this intelligent function concept is observed to be well taken by consumers.

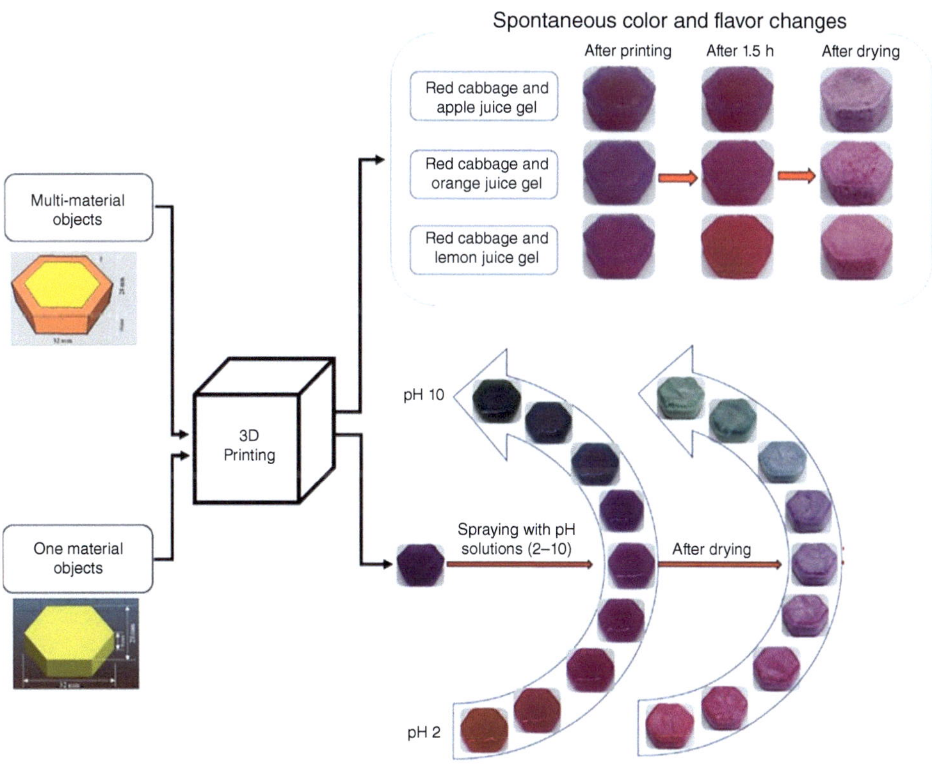

Figure 6.11 Multi-material 4D food printing – color changes due to internal (pH) and external stimuli (microwave). *Source*: Ghazal et al. 2021 / with permission of Elsevier.

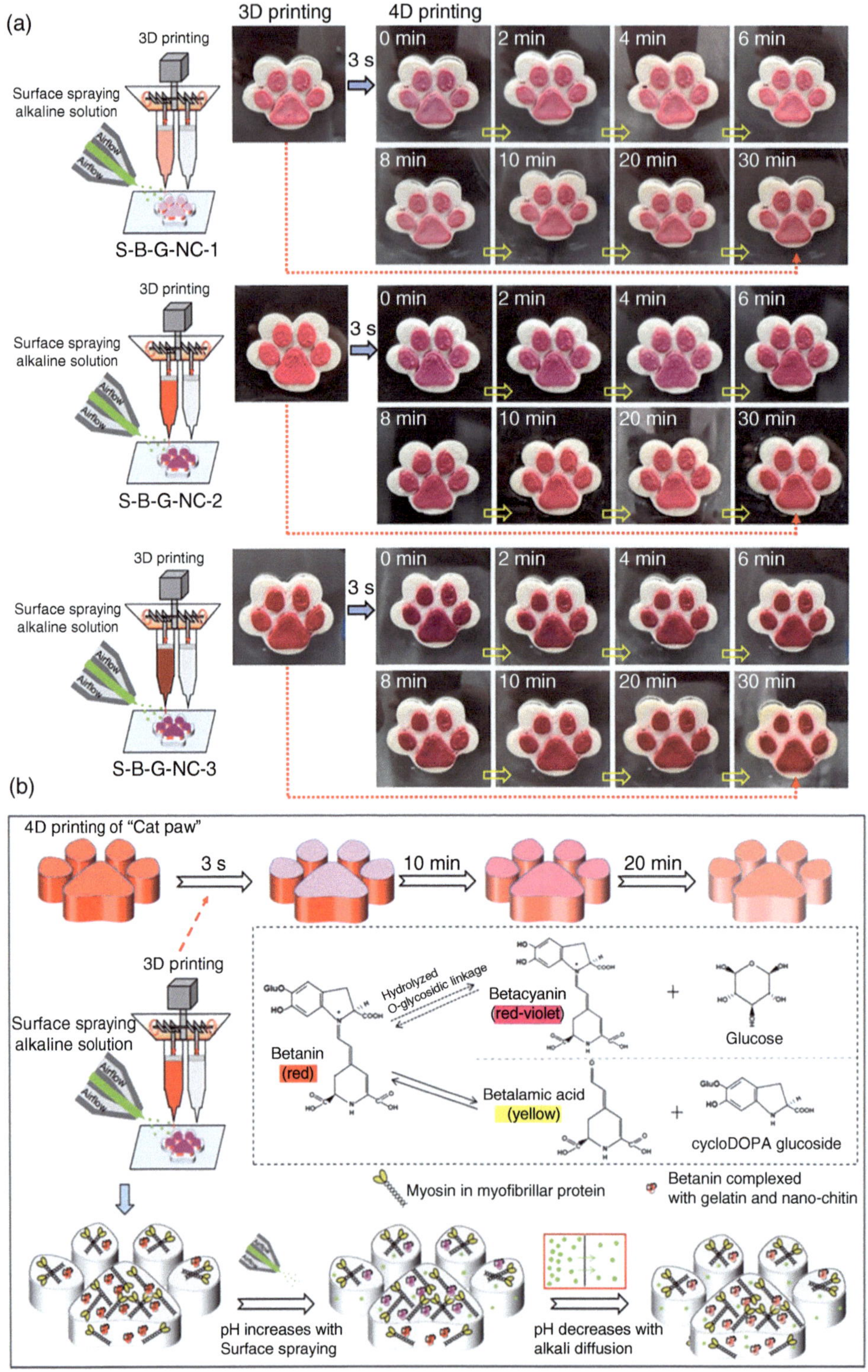

Figure 6.12 (a) 4D-printed betanin complexes – functionalized surimi (S-B-G-NC = Surimi- Betanin-Gelatin- Nano-chitin), (b) underlying mechanism of color changes, *Source*: Zhu et al. 2024 / with permission of Elsevier.

6.6.3 Nutrients

Food printing offers the possibility to alter the nutritional value of food according to consumer preferences. This can be accomplished by integrating beneficial elements such as phytochemicals macro- and micro-nutrients while reducing detrimental ones such as allergens and antinutritional factors. The nutritional composition of printed food can be regulated by the deliberate organization of nutrients inside inks or by employing diverse printing techniques. The selection of printing materials determines the nutritional quality of 4D-printed food, the printing methodology, and the postprocessing processes. Also, components utilized in printing ink significantly influence the nutritional content of the final printed food. For instance, whole grains, fruits, and vegetables provide fiber, vitamins, and minerals, rendering the end product superior. The printing process and postprinting conditions must support nutritional retention, as temperature, speed, and light exposure affect heat-sensitive vitamins. Excessive heat can deteriorate some vitamins, while prolonged exposure to light can reduce vitamin C levels in specific foods (Shen et al. 2023b). Postprocessing techniques, including drying, heating, and freezing, significantly affect the nutritional composition of the printed foods. For example, the authors reported the different postprocessing impacts on printed foods' nutritional compositions (Santhoshkumar et al. 2024b; Theagarajan et al. 2021). Although drying might enhance nutritional concentration, it may also result in the degradation of heat-sensitive vitamins.

Food ingredients from various sources can enrich printing inks with high-nutrition profiles, vitamins, polysaccharides, and other valuable compounds. 4D printing, which includes bioprinting, can improve the nutritional qualities of printed foods. By utilizing food-grade printing inks that can be influenced by external stimuli such as temperature, humidity, and pH, the texture of flesh can be replicated whilst assuring the high nutritional value of the final printed food. When treated to ideal conditions, the printed foods can induce the formation of tissue that replicates natural structures, perhaps carrying more nutrient profiles (Aghababaei et al. 2025).

Recently, microflora integration in food printing inks has improved nutritional quality, as the inclusion of ingredients such as microalgae contributed essential nutrients such as proteins, sterols, vitamins, and fatty acids (Uribe-Wandurraga et al. 2020). In addition, the selection of post-treatment plays a critical role in 4D foods. For instance, ultraviolet (UV) treatment considerably enhanced the vitamin D2 concentration in 4D-printed purple sweet potato pastes containing ergosterol, particularly in the treated regions with stimuli (Chen et al. 2021d). Similarly, mushroom-based studies have been reported on such aspects, and the mechanisms involved are summarized in Figure 6.13 (Chen et al. 2022a). Studies have also shown that microwave-stimulated oleogel powder ink from purple potatoes could produce low-fat printed foods (Shi et al. 2022).

4D food printing facilitates personalization and an improved culinary experience for consumers. It is essential to carefully assess potential changes in nutritional content during printing to ensure that the printed food remains nutritious and contributes to a balanced diet. Strategies such as selecting nutrient-rich materials, optimizing printing parameters, employing protective packaging on printed foods, and reducing postprocessing exposure time can help retain the nutritional integrity of 4D-printed foods (Liu et al. 2021). The recent study focuses on enriching the nutrient material as printing inks to develop 4D

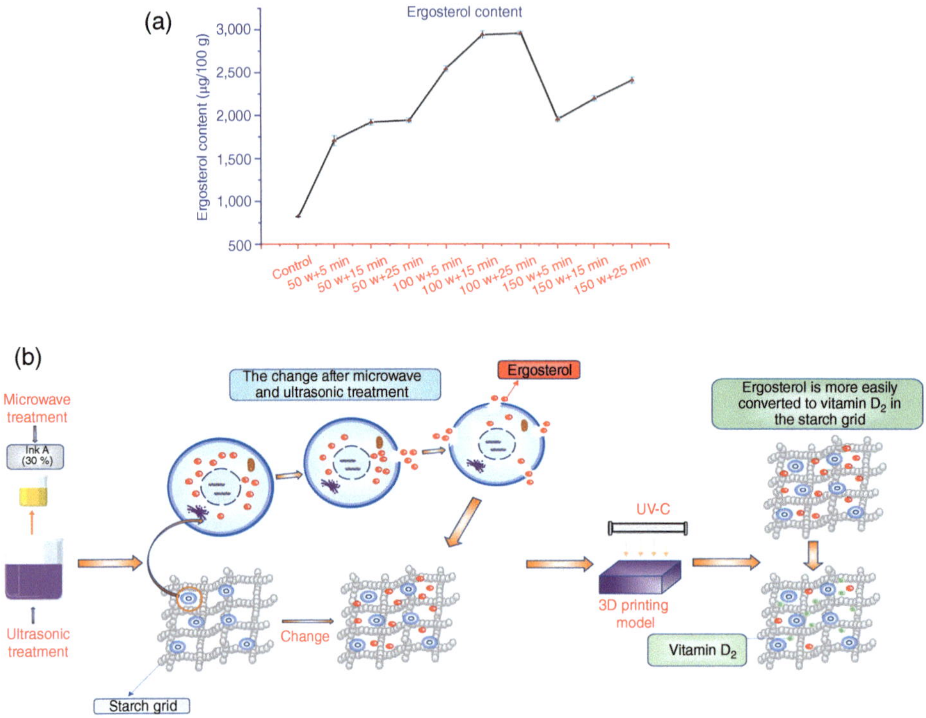

Figure 6.13 Nutritional changes in 4D food printing due to microwave stimuli. (a) Vitamin content and, (b) Underlying mechanism involved in vitamin conversion. *Source*: Chen et al. 2022a) / with permission of Elsevier.

foods by incorporating probiotics to improve their biological value. This is achieved through the growth of probiotics in a starch gel or a suitable medium, leading to an overall increase in functional benefits (Chen et al. 2023a).

6.6.4 Texture

Consumer acceptability of food relies on its visual appearance, taste, and texture of the printed foods. Textural modifications have been made to starch-based printed foods using a three-head extruder with a near-infrared (IR) heater. These modifications are justified by the fact that the calculated offsets in Cartesian coordinates between the two custom extruders did not significantly impact the textural features within the elastic range and slightly affected the variations in layer height along the z-axis, in addition to the effects of slicing parameters and printing settings on textural properties of printed items (Fahmy et al. 2020). Researchers have also successfully developed a 4D food of mashed potatoes with diverse interior configurations, indicating that infill concentrations and perimeter settings significantly influenced the hardness and gumminess of the printed constructs (Liu et al. 2018). Recent studies, such as the one reported by Raja et al. (2023), show the impact of infill density on printed fermented products. Similarly, in another study, the authors reported the impact of infills on textural attributes of printed foods (Oral et al. 2021).

A study on a 4D deformable structure composed of pumpkin and paper highlighted the significance of anisotropy and the directional properties of the planar layer; scientists noted an exponential increase in hardness over time, indicating a negative connection with time (Chen et al. 2021c). Similarly, mechanical qualities, such as hardness, gumminess, springiness, and cohesiveness, are known to be markedly affected by moisture content (Santhoshkumar et al. 2024a). Additionally, treatments such as microwave stimulation improved the characteristics of 4D-printed lotus root powder gel, increasing hardness due to moisture loss and increasingly dense structure (Chen et al. 2021a). In the context of texture-modified foods, through AM, foods suitable for dysphagic patients can be developed. A recent study in 4D printing, which involves color changes, suggests that the intelligent properties of 3D-printed materials lead to these color variations (thus classifying them as 4D). It also found that the textural attributes remain unchanged, indicating that 3D- and 4D-printed foods exhibit similar textures. This was validated by International Dysphagia Diet Standardization Initiative (IDDSI) tests specifically for dysphagia foods (Zhu et al. 2024).

Further, stimuli such as pH were reported to induce alterations in the textural characteristics of 3D-printed soy protein isolate, pumpkin, and beetroot composites. A spray of a lower pH solution (pH - 3) enhanced hardness and gumminess, but elevated pH levels amplified springiness and cohesiveness due to the aggregation of protein molecules (Phuhongsung et al., 2020). In addition, different pH solutions in a mixture of 3D-printed red cabbage juice, vanillin powder, and potato starch were found to adversely impact textural qualities (Ghazal et al. 2021). The detrimental effects of surface treatment with pH solutions have been ascribed to moisture absorption, potentially diminishing the hardness of the printed items (Navaf et al. 2022).

6.6.5 Taste

Food printing technology can be used to create distinct tastes by integrating diverse compounds to facilitate advantageous reactions. Also, layers can be engineered/structured in a manner so as to release different molecules responsible for the taste with different release behaviors. The taste profile of printed food samples can be assessed through E-tongue (Electronic tongue) technology, which typically measures eight sensory characteristics aligned with human taste perceptions: sourness, bitterness, astringency, aftertaste-A, aftertaste-B, umami, richness, and saltiness. The radar chart generated from the e-tongue analysis can depict the taste profile of the samples. For instance, pH-stimulated alterations in the taste properties of multismart materials within 4D-printed foods were assessed (Ghazal et al. 2021). Saltiness and umami flavors diminished as the pH value decreased (from 2 to 10), whereas sourness intensified with lower pH levels (Figure 6.14a). Another study revealed elevated saltiness at a pH of 10 in a mixture comprising 3D-printed soy protein isolate, pumpkin, and beetroot, while increased astringency was noted at a pH of 4. The increase in hydrogen ions at decreased pH levels lowers the pH of the sprayed solution (Phuhongsung et al., 2020). The diminished umami flavor in acidic conditions arises from the suppression of umami taste at lower pH levels, as acidic environments hinder the binding of phenolic compounds to proteins. This leads to a higher concentration of free phenolic compounds and increased astringency at lower pH (Navaf et al. 2022; Zdunic et al. 2016).

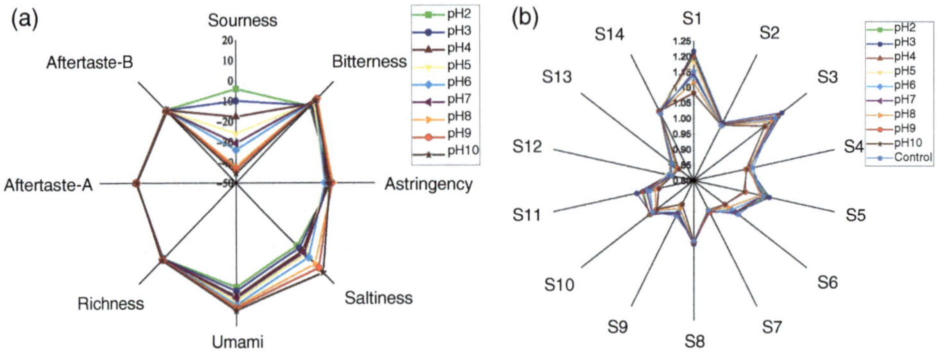

Figure 6.14 (a) E-tongue and, (b) E-nose observations of 4D-printed samples (red cabbage juice and vanillin powder composites) under pH stimuli. *Source*: Ghazal et al. (2021) / with permission of Elsevier.

6.6.6 Flavor/Aroma

The flavor or aroma of food is a vital attribute and a significant issue for consumer attitudes toward printed foods. Particularly, 4D food printing, the formulation of inks, and the alignment of layers can create the ideal conditions for developing distinct flavors in a single printed food. The printing ink formulations enhance the chemical interactions of taste components to ensure their uniform distribution among the product. High-tech instrumental techniques such as E-nose and gas chromatography-mass spectrometry (GC-MS) can help understand/quantify the flavor changes in printed foods. For instance, vanillin is a crucial flavoring component in food, and its flavor strength is significantly influenced by factors such as pH, temperature, and moisture present in the material compositions (Weerawatanakorn et al. 2015). Researchers studied using vanillin as a stimulus-responsive material in 4D food printing (Phuhongsung et al., 2020). This researchers also studied the impact of microwave irradiation on 4D-printed food incorporating vanilla flavor in the printing inks.

Another study by Guo et al. (2021) examined the changes in flavor compounds from microwave-treated 3D-printed foods, particularly those involving gelatin-gum and arabic-oil complex coacervates as stimulus-responsive materials; the authors reported an increase in volatile taste compounds such as 1-hexanol, hexanal, and benzaldehyde, which are associated with yellow peaches. After four minutes of microwave treatment, there was a two-fold increase in E-cinnamaldehyde. The intensified aroma of microwave-treated items arises from the rupture of microcapsules and the ensuing diffusion of aromatic chemicals owing to thermal exposure (Guo et al. 2021). Marked variations in scent were detected in 4D-printed constructs composed of red cabbage juice, vanillin powder, and potato starch at different pH levels, as assessed by an E-nose (Electronic nose) (Figure 6.14b). Samples subjected to elevated pH values (9 and 10) demonstrated a diminished scent reaction, while those at reduced pH levels indicated a more pronounced response. The reduced odor intensity at elevated pH levels was ascribed to the ionization of vanillin's phenolic group, possessing a pKa of 7.38, resulting in diminished volatility under basic conditions (Ghazal et al. 2021). Similarly, another study on red cabbage juice with the olive oil composite was reported for pasta printing with different flavored water (Ghazal et al. 2024). The impact of varying pH values on a 4D-printed soy protein complex combined with pumpkin and beetroot powders was also observed. Following five hours of spraying, alterations in flavor suggested the synthesis of novel fragrance compounds at elevated pH settings (pH 8 and 10) (Phuhongsung et al., 2020). Further, microcapsules were developed from yam gel utilizing 4D printing approaches to enhance the flavor by inducing microwave heat (Chen et al. 2024b).

6.6.7 Live Cell Cultures

4D printing can also involve processes by which texture, flavor, and nutritional properties evolve by culturing plant or animal cells within structured scaffolds. While 3D printing in cell culturing is not usually categorized as 4D printing, it may be regarded as such when the printed structures demonstrate additional functionalities, such as biological growth, that develop over time, an aspect absent in traditional 3D-printed foods. Cell culturing in printing applications typically entails fabricating scaffold structures that incorporate plant or animal cells within a medium conducive to cellular growth alongside a scaffold design that promotes cell proliferation. Through these living cell cultures, the printed food obtained is called "simulated foods"; depending on their sources, they can be categorized as plant-based simulated foods and animal-based simulated foods (Shi et al. 2024). Plant-based simulated foods rely on the potential of plant tissue cells, enabling successful 4D printing in combination with plant cell culture technology. The substrate used for cell-based 3D printing must be biocompatible to support cellular proliferation and exhibit sufficient printability and rigidity. Printed materials can be combined with plant tissue culturing postprinting to closely mimic the structure and nutritional characteristics of natural plants. Traditionally, plant cell culture methods involve breaking down cells in a reactor to extract metabolites (Davies and Deroles 2014). However, the introduction of 4D printing has brought about greatly enhanced control over callus morphology and functional design in early development (Shi et al. 2024).

The environmental impact of traditional animal husbandry, such as pollution, greenhouse gas emissions, food security concerns, and animal welfare issues, has led to increased interest in 4D printing and related technologies for creating animal-based simulated foods, commonly referred to as artificial meat or meat analogs (Kang et al. 2021; Ramachandraiah 2021). Artificial meat can be categorized into three types (Shi et al. 2024): (i) vegetarian meat, which consists of plant proteins – soy, wheat, or pea protein and plant-derived hemoglobin synthesized by yeast (this type mimics the properties of conventional meat without containing any animal tissue), (ii) animal-derived meat, grown *in vitro* from animal cells or reconstructed using 3D/4D printing, referred to as bioprinting, where actual animal meat is produced, and (iii) combining plant tissue and animal protein, resulting in hybrid meats, which are considered a next-generation meat analog with superior nutrition compared to pure plant-based alternatives. A range of other protein sources, such as insect and fungal proteins, are also being explored for meat analogs.

Two printing methodologies have been investigated before applying 4D printing in food systems utilizing plant or animal cells. The initial approach, termed "Edible Food Growth," was launched by a food designer in 2014 and involved the creation of 3D-printed food structures that facilitate biological growth (Rutzerveld 2014). Offered dynamic sensory experiences and nutrition over time; however, it failed to satisfy the 4D food printing criteria, as the product was not 3D printed. Then, the second approach established the foundation for 4D food printing utilizing plant cells by encapsulating these cells within a food material system. Vancauwenberghe et al. (2017) demonstrated textural changes and cell viability in low-methoxylated pectin gels with isolated lettuce cells. Printability and mechanical strength vary based on pectin concentration and cell content. This study did not investigate the time-dependent growth of the cell culture and, therefore, was termed 3D printing instead of 4D printing. Following this, a 3D-printed plant cell-laden hydrogel scaffold to demonstrate time-dependent textural changes was developed, representing 4D food printing. The researchers examined carrot callus cells embedded in alginate, analyzing the correlation between printability, cell proliferation, and temporal texture alterations. Their findings indicated that cell growth reduced hardness, which was dependent on the initial cell concentrations (Park et al. 2020). X. Guo et al. (2024) documented cultured meat applications through printing approaches. Their study indicated potential time-dependent changes; however, additional research is necessary to enhance mechanical strength and sensory perception via 4D printing.

6.7 Application Range – Dynamic/Intelligent Structure Changes

The practice of food preparation has significantly improved life satisfaction, while 4D food printing automation offers a viable solution for accommodating individualized meal preferences through modifications in the color, flavor, and shape of food items. Microcapsules containing a gelatin-gum and arabic oil compound were integrated into imprinted buckwheat flatbread with yellow-fleshed peach. Upon exposure to microwave heating, the microparticles progressively dissolved, liberating the encapsulated oil, which permeated the dough, altering its color and fragrance (Guo et al. 2021). In another instance, bilayered 3D-printed sweet potato purees underwent dehydration via microwave irradiation, resulting in shape distortion. The modification of flavor via microwave heating was also evidenced in printed foods comprising carrageenan, vanilla flavor, and soy protein isolate. Four newly synthesized flavor compounds were combined with vanilla, and their flavors were characterized (Phuhongsung et al., 2020). Further, the nutrient development and color through probiotic growth (Chen et al. 2023a) and live cell-based culture to develop artificial meat and plant tissues (Shi et al. 2024). In recent days, multi-intelligence can be introduced in 4D food printing; for instance, single stimuli can induce change in color, aroma, and shape of the printed products (Chen et al. 2024b). These examples underscore the significance of stimulus-responsive food additives as essential elements of 4D food printing, a field poised for ongoing research.

A significant application of 4D-printed structures lies in digital programming, where various material distortion algorithms are incorporated with 4D printing technology. This augments the diversity of structural outcomes, enhancing visual effects and enabling artists to express themselves more innovatively. It provides designers and artists with boundless opportunities for patterns, encompassing diverse digital geometric forms, artworks, and ceramics. Instances of programmable structures comprise an expanding reentrant honeycomb (150% area augmentation), a shrinkable reentrant honeycomb (200% area reduction), and an expandable antitetrarchical honeycomb (62% area augmentation) (Singh et al. 2024).

Recently, Singh et al. (2024) proposed actuation hinges, which were reported by Ahmed et al. (2021), designed in different shapes such as squares, zigzags, and honeycombs, allowing varying degrees of bending. When exposed to IR light, these hinges, made from chitosan hydrogel ink infused with black carbon particles, enabled the deformation of a polystyrene polymer substrate. Such designs offer potential applications in the food industry, particularly confectionery and baking.

6.8 Challenges and Future Aspects in 4D Food Printing

4D printing advances from A methods such as 3D printing by enabling printed items to undergo spontaneous transformations in reaction to environmental stimuli. This has generated increasing interest in industrial and academic sectors, especially in 4D food printing. Although contemporary 4D printing is based on conventional 3D printing technologies, forthcoming advancements are expected to improve printing devices and settings with superior functionalities.

The progression of 4D printing technology will depend significantly on innovations in printing techniques, intelligent materials, structural design, and the creation of sophisticated modeling tools and software. As these technologies advance, their applications will proliferate across diverse domains, necessitating significant research and development endeavors. To augment this technology, several critical areas require focus. Initially, it is imperative to standardize experimental apparatus and create high-precision printers. Currently, most printing devices are bespoke creations by individual laboratories with restricted functionalities (Pei and Loh 2018; Wan et al. 2024). Engaging professional manufacturers will expedite advancement and enhance the quality of the technology.

Given that 4D food printing is in its nascent stages, there exist ample prospects for additional investigation and development. Particularly, notwithstanding its potential, 4D food printing continues to encounter numerous hurdles. Contemporary 4D printers have challenges that include the need to eliminate support structures, simultaneous printing of several materials, limited availability of cost-effective materials, and prolonged printing durations. Moreover, technological advancements are necessary to modify 3D printing techniques initially intended for molten plastics and metals to handle "sensitive" food materials for 4D printing applications. Challenges related to mechanical characteristics, sluggish actuation, and insufficient availability of materials persist. Interestingly, current research is investigating methods to regulate moisture absorption in hydrogels and boost stimulus application, which may improve actuation precision. The advancement of five-axis 4D food printing offers prospective solutions to some difficulties; however, additional progress in printing technology and material innovation is crucial for achieving the complete capabilities of 4D printing. Nevertheless, practical constraints remain, such as challenges in developing smart materials with exact shape-memory properties, enhancing accuracy for intricate designs, and attaining scalability for large-scale production. Moreover, although 4D printing reduces postprocessing, certain goods may require further treatment.

Most food substances lack the self-healing capabilities and flexibility characteristic of sophisticated smart materials, and their stability and reactivity to stimuli are much worse. Moreover, food structures may incur degradation during postprocessing. The complex internal composition of food can be influenced by interactions among its components, potentially resulting in less adaptability in printing or diminished quality of 4D-printed products. A thorough comprehension of the attributes of stimuli-responsive food components is crucial, alongside the refinement of material selection and formulations and the enhancement of the relationships between the internal structures of food materials and stimuli.

Establishing a correlation model that connects material, process, performance, and function is essential for attaining the requisite deformation, denaturation, and functional alterations in food materials. As of now, multiphysics and simulation methodologies within the 4D food printing paradigm remain underdeveloped. Precisely assessing the correlation between a model's internal architecture and external inputs is complex, complicating the prediction of how food structure and qualities will develop over time. Future developments in predictive modeling and multiphysics analysis may offer insights for design optimization. This would facilitate the regulation of flow, layer adhesion, thermal transfer, and structural integrity throughout the printing process, guaranteeing that the printed food satisfies both quality and functional criteria.

As 4D printing technology advances, it is imperative to develop precise predictive models for material behavior, especially for hydrogel constructs exhibiting anisotropic features, where forecasting nonlinear behavior and anisotropic expansion is vital. To improve understanding of the 4D transformation of food materials, it is essential to build finite element models and algorithms such as these that accurately predict experimental results, hence minimizing research and development costs and duration. Further, machine learning and data-driven methodologies are emerging as potent solutions for addressing design issues in 4D printing and are anticipated to significantly contribute to the development of intricate food structures and deformation processes in the future.

4D food printing can alter food characteristics, including texture, flavor, color, and shape. Contemporary research predominantly examines materials such as starch hydrogel, soy protein, and potato, explores the effects of stimuli such as microwave cooking and pH variations on the characteristics of food. For instance, these stimuli can modify the color of printed food by ingredient denaturation and distortion; however, research on other categories of stimuli-responsive materials remains limited. In the future, 4D food printing may explore novel stimuli, such as light and changes in ionic content, to induce transformations. Additionally, the potential to revolutionize food manufacturing with 4D printing can be strengthened by integrating innovative compounds such as diacetyl and vanillin, which simultaneously modify numerous culinary attributes through stimuli-responsive materials. Such aspects can be achieved by implementing a surveillance and regulation system for these modifications. The inclusion of a variety of responsive elements in food inks may improve our understanding of the reactions of 4D-printed food, which is why it is recommended to conduct further research on the impact of stimuli on the texture and shape of foods.

Simulation software must incorporate smart material characteristics to forecast the AM trajectory and simulate how printed objects autonomously adapt to stimuli. Also, modeling and slicing tools must offer extensive functionalities for creating accurate, intricate products equipped with tools that oversee the complete process from design to output. Considering that 4D printing frequently utilizes various smart materials, these approaches must integrate sophisticated algorithms to manage material interactions proficiently. Current monitoring software primarily focuses on tracking printer status and automating supply replacements. Future monitoring solutions must extend scopes beyond this by tracking the morphological changes of printed products over time to guarantee optimal functionality. Specialized printing management systems will also prove beneficial. In contrast to the traditional management system, which emphasizes printer usage restrictions and cost management, a 4D printing management system must synchronize every phase of the printing process, facilitating uninterrupted design, production, and transformation oversight (Chung et al. 2017). In addition, the artificial intelligence incorporation enhances printability through intelligent actions. These aspects are detailed in upcoming chapters. Overall, developing specialized software solutions will not only simplify the field of 4D printing but also help unlock its full potential for widespread applications in diverse sectors.

Currently, the application of 4D food printing is primarily limited to laboratory research rather than commercial applications. Future initiatives should prioritize the commercial viability of 4D food printing, enhance the quality of customized nutrition, and meet the unique needs of consumers. To ensure the sustainable advancement and safety of 4D food printing, it will be essential to establish cohesive technical standards, regulatory frameworks, policies, and human training as this domain develops.

6.9 Conclusion

Presently, 4D-printed foods are employed in various food categories, including simulated meals, primary foods, fermented products, and personalized diets for clients with specific needs. Emerging applications include esthetically appealing appearances, superior nutritional value, enhanced microstructures, improved printability, and customizable textures. This technology offers a variety of benefits, such as enhanced functionality and innovative culinary designs. However, some challenges require further research and development to address their constraints. Numerous opportunities exist for the 4D printing of structurally designed edibles, particularly when combined with advanced food processing technologies. In addition to naturally occurring materials, synthetic components, including hydrogels, polysaccharide-based substances, and intelligent materials, can be explored for novel opportunities.

References

Aghababaei, F., McClements, D. J., Pignitter, M., & Hadidi, M. (2025). Plant protein edible inks: Upgrading from 3D to 4D food printing. *Food Chemistry: X*, 102280. doi:10.1016/j.fochx.2025.102280

Ahmed, A., Arya, S., Gupta, V., Furukawa, H., & Khosla, A. (2021). 4D printing: Fundamentals, materials, applications and challenges. *Polymer*, *228*, 123926. doi:10.1016/j.polymer.2021.123926

Anandharamakrishnan, C., Moses, J. A., & Anukiruthika, T. (2022a). *3D printing of foods*. John Wiley & Sons Ltd.

Anandharamakrishnan, C., Moses, J. A., & Anukiruthika, T. (2022b). 4D Printing Technology. In *3D Printing of Foods* (pp. 310–354). Wiley. https://doi.org/10.1002/9781119669838.ch10

Anukiruthika, T., Moses, J. A., & Anandharamakrishnan, C. (2020). 3D printing of egg yolk and white with rice flour blends. *Journal of Food Engineering, 265*. doi:10.1016/j.jfoodeng.2019.109691

Benny, A., Pandi, K., & Upadhyay, R. (2022). Techniques, challenges and future prospects for cell-based meat. *Food Science and Biotechnology, 31*(10), 1225–1242. doi:10.1007/s10068-022-01136-6

Blutinger, J. D., Cooper, C. C., Karthik, S., Tsai, A., Samarelli, N., Storvick, E., Seymour, G., Liu, E., Meijers, Y., & Lipson, H. (2023). The future of software-controlled cooking. *NPJ Science of Food, 7*(1), 6. doi:10.1038/s41538-023-00182-6

Blutinger, J. D., Tsai, A., Storvick, E., Seymour, G., Liu, E., Samarelli, N., Karthik, S., Meijers, Y., & Lipson, H. (2021). Precision cooking for printed foods via multiwavelength lasers. *NPJ Science of Food, 5*(1), 24. doi:10.1038/s41538-021-00107-1

Cen, S., Li, Z., Guo, Z., Li, H., Shi, J., Huang, X., Zou, X., & Holmes, M. (2022). 4D printing of a citrus pectin/β-CD Pickering emulsion: A study on temperature induced color transformation. *Additive Manufacturing, 56*, 102925. doi:10.1016/j.addma.2022.102925

Cen, S., & Meng, Z. (2024). 4D printing: A novel application for structuring oils with fat-analog characteristics. *Trends in Food Science and Technology, 153*, 104729. doi:10.1016/j.tifs.2024.104729

Chen, C., Zhang, M., Guo, C., & Chen, H. (2021a). 4D printing of lotus root powder gel: Color change induced by microwave. *Innovative Food Science and Emerging Technologies, 68*, 102605. doi:10.1016/j.ifset.2021.102605

Chen, C., Zhang, M., Mujumdar, A. S., & Phuhongsung, P. (2021b). Investigation of 4D printing of lotus root-compound pigment gel: Effect of pH on rapid colour change. *Food Research International, 148*, 110630. doi:10.1016/j.foodres.2021.110630

Chen, F., Zhang, M., Liu, Z., & Bhandari, B. (2021c). 4D deformation based on double-layer structure of the pumpkin/paper. *Food Structure, 27*, 100168. doi:10.1016/j.foostr.2020.100168

Chen, J., Teng, X., Zhang, M., Bhandari, B., Adhikari, B., & Yu, D. (2023a). 5D food printing with color change induced by probiotic growth in a starch-protein-based gel system. *Food and Bioprocess Technology, 16*(10), 2304–2314. doi:10.1007/s11947-023-03064-7

Chen, J., Zhang, M., & Devahastin, S. (2021d). UV-C irradiation-triggered nutritional change of 4D printed ergosterol-incorporated purple sweet potato pastes: Conversion of ergosterol into vitamin D2. *Lebensmittel-Wissenschaft and Technologie, 150*, 111944. doi:10.1016/j.lwt.2021.111944

Chen, J., Zhang, M., Mujumdar, A. S., & Phuhongsunge, P. (2022a). 4D printing induced by microwave and ultrasound for mushroom mixtures: Efficient conversion of ergosterol into vitamin D2. *Food Chemistry, 387*, 132840. doi:10.1016/j.foodchem.2022.132840

Chen, X., Teng, X., Zhang, M., Mujumdar, A. S., & Li, J. (2024a). Functional changes in 5D printed starch-based gel systems caused by spontaneous growth of probiotics. *Food and Bioprocess Technology, 17*(8), 2470–2482. doi:10.1007/s11947-023-03267-y

Chen, X., Zhang, M., & Tang, T. (2023b). Microwave-induced rapid shape change of 4D printed vegetable-based food. *Food, 12*(11), 2158. doi:10.3390/foods12112158

Chen, X., Zhang, M., Tang, T., & Yu, D. (2024b). The color/shape/flavor of yam gel with double emulsified microcapsules changed synchronously in 4D printing induced by microwave. *International Journal of Biological Macromolecules*, *260*, 129631. doi:10.1016/j.ijbiomac.2024.129631

Chen, Y., Zhang, M., Sun, Y., & Phuhongsung, P. (2022b). Improving 3D/4D printing characteristics of natural food gels by novel additives: A review. *Food Hydrocolloids*, *123*, 107160. doi:10.1016/j.foodhyd.2021.107160

Cheng, Y., Fu, Y., Ma, L., Yap, P. L., Losic, D., Wang, H., & Zhang, Y. (2022). Rheology of edible food inks from 2D/3D/4D printing, and its role in future 5D/6D printing. *Food Hydrocolloids*, *132*, 107855. doi:10.1016/j.foodhyd.2022.107855

Chung, S., Song, S. E., & Cho, Y. T. (2017). Effective software solutions for 4D printing: A review and proposal. *International Journal of Precision Engineering and Manufacturing-Green Technology*, *4*(3), 359–371. doi:10.1007/s40684-017-0041-y

Cui, L., Guo, J., & Meng, Z. (2023). A review on food-grade-polymer-based O/W emulsion gels: Stabilization mechanism and 3D printing application. *Food Hydrocolloids*, *139*, 108588. doi:10.1016/j.foodhyd.2023.108588

Davies, K. M., & Deroles, S. C. (2014). Prospects for the use of plant cell cultures in food biotechnology. *Current Opinion in Biotechnology*, *26*, 133–140. doi:10.1016/j.copbio.2013.12.010

Derossi, A., Spence, C., Corradini, M. G., Jekle, M., Fahmy, A. R., Caporizzi, R., Devahastin, S., Moses, J. A., Le-Bail, A., Zhou, W., Zhang, M., Bhandari, B., & Severini, C. (2024). Personalized, digitally designed 3D printed food towards the reshaping of food manufacturing and consumption. *NPJ Science of Food*, *8*(1), 54. doi:10.1038/s41538-024-00296-5

Dick, A., Bhandari, B., & Prakash, S. (2019). Post-processing feasibility of composite-layer 3D printed beef. *Meat Science*, *153*, 9–18. doi:10.1016/j.meatsci.2019.02.024

Dick, A., Gao, Y., Bhandari, B., & Prakash, S. (2021). Influence of drying method and 3D design on the 4D morphing of beef products. *Applied Food Research*, *1*(2), 100017. doi:10.1016/j.afres.2021.100017

Domżalska, Z., & Jakubczyk, E. (2025). Characteristics of food printing inks and their impact on selected product properties. *Food*, *14*(3), 393. doi:10.3390/foods14030393

Donn, P., Prieto, M. A., Mejuto, J. C., Cao, H., & Simal-Gandara, J. (2022). Functional foods based on the recovery of bioactive ingredients from food and algae by-products by emerging extraction technologies and 3D printing. *Food Bioscience*, *49*, 101853. doi:10.1016/j.fbio.2022.101853

Fahmy, A. R., Becker, T., & Jekle, M. (2020). 3D printing and additive manufacturing of cereal-based materials: Quality analysis of starch-based systems using a camera-based morphological approach. *Innovative Food Science and Emerging Technologies*, *63*, 102384. doi:10.1016/j.ifset.2020.102384

Fahmy, A. R., Derossi, A., & Jekle, M. (2023). Four-dimensional (4D) printing of dynamic foods—definitions, considerations, and current scientific status. *Food*, *12*(18), 3410. doi:10.3390/foods12183410

Ghazal, A. F., Zhang, M., Bhandari, B., & Chen, H. (2021). Investigation on spontaneous 4D changes in color and flavor of healthy 3D printed food materials over time in response to external or internal pH stimulus. *Food Research International*, *142*, 110215. doi:10.1016/j.foodres.2021.110215

Ghazal, A. F., Zhang, M., & Guo, Z. (2023). Microwave-induced rapid 4D change in color of 3D printed apple/potato starch gel with red cabbage juice-loaded WPI/GA mixture. *Food Research International*, *172*, 113138. doi:10.1016/j.foodres.2023.113138

Ghazal, A. F., Zhang, M., & Li, C. (2024). 4D color change of 3D-printed pasta cooked with different-flavored waters and its potential use as a cooking indicator. *Food and Bioprocess Technology*. doi:10.1007/s11947-024-03409-w

Ghazal, A. F., Zhang, M., & Liu, Z. (2019). Spontaneous color change of 3D printed healthy food product over time after printing as a novel application for 4D food printing. *Food and Bioprocess Technology*, *12*(10), 1627–1645. doi:10.1007/s11947-019-02327-6

Guo, C., Zhang, M., & Bhandari, B. (2019). Model building and slicing in food 3D printing processes: A review. *Comprehensive Reviews in Food Science and Food Safety*, *18*(4), 1052–1069. doi:10.1111/1541-4337.12443

Guo, C., Zhang, M., Bhandari, B., & Devahastin, S. (2022). Investigation on simultaneous change of deformation, color and aroma of 4D printed starch-based pastes from fruit and vegetable as induced by microwave. *Food Research International*, *157*, 111214. doi:10.1016/j.foodres.2022.111214

Guo, C., Zhang, M., & Devahastin, S. (2021). Color/aroma changes of 3D-Printed buckwheat dough with yellow flesh peach as triggered by microwave heating of gelatin-gum Arabic complex coacervates. *Food Hydrocolloids*, *112*, 106358. doi:10.1016/j.foodhyd.2020.106358

Guo, J., Zhang, M., Li, J., & Fang, Z. (2023). Using soy protein isolate to improve the deformation properties of 4D-printed oat flour butterfly. *Food and Bioprocess Technology*, *16*(5), 1165–1176. doi:10.1007/s11947-023-02994-6

Guo, X., Wang, D., He, B., Hu, L., & Jiang, G. (2024). 3D bioprinting of cultured meat: A promising avenue of meat production. *Food and Bioprocess Technology*, *17*(7), 1659–1680. doi:10.1007/s11947-023-03195-x

Guo, Z., Yang, B., Liang, N., Huang, X., Shi, J., Li, Z., Paximada, P., & Xiaobo, Z. (2025). 4D printing of Pickering emulsion: Temperature-driven color changes. *Journal of Food Engineering*, *386*, 112258. doi:10.1016/j.jfoodeng.2024.112258

He, C., Zhang, M., & Devahastin, S. (2020a). Investigation on spontaneous shape change of 4D printed starch-based purees from purple sweet potatoes as induced by microwave dehydration. *ACS Applied Materials and Interfaces*, *12*(34), 37896–37905. doi:10.1021/acsami.0c10899

He, C., Zhang, M., & Devahastin, S. (2021). Microwave-induced deformation behaviors of 4D printed starch-based food products as affected by edible salt and butter content. *Innovative Food Science and Emerging Technologies*, *70*, 102699. doi:10.1016/j.ifset.2021.102699

He, C., Zhang, M., & Fang, Z. (2020b). 3D printing of food: pretreatment and post-treatment of materials. *Critical Reviews in Food Science and Nutrition*, *60*(14), 2379–2392. doi:10.1080/10408398.2019.1641065

He, C., Zhang, M., & Guo, C. (2020c). 4D printing of mashed potato/purple sweet potato puree with spontaneous color change. *Innovative Food Science and Emerging Technologies*, *59*, 102250. doi:10.1016/j.ifset.2019.102250

Hertafeld, E., Zhang, C., Jin, Z., Jakub, A., Russell, K., Lakehal, Y., Andreyeva, K., Bangalore, S. N., Mezquita, J., Blutinger, J., & Lipson, H. (2019). Multi-material three-dimensional food printing with simultaneous infrared cooking. *3D Printing and Additive Manufacturing*, *6*(1), 13–19. doi:10.1089/3dp.2018.0042

Huang, Y., Zhang, M., Pattarapon, P., & Mujumdar, A. S. (2022). 4D printing of mixed vegetable gel based on deformation and discoloration induced by acidification and dehydration. *Journal of Food Process Engineering*, *45*(12). doi:10.1111/jfpe.14172

Jiang, Q., Binks, B. P., & Meng, Z. (2022). Double scaffold networks regulate edible Pickering emulsion gel for designing thermally actuated 4D printing. *Food Hydrocolloids, 133*, 107969. doi:10.1016/j.foodhyd.2022.107969

Kang, D.-H., Louis, F., Liu, H., Shimoda, H., Nishiyama, Y., Nozawa, H., Kakitani, M., Takagi, D., Kasa, D., Nagamori, E., Irie, S., Kitano, S., & Matsusaki, M. (2021). Engineered whole cut meat-like tissue by the assembly of cell fibers using tendon-gel integrated bioprinting. *Nature Communications, 12*(1), 5059. doi:10.1038/s41467-021-25236-9

Kang, Z., Wang, Z., Zhang, W., & Fang, Z. (2023). 4D printing of edible insect snacks: Color changes with external pH stimulation. *Food Bioscience, 55*, 102971. doi:10.1016/j.fbio.2023.102971

Kavimughil, M., Leena, M. M., Moses, J. A., & Anandharamakrishnan, C. (2022). 3D printed MCT oleogel as a co-delivery carrier for curcumin and resveratrol. *Biomaterials, 287*, 121616. doi:10.1016/j.biomaterials.2022.121616

Keerthy, M. A., Dhanaselvam, K. R., Santhoshkumar, P., Ravikrishnan, V., & Moses, J. A. (2025). Improving the printability of watermelon rind using pectin: 3D printing optimization for the development of gummies. *Food Biomacromolecules, 2*(2), 143–156. doi:10.1002/fob2.70008

Koirala, S., Prakash, S., Karim, A., & Bhandari, B. (2023). Shape morphing of foods: Mechanism, strategies, and applications. *Trends in Food Science and Technology, 139*, 104135. doi:10.1016/j.tifs.2023.104135

Koirala, S., Prakash, S., Karim, A., & Bhandari, B. (2025). 4D shape morphing in 3D-printed pea protein structures through designed surface grooves under drying and frying conditions. *Innovative Food Science and Emerging Technologies, 100*, 103924. doi:10.1016/j.ifset.2025.103924

Kokane, S. B., Anjaly, P., Thangalakshmi, S., & Arora, V. K. (2024a). Current trends in additive manufacturing based 4D food printing technology: A review. *Future Foods, 10*, 100450. doi:10.1016/j.fufo.2024.100450

Kokane, S. B., Arora, V. K., & Thangalakshmi, S. (2024b). Potential of bioactive ingredients for healthy 4D food printing: Applications and challenges. *Food Reviews International, 1–24*. doi:10.1080/87559129.2024.2374808

Lee, S. H., Kim, H. W., & Park, H. J. (2023). Integrated design of micro-fibrous food with multi-materials fabricated by uniaxial 3D printing. *Food Research International, 165*, 112529. doi:10.1016/j.foodres.2023.112529

Li, G., Zhan, J., Huang, J., Xu, E., Yuan, C., Chen, J., Yao, Q., & Hu, Y. (2023a). Enhanced fresh-keeping capacity of printed surimi by Ca2+-nano starch-lutein and its nondestructive freshness monitoring based on 4D printed anthocyanin. *International Journal of Biological Macromolecules, 252*, 126543. doi:10.1016/j.ijbiomac.2023.126543

Li, X., Fan, L., Li, R., Han, Y., & Li, J. (2023b). 3D/4d printing of β-cyclodextrin-based high internal phase emulsions. *Journal of Food Engineering, 348*, 111455. doi:10.1016/j.jfoodeng.2023.111455

Lin, Q., Hu, Y., Qiu, C., Li, X., Sang, S., McClements, D. J., Chen, L., Long, J., Xu, X., Wang, J., & Jin, Z. (2023). Peanut protein-polysaccharide hydrogels based on semi-interpenetrating networks used for 3D/4D printing. *Food Hydrocolloids, 137*, 108332. doi:10.1016/j.foodhyd.2022.108332

Liu, Z., He, C., Guo, C., Chen, F., Bhandari, B., & Zhang, M. (2021). Dehydration-triggered shape transformation of 4D printed edible gel structure affected by material property and heating mechanism. *Food Hydrocolloids, 115*, 106608. doi:10.1016/j.foodhyd.2021.106608

Liu, Z., Zhang, M., & Bhandari, B. (2018). Effect of gums on the rheological, microstructural and extrusion printing characteristics of mashed potatoes. *International Journal of Biological Macromolecules*, *117*, 1179–1187. doi:10.1016/j.ijbiomac.2018.06.048

Lueyot, A., Rungsardthong, V., Vatanyoopaisarn, S., Hutangura, P., Wonganu, B., Wongsa-Ngasri, P., Charoenlappanit, S., Roytrakul, S., & Thumthanaruk, B. (2021). Influence of collagen and some proteins on gel properties of jellyfish gelatin. *PLOS One*, *16*(6), e0253254. doi:10.1371/journal.pone.0253254

Lv, Y., Wang, B., Cheng, Y., Lv, W., Zeng, S., & Xiao, H. (2024). Printing characteristics and microwave infrared-induced 4D printing of chestnut powder composite paste. *Journal of Food Engineering*, *382*, 112197. doi:10.1016/j.jfoodeng.2024.112197

Miller, O., Scarlett, C. J., & Akanbi, T. O. (2024). Plant-based meat analogues and consumer interest in 3D-printed products: A mini-review. *Food*, *13*(15), 2314. doi:10.3390/foods13152314

Momeni, F., Mehdi Hassani, M., & N, S., Liu, X., & Ni, J. (2017). A review of 4D printing. *Materials and Design*, *122*, 42–79. doi:10.1016/j.matdes.2017.02.068

Momeni, F., & Ni, J. (2020). Laws of 4D printing. *Engineering*, *6*(9), 1035–1055. doi:10.1016/j.eng.2020.01.015

Nachal, N., Moses, J. A., Karthik, P., & Anandharamakrishnan, C. (2019). Applications of 3D printing in food processing. *Food Engineering Reviews*, *11*(3), 123–141. doi:10.1007/s12393-019-09199-8

Navaf, M., Sunooj, K. V., Aaliya, B., Akhila, P. P., Sudheesh, C., Mir, S. A., & George, J. (2022). 4D printing: A new approach for food printing; effect of various stimuli on 4D printed food properties. A comprehensive review. *Applied Food Research*, *2*(2), 100150. doi:10.1016/j.afres.2022.100150

Nida, S., Moses, J. A., & Anandharamakrishnan, C. (2022). Emerging applications of 5D and 6D printing in the food industry. *Journal of Agriculture and Food Research*, *10*, 100392. doi:10.1016/j.jafr.2022.100392

Niu, D., Zhang, M., Mujumdar, A. S., & Li, J. (2023). Research on microwave-induced bidirectional deformation of coix seed compound materials in 4D printing. *Food and Bioprocess Technology*, *16*(11), 2400–2415. doi:10.1007/s11947-023-03078-1

Niu, D., Zhang, M., Mujumdar, A. S., & Yu, D. (2025). Investigation of 5D printing of children food based on anthocyanin microcapsules and spontaneous changes in probiotics. *Innovative Food Science and Emerging Technologies*, *100*, 103921. doi:10.1016/j.ifset.2025.103921

Noree, S., Pinyakit, Y., Tungkijanansin, N., Kulsing, C., & Hoven, V. P. (2023). Shape transformation of 4D printed edible insects triggered by thermal dehydration. *Journal of Food Engineering*, *358*, 111666. doi:10.1016/j.jfoodeng.2023.111666

Oral, M. O., Derossi, A., Caporizzi, R., & Severini, C. (2021). Analyzing the most promising innovations in food printing. Programmable food texture and 4D foods. *Future Foods*, *4*, 100093. doi:10.1016/j.fufo.2021.100093

Park, S., Shou, W., Makatura, L., Matusik, W., Fu, K., & (Kelvin). (2022). 3D printing of polymer composites: Materials, processes, and applications. *Matter*, *5*(1), 43–76. doi:10.1016/j.matt.2021.10.018

Park, S. M., Kim, H. W., & Park, H. J. (2020). Callus-based 3D printing for food exemplified with carrot tissues and its potential for innovative food production. *Journal of Food Engineering*, *271*, 109781. doi:10.1016/j.jfoodeng.2019.109781

Patrocinio, D., Galván-Chacón, V., Gómez-Blanco, J. C., Miguel, S. P., Loureiro, J., Ribeiro, M. P., Coutinho, P., Pagador, J. B., & Sanchez-Margallo, F. M. (2023). Biopolymers for tissue engineering: Crosslinking, printing techniques, and applications. *Gels*, *9*(11), 890. doi:10.3390/gels9110890

Pei, E., & Loh, G. H. (2018). Technological considerations for 4D printing: an overview. *Progress in Additive Manufacturing*, *3*(1–2), 95–107. doi:10.1007/s40964-018-0047-1

Phuhongsung, P., Zhang, M., & Bhandari, B. (2020a). 4D printing of products based on soy protein isolate via microwave heating for flavor development. *Food Research International*, *137*, 109605. doi:10.1016/j.foodres.2020.109605

Phuhongsung, P., Zhang, M., & Devahastin, S. (2020b). Influence of surface pH on color, texture and flavor of 3D printed composite mixture of soy protein isolate, pumpkin, and beetroot. *Food and Bioprocess Technology*, *13*(9), 1600–1610. doi:10.1007/s11947-020-02497-8

Phuhongsung, P., Zhang, M., Devahastin, S., & Mujumdar, A. S. (2022). Defects in 3D/4D food printing and their possible solutions: A comprehensive review. *Comprehensive Reviews in Food Science and Food Safety*, *21*(4), 3455–3479. doi:10.1111/1541-4337.12984

Pulatsu, E., Su, J.-W., Lin, J., & Lin, M. (2022). Utilization of ethyl cellulose in the osmotically-Driven and Anisotropically-Actuated 4D Printing Concept of Edible Food Composites. *Carbohydrate Polymer Technologies and Applications*, *3*, 100183. doi:10.1016/j.carpta.2022.100183

Qin, Z., Li, Z., Huang, X., Du, L., Li, W., Gao, P., Chen, Z., Zhang, J., Guo, Z., Li, Z., Liu, B., & Shen, T. (2025). Advances in 3D and 4D printing of gel-based foods: mechanisms, applications, and future directions. *Gels*, *11*(2), 94. doi:10.3390/gels11020094

Qiu, R., Liu, X., Tian, H., Hu, Z., Wang, K., & Zhao, L. (2024). Exploration of bigel 4D printing with spontaneous colour change for monitoring bio-actives kinetic behaviour based on the dual-units 3D printer. *Journal of Food Engineering*, *367*, 111861. doi:10.1016/j.jfoodeng.2023.111861

Raja, V., Moses, J. A., & Anandharamakrishnan, C. (2023). Effect of 3D printing conditions and post-printing fermentation on pearl millet fortified idli. *Journal of the Science of Food and Agriculture*, *103*(5), 2401–2412. doi:10.1002/jsfa.12410

Ramachandraiah, K. (2021). Potential development of sustainable 3D-printed meat analogues: A review. *Sustainability*, *13*(2), 938. doi:10.3390/su13020938

Rashwan, A. K., Younis, H. A., Abdelshafy, A. M., Osman, A. I., Eletmany, M. R., Hafouda, M. A., & Chen, W. (2024). Plant starch extraction, modification, and green applications: a review. *Environmental Chemistry Letters*, *22*(5), 2483–2530. doi:10.1007/s10311-024-01753-z

Reuben, S. J., Chandra, S., Praveena, B., Santhoshkumar, P., & Moses, J. A. (2025). 3D printing of alternative proteins: Approaches, challenges and advances. *Sustainable Materials and Technologies*, e01285. doi:10.1016/j.susmat.2025.e01285

Rubio, N. R., Xiang, N., & Kaplan, D. L. (2020). Plant-based and cell-based approaches to meat production. *Nature Communications*, *11*(1), 6276. doi:10.1038/s41467-020-20061-y

Rutzerveld, C. (2014). Edible Growth–The use of additive manufacturing technologies to create an edible ecosystem. *Food Futurist*. https://www.chloerutzerveld.com/edible-growth

Said, N. S., Olawuyi, I. F., & Lee, W. Y. (2023). Pectin hydrogels: gel-forming behaviors, mechanisms, and food applications. *Gels*, *9*(9), 732. doi:10.3390/gels9090732

Santhoshkumar, P., & Moses, J. A. (2025). Impact of konjac glucomannan on hot extrusion 3D printability of cake gel. *International Journal of Biological Macromolecules*, *305*, 141162. doi:10.1016/j.ijbiomac.2025.141162

Santhoshkumar, P., Negi, A., & Moses, J. A. (2023). 3D printing for space food applications: Advancements, challenges, and prospects. *Life Sciences in Space Research*. doi:10.1016/j.lssr.2023.08.002

Santhoshkumar, P., Raja, V., Priyadarshini, S. R., & Moses, J. A. (2024a). Evaluating the 3D printability of pearl millet flour with banana pulp blends. *Journal of the Science of Food and Agriculture*. doi:10.1002/jsfa.13389

Santhoshkumar, P., Raja, V., Priyadarshini, S. R., & Moses, J. A. (2024b). Impact of Baking and Steaming of 3D printed Pearl millet–Banana pulp blends on physicochemical and sensory attributes. *International Journal of Gastronomy and Food Science*, *37*, 100962. doi:10.1016/j.ijgfs.2024.100962

Santhoshkumar, P., Swathika, B. S., Negi, A., & Moses, J. A. (2025). Improving gel performance and hot extrusion three-dimensional printability of blueberry gel with guar gum. *Food Biophysics*, *20*(2), 1–15. doi:10.1007/s11483-025-09964-9

Shabir, I., Dar, A. H., Manzoor, S., Pandey, V. K., Srivastava, S., Shams, R., Dash, K. K., Bashir, I., Fayaz, U., Prithviraj, V., Singh, P., Rustagi, S., Ramniwas, S., & Pandiselvam, R. (2024). Advancements in food printing technologies and their potential culinary applications: A contemporary exploration. *Journal of Food Processing and Preservation*, *2024*, 1–17. doi:10.1155/2024/6621344

Shahbazi, M., Jäger, H., Ettelaie, R., Chen, J., Mohammadi, A., & Kashi, P. A. (2024a). 4D printing of poly-Pickering high internal phase emulsions to assembly a thermo-responsive shape-memory hierarchical macroporous scaffold. *Additive Manufacturing*, *81*, 103977. doi:10.1016/j.addma.2024.103977

Shahbazi, M., Jäger, H., Ettelaie, R., Chen, J., Mohammadi, A., Kashi, P. A., & Ulbrich, M. (2024b). A smart thermoresponsive macroporous 4D structure by 4D printing of Pickering-high internal phase emulsions stabilized by plasma-functionalized starch nanomaterials for a possible delivery system. *Current Research in Food Science*, *8*, 100686. doi:10.1016/j.crfs.2024.100686

Shanthamma, S., Preethi, R., Moses, J. A., & Anandharamakrishnan, C. (2021). 4D printing of sago starch with turmeric blends: A study on pH-triggered spontaneous color transformation. *ACS Food Science and Technology*, *1*(4), 669–679. doi:10.1021/acsfoodscitech.0c00151

Shen, C., Chen, W., Li, C., Chen, X., Cui, H., & Lin, L. (2023a). 4D printing system stimulated by curcumin/whey protein isolate nanoparticles: A comparative study of sensitive color change and post-processing. *Journal of Food Engineering*, *342*, 111357. doi:10.1016/j.jfoodeng.2022.111357

Shen, D., Zhang, M., Mujumdar, A. S., & Li, J. (2023b). Advances and application of efficient physical fields in extrusion based 3D food printing technology. *Trends in Food Science and Technology*, *131*, 104–117. doi:10.1016/j.tifs.2022.11.017

Shi, H., Zhang, M., & Mujumdar, A. S. (2024). 3D/4D printed super reconstructed foods: Characteristics, research progress, and prospects. *Comprehensive Reviews in Food Science and Food Safety*, *23*(2). doi:10.1111/1541-4337.13310

Shi, Y., Zhang, M., & Phuhongsung, P. (2022). Microwave-induced spontaneous deformation of purple potato puree and oleogel in 4D printing. *Journal of Food Engineering*, *313*, 110757. doi:10.1016/j.jfoodeng.2021.110757

Silva, R., Silva, B., Fernandes, C., Morouço, P., Alves, N., & Veloso, A. (2024). A review on 3D scanners studies for producing customized orthoses. *Sensors*, *24*(5), 1373. doi:10.3390/s24051373

Singh, P., Pandey, V. K., Tripathi, A., Singh, R., Net, T. S. S., Ramniwas, S., & Pandiselvam, R. (2024). 4D food printing technology: Structural changes to culinary art and beyond. *Journal of Food Process Engineering*, *47*(1). doi:10.1111/jfpe.14535

Sinha, S., Lynch, S. P., & Meisel, N. A. (2021). Heat transfer simulation of material extrusion additive manufacturing to predict weld strength between layers. *Additive Manufacturing*, *46*, 102117. doi:10.1016/j.addma.2021.102117

Srivastava, S., Pandey, V. K., Singh, R., & Dar, A. H. (2023). Recent insights on advancements and substantial transformations in food printing technology from 3 to 7D. *Food Science and Biotechnology*, *32*(13), 1783–1804. doi:10.1007/s10068-023-01352-8

Su, C., Wu, Y., Vardhanabhuti, B., & Lin, M. (2025). 4D printing of dysphagia foods using pea protein and purple sweet potato flour. *Food Bioscience*, *64*, 105887. doi:10.1016/j.fbio.2025.105887

Subhasri, D., Dutta, S., Leena, M. M., Moses, J. A., & Anandharamakrishnan, C. (2022). Gastronomy: An extended platform for customized nutrition. *Future Foods*, *5*, 100147. doi:10.1016/j.fufo.2022.100147

Sun, X., Yue, L., Yu, L., Forte, C. T., Armstrong, C. D., Zhou, K., Demoly, F., Zhao, R. R., & Qi, H. J. (2024). Machine learning-enabled forward prediction and inverse design of 4D-printed active plates. *Nature Communications*, *15*(1), 5509. doi:10.1038/s41467-024-49775-z

Swathika, B. S., Santhoshkumar, P., Moses, J. A., & Sinija, V. R. (2025). Novel three-dimensional-printed ι-carrageenan–carnauba wax bigel formulation for codelivery applications. *ACS Food Science & Technology*, *5*(3), 1183–1192. doi:10.1021/acsfoodscitech.5c00004

Tambe, S., Jain, D., Amin, P., Mali, S. N., & Cruz, J. N. (2023). Advances in the development of a 3D-printed nutraceutical delivery platform. In *Nutraceuticals* (pp. 193–222). Elsevier. doi:10.1016/B978-0-443-19193-0.00008-3

Tang, T., Zhang, M., Mujumdar, A. S., & Teng, X. (2022). 3D printed white radish/potato gel with microcapsules: Color/flavor change induced by microwave-infrared heating. *Food Research International*, *158*, 111496. doi:10.1016/j.foodres.2022.111496

Tao, Y., Do, Y., Yang, H., Lee, Y.-C., Wang, G., Mondoa, C., Cui, J., Wang, W., & Yao, L. (2019). Morphlour: Personalized flour-based morphing food induced by dehydration or hydration method. *Proceedings of the 32nd Annual ACM Symposium on User Interface Software and Technology*, 329–340. https://doi.org/10.1145/3332165.3347949

Teng, X., Zhang, M., & Mujumdar, A. S. (2021). 4D printing: Recent advances and proposals in the food sector. *Trends in Food Science and Technology*, *110*, 349–363. doi:10.1016/j.tifs.2021.01.076

Theagarajan, R., Nimbkar, S., Moses, J. A., & Anandharamakrishnan, C. (2021). Effect of post-processing treatments on the quality of three-dimensional printed rice starch constructs. *Journal of Food Process Engineering*, *44*(9), e13772. doi:10.1111/jfpe.13772

Tong, Q., Jiang, Y., Xiao, S., Meng, Y., & Dong, X. (2024). Research on improving the structural stability of surimi 3D printing through laser cooking techniques. *Journal of Food Engineering*, *375*, 112075. doi:10.1016/j.jfoodeng.2024.112075

Uribe-Wandurraga, Z. N., Zhang, L., Noort, M. W. J., Schutyser, M. A. I., García-Segovia, P., & Martínez-Monzó, J. (2020). Printability and physicochemical properties of microalgae-enriched 3D-printed snacks. *Food and Bioprocess Technology*, *13*(11), 2029–2042. doi:10.1007/s11947-020-02544-4

Vancauwenberghe, V., Katalagarianakis, L., Wang, Z., Meerts, M., Hertog, M., Verboven, P., Moldenaers, P., Hendrickx, M. E., Lammertyn, J., & Nicolaï, B. (2017). Pectin based food-ink formulations for 3-D printing of customizable porous food simulants. *Innovative Food Science and Emerging Technologies*, *42*, 138–150. doi:10.1016/j.ifset.2017.06.011

Wan, X., Xiao, Z., Tian, Y., Chen, M., Liu, F., Wang, D., Liu, Y., Bartolo, P. J. D. S., Yan, C., Shi, Y., Zhao, R. R., Qi, H. J., & Zhou, K. (2024). Recent advances in 4D printing of advanced materials and structures for functional applications. *Advanced Materials*, *36*(34). doi:10.1002/adma.202312263

Wang, N., Li, R., Wang, X., & Yang, X. (2024). 4D food printing: Key factors and optimization strategies. *Trends in Food Science and Technology*, *145*, 104380. doi:10.1016/j.tifs.2024.104380

Wang, W., Yao, L., Zhang, T., Cheng, C.-Y., Levine, D., & Ishii, H. (2017). Transformative appetite: Shape-changing food transforms from 2D to 3D by water interaction through cooking. *Proceedings of the 2017 CHI Conference on Human Factors in Computing Systems*, 6123–6132. https://doi.org/10.1145/3025453.3026019

Waseem, M., Tahir, A. U., & Majeed, Y. (2024). Printing the future of food: The physics perspective on 3D food printing. *Food Physics*, *1*, 100003. doi:10.1016/j.foodp.2023.100003

Weerawatanakorn, M., Wu, J.-C., Pan, M.-H., & Ho, C.-T. (2015). Reactivity and stability of selected flavor compounds. *Journal of Food and Drug Analysis*, *23*(2), 176–190. doi:10.1016/j.jfda.2015.02.001

Xiao, K., Zhang, Y., Pan, L., & Tu, K. (2024). Study on color and flavor changes of 4D printed white mushroom gel with microcapsules containing gelatin/β-cyclodextrin induced by microwave heating. *International Journal of Biological Macromolecules*, *279*, 135365. doi:10.1016/j.ijbiomac.2024.135365

Xie, R., Zhang, S., Liu, S., Gu, Y., Zhang, J., Guo, Y., Tang, X., Jiang, L., Guo, Z., & Wang, Z. (2025). Stabilization of oil-in-water high internal phase Pickering emulsion by using 7S globulin/konjac glucomannan complexes: Rheology properties, stability properties and 3D/4D printing performance. *Food Research International*, *203*, 115878. doi:10.1016/j.foodres.2025.115878

Yoha, K. S., Anukiruthika, T., Anila, W., Moses, J. A., & Anandharamakrishnan, C. (2021). 3D printing of encapsulated probiotics: Effect of different post-processing methods on the stability of Lactiplantibacillus plantarum (NCIM 2083) under static in vitro digestion conditions and during storage. *Lebensmittel-Wissenschaft and Technologie*, *146*, 111461. doi:10.1016/j.lwt.2021.111461

Zdunic, G., Menkovic, N., Jadranin, M., Novakovic, M., Savikin, K., & Zivkovic, J. (2016). Phenolic compounds and carotenoids in pumpkin fruit and related traditional products. *Hemijska Industrija*, *70*(4), 429–433. doi:10.2298/HEMIND150219049Z

Zhan, J., Fu, J., Jin, D., Yuan, Y., Shen, S., Li, G., & Chen, Y. (2023). Surimi freshness monitoring of 4D printing material with anthocyanin. *Journal of Food Engineering*, *358*, 111678. doi:10.1016/j.jfoodeng.2023.111678

Zhu, J., Yang, Y., Qiao, S., Dai, H., Chen, H., Fu, Y., Ma, L., Wang, H., & Zhang, Y. (2024). 4D printing of betanin/gelatin/nano-chitin complexes-functionalized surimi via disulfide bonds, and its applicability in dysphagia diets. *Food Hydrocolloids*, *152*, 109891. doi:10.1016/j.foodhyd.2024.109891

Zhu, Y., Liu, Z., Zhang, X., He, C., & Zhang, X. (2023). Microwave-triggered 4D automatic color change in 3D-printed food materials incorporating natural pigments. *Food*, *12*(10), 2055. doi:10.3390/foods12102055

7

Stimuli in 4D Printing of Foods

CHAPTER MENU

7.1 Introduction

4D food printing is noteworthy because of its distinctive and beneficial capabilities. The emergence of smart materials, particularly shape memory alloys, has markedly enhanced the utilization of 4D printing technology in diverse sectors such as aerospace, automotive, soft robotics, and biomedicine (Sajjad et al. 2024). Conversely, food substances utilized in printing inks are generally nonintelligent compounds. At present, 4D printing in the culinary domain is accomplished by combining various food substances (recombined foods), considering precise formulas and structural designs. The characteristics of printed products, namely, shape, color, flavor, texture, and nutritional content, may experience predictable alterations when subjected to specific stimuli, including water, heat, magnetic

4D Printing of Foods: Fundamentals, Applications and Prospects, First Edition.
C. Anandharamakrishnan, Jeyan A. Moses, and P. Santhoshkumar.

fields, light, and pH (Anandharamakrishnan et al. 2022). Although 3D food printing has been famous for its capabilities to provide unique esthetics, nutritional advantages, textural modifications, and other benefits, 4D printing enhances these functionalities by the action of different stimuli. Also, while 3D printing may create complex geometries, these typically necessitate support owing to the nature of the additive manufacturing process. Conversely, 4D printing can produce hanging structures, such as blossoming flowers (Li et al. 2023b). Figure 7.1 represents a flower blooming concept achieved through biomimetic anisotropic self-sensing actuators developed from hydrogels.

The transition from closed to open form can enhance dining experiences, particularly attracting discerning eaters such as youngsters (Xia et al. 2024). 4D-printed foods promote participatory dining experiences, strengthening the connection between consumers and their meals (Burke 2024). As an example of an inspired concept from 3D food printing, Figure 7.2 illustrates the "Print A Drink" concept of KUKA's LBR iiwa robots to produce elaborate 3D designs within a cocktail. The robots accurately injected microliters of liquid into the beverage, creating suspended patterns. This robotic methodology facilitated the development of patterns and models, almost unattainable by traditional methods. The approach was formulated at the University of Arts and Design in Linz (Burke 2024).

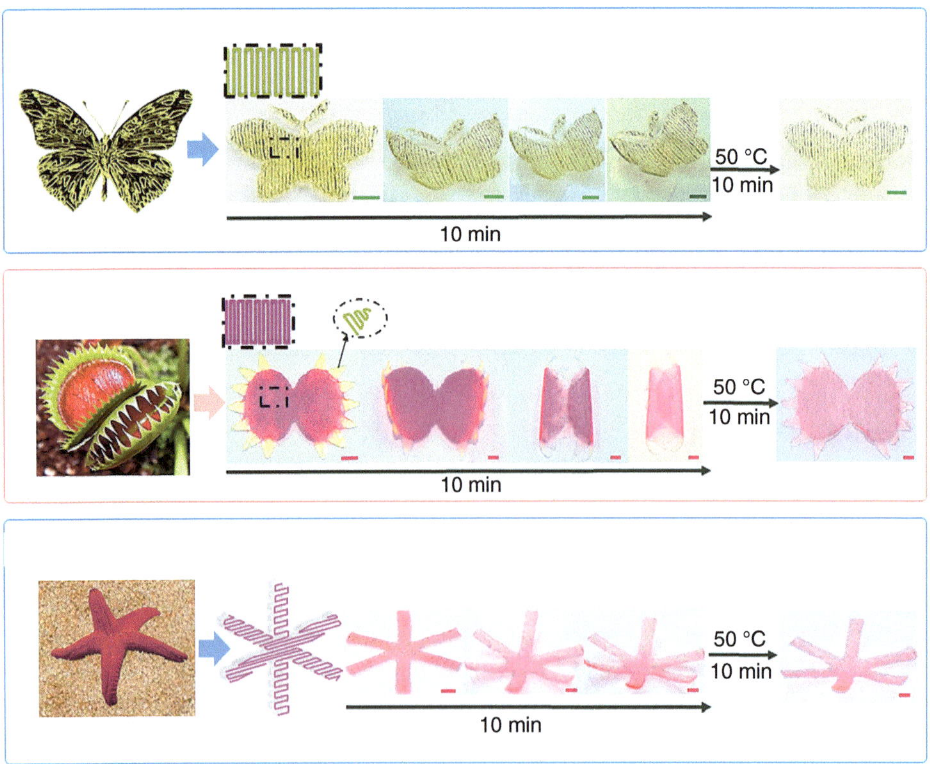

Figure 7.1 4D-printed hydrogels based on nature-inspired concepts. *Source:* Li et al. (2023) / with permission of Elsevier.

Figure 7.2 "Print a drink" patterns in a cocktail. *Source:* Burke (2024) / Elsevier / CC BY 4.0.

A significant benefit of 4D printing is its capacity to release flavors, nutrients, or colors during consumption rather than during storage. Particularly, the sensory attributes of 3D-printed foods generally decline with time. 3D-printed foods frequently encounter challenges with drying and baking since it may result in undesirable deformation that undermines their structural integrity (Derossi et al. 2024). Nonetheless, 4D printing facilitates regulated structural modifications throughout the drying process, which is advantageous for producing healthier snacks. For instance, the preferred crispiness of potato chips is affected by their curvature, and the integration of 4D printing and drying can yield snacks with an attractive texture (Koirala et al. 2023). As printing technology progresses and becomes more established, it may be incorporated into production lines for automated manufacturing. 4D food printing is consistent with the "flat packaging" concept popularized by Swedish furniture makers, enabling consumers to construct 3D objects from 2D components at home (Teng et al. 2021). This method is optimal for distorted foods produced via 4D printing, as it reduces transit expenses and conserves storage capacity.

The previous chapter summarized the fundamental concepts, materials, and approaches to 4D food printing, also covering the wide range of food applications. This chapter focuses on the different types of stimuli used in 4D food printing technology. There are various types of stimuli based on their mode of application on printed products, called internal and external stimuli (Fahmy et al. 2023). In the context of additive manufacturing, "stimuli" refers to external factors such as temperature, light, pH, magnetic fields, humidity that can trigger a change in the properties of a "stimuli-responsive" material, allowing a 3D-printed object to alter its shape or function dynamically when exposed to these specific conditions, which is particularly relevant in advanced applications such as 4D printing where the object can change over time based on the applied stimulus. These stimuli are typically classified as physical, chemical, and biological stimuli, and in some cases, combinations of different stimuli exist. Accordingly, in this chapter, the underlying concepts and the interaction mechanism between these stimuli and their corresponding responsive material are discussed with examples from scientific studies. Finally, the chapter presents concerns that must be considered for future explorations on the 4D printing of foods.

7.2 Engineering Structures Through Physical Stimuli

Physical stimuli such as temperature, humidity, light, magnetic fields, and electrical stimulation can modify the chain dynamics or internal atomic arrangement of stimulus-responsive materials, resulting in the shape, color, texture, nutrition, and sensory behavior of the printed products. The inherent ability of these smart materials to respond uniquely to stimuli positions them as transformative innovations for food manufacturing applications.

7.2.1 Temperature

Most experimental studies have focused on employing temperature as the external stimulus for fabricating 4D food-printed constructs. Shape-memory materials are generally produced or reacted under designated circumstances. In shape morphing, mechanical stress is applied and sustained above a certain temperature, such as the glass transition temperature, and subsequently cooled to achieve a metastable temporary form in polymer-based materials (Rabaux and Jérôme 2025). In foods, major mechanical stress is typically absent, and only temperature-stimulated shape changes and other intelligent functions are attainable. In some cases, this alteration is irrevocable or irreversible. Shape-changing materials are preferred over conventional shape-memory materials for developing 4D-printed structures that endure reversible changes. Typically, these materials endure shape alteration through fundamental expansion or contraction in response to a stimulus and revert to their original form upon removal (Mahmood et al. 2023). This reversible property can potentially have significant implications in drug delivery applications and cell encapsulation (Cui et al. 2020). The development of biocompatible substances with reduced glass transition or materials that exhibit multiresponsiveness, which allows them to respond to multiple stimuli simultaneously, may be necessary to increase the variety of these materials (Wang et al. 2019).

Temperature is the primary stimulus utilized in 4D food printing. Multiple studies in this domain have employed this stimulus. In 4D food printing, materials modify their structure due to temperature variations (Chen et al. 2021a; Guo et al. 2022; He et al. 2020). Food gels made from edible substances modify their texture with temperature changes, enabling foods to change texture during preparation or presentation, hence enhancing the sensory experience (Liu et al. 2021). The glass transition temperature and melting temperature of the food ink composition are also associated with temperature-induced changes (Joardder et al. 2024; Kashiwakura et al. 2022).

In food hydrogels, temperature variations can induce dehydration, resulting in the shrinking or contraction of the structure. These alterations are notably reversible (Xu et al. 2024). Microwave heating is a noncontact thermal technique used to induce thermal changes in 4D-printed edibles. It is advantageous as a postprinting procedure due to its rapid heating, improved energy efficiency, precise control, and hygienic operation (Guo et al. 2021; Guo et al. 2022). Microwave-induced dehydration induces morphological alterations, including bending or shrinking (Chen et al. 2023b; He et al., 2021; Liu et al. 2021). Furthermore, it can emit flavors and colors by rupturing microcapsules incorporated in 4D-printed materials (Xiao et al. 2024). Utilizing this stimulus, many studies have been reported for food applications. For instance, multiresponsiveness (color/aroma/

shape) through microwave treatment has been observed (Chen et al. 2024; Guo et al. 2022; Xiao et al. 2024). Other thermal treatments, such as infrared and hot air drying, are also utilized as stimuli in 4D food printing intelligent functions, and key highlights are documented in Table 7.1.

Subsequently, an infrared imaging system was used to identify the temperature distribution of printed products. Microwave, infrared, and hot air drying are prevalent techniques for food dehydration, each influencing the deformation of printed samples (Shen et al. 2023a). Hot air drying typically produces a more pronounced bending angle among these approaches due to its inadequate heat penetration. The dehydration rate is crucial in

Table 7.1 Temperature stimulus responses in 4D food printing.

Smart materials	Temperature-based stimuli	Responses	Exposure ranges	References
Starch-based pastes from fruits and vegetables	Microwave	Shape, color, aroma	Power level – 700 W Time – 15 s	Guo et al. (2022)
Vegetable gels (pumpkin and spinach)	Microwave	Shape	Power level – 700 W Time – 30 s	(Chen et al., 2023)
Purple potato puree and oleogel	Microwave	Shape	Power level – 140 W Time – 78 s	Shi et al. (2022)
Starch-based food with salt and butter	Microwave	Shape	Microwave power – 2.0 W/g	(He et al., 2021)
Coix seeds and purple potato powders	Microwave	Shape (bi-directional deformation)	Power level – 700 W Time – 15 s	Niu et al. (2023)
Soy protein isolate	Microwave	Flavor	Power level – 110 W Time – 5 min	Phuhongsung et al. (2020)
Lotus root powder gel	Microwave	Color	Power level – 280 W Time – 1-3 min	(Chen et al., 2021)
Yam gel with double-emulsified microcapsules	Microwave	Color, shape, flavor	Power level – 700 W Time – 45 s	Chen et al. (2024)
Chestnut powder composite paste	Microwave and infrared	Shape	Power level – 200 W Time – 8 min	(Lv et al. 2024)
Purple sweet potato	Microwave	Shape	Power level – 2 W/g Time – 54 min	He et al. (2020)
White mushroom gel with microcapsules containing gelatin/β-cyclodextrin	Microwave	Color and flavor	Power level – 280 W Time – 3 min	Xiao et al. (2024)

(Continued)

Table 7.1 (Continued)

Smart materials	Temperature-based stimuli	Responses	Exposure ranges	References
Natural pigments	Microwave	Color	Power level – 119 W Time – 3 min	Zhu et al. (2023)
Apple and potato starch gel with red cabbage juice	Microwave	Color	Power level – 120 W Time – 80 s	Ghazal et al. (2023)
Oat flour	Microwave	Shape	Power level – 230 W Time – 30 s	Guo et al. (2023)
Buckwheat dough with yellow flesh peach, gelatin-gum arabic complex coacervates	Microwave	Color and aroma	Power level – 200 W Time – 4 min	Guo et al. (2021)
Potato starch gel	Microwave	Shape	Power level – 1.0 W/g Time – 16 min	Liu et al. (2021)
Radish and potato gel	Microwave-infrared	Color and flavor	Power level – 200 W Time – 40 min	Tang et al. (2022)
Curcumin and whey protein isolate nanoparticles	Catalytic infrared drying	Color	Temperature – 70°C, time – 2 min	Shen et al., (2023)
Mixed vegetable gel	Oven drying	Shape and color	Temperature – 45°, time – 90 min	Huang et al. (2022)
Insects	Oven drying	Color	Temperature – 80°C, time – 15 min	Noree et al. (2023)

the deformation caused by dewatering. A higher dehydration rate increases the surface hardness of the product, thereby delaying deformation and reducing the bending angle. Insufficient water evaporation can lead to cracks and voids within the material. The internal structural configuration of the printed samples, including filling rate and filling angle, significantly influences the extent of deformation due to dehydration. An increase in the filling angle decreased the maximum bending or torsion degree (He et al. 2020). Similarly, Yang et al. (2019) found that greater filling density hindered the release of water vapor, resulting in undesirable sample expansion.

7.2.2 Liquid/Water

Liquid, or water, is recognized as a promising catalyst for facilitating intelligent material transformations. It is also crucial for inducing deformations in 4D food printing. The active material's geometry is altered by the forces it generates as it expands. Complex deformation behaviors can be obtained when these materials are arranged effectively (Wang et al. 2017).

However, in food printing, without external forces, it can be expandable due to its inherent chemical properties (e.g., hydrogel). The transformation is intended to enable differential enlargement in various compartments of liquid-responsive materials, contingent upon spatial and temporal factors. These materials provide various beneficial functions, such as the reversible actuation of smart valves, controlled drug delivery, and cell encapsulation. Nevertheless, it is imperative to consider factors such as potential degradation or hydrolysis after multiple swelling and de-swelling cycles, weakened mechanical properties post expansion, and sluggish response times when evaluating the material's lifespan (Billiet et al. 2012; Ionov 2014). Additionally, the authors have also detailed concepts and applications of liquid-responsive materials.

As they incorporate water, most food gels undergo volume changes and hydration due to intermolecular hydrophilic interactions. Insoluble fibers, such as ethyl cellulose, have a dual function in regulating the shape changes of the entire structure. The cellulose layer functions as a mechanical constraint that regulates the structure's bending deformation and modifies the expansive behavior of the underlying gel by acting as a barrier to water diffusion (Gauss et al. 2021). By incorporating anisotropic expansion behaviors into the recombined food structure, this insight is essential for the design of shape transformations (Ghazal et al. 2023; Pulatsu et al. 2022; Sydney Gladman et al. 2016).

In an interesting study on 4D printing through liquid transformation, a printer with temperature-controlled double nozzles (one for gelatin and the other for ethyl cellulose) and a cooling platform to consolidate the gelatin to create gelatin films in different geometry (e.g., disc-shaped, ellipse-shaped, s-shaped) was used. The characteristics of the shape constraint (e.g., cellulose density, thickness, line gap, total coverage) were manipulated to adjust the stiffness and water diffusivity of the constraint (Wang et al. 2017). Ethyl cellulose was printed onto the gel film. These thin films are widely used in the food industry, facilitating the customization of food material fragmentation, shaping, and textured interactions (Koirala et al. 2023). Examples include self-watering sushi, cold pasta salad design, and self-wrapping cannoli.

Food hydrogels and moisture-sensitive materials can expand or contract in response to humidity levels, allowing modifications in the texture and shape of printed foods (Pulatsu et al. 2022). This characteristic enables the development of a dynamic culinary experience by allowing foods to endure physicochemical changes during storage or consumption. The utilization of hydrocolloids, proteins, natural pigments, starches, and bioactive compounds is intended to optimize nutritional and esthetic values while simultaneously attaining the desired sensory and structural properties (Kokane et al. 2024). The potential of 4D food printing should be expanded through the optimization of formulations and the discovery of novel stimuli-responsive materials in future research. This technology can potentially revolutionize the culinary industry by providing a more personalized, engaging, and interactive experience.

7.2.3 Radiation and Sound-Assisted Stimuli

Recently, radiation and sound-assisted technologies have gained prominence in food processing, including techniques such as ultrasound, pulsed light, radio frequency, pulsed electric fields, and irradiation (Chen et al., 2021c). In food printing, these technologies are often utilized as pretreatment processes to modify the properties of food components. Gao et al. (2023) reported that ultrasound treatment enhanced the printability of blueberry gel composites. Similarly, many studies reported that utilizing ultrasound improves the printability of the materials (Jeong et al. 2024; Ma et al. 2024; Maleki et al. 2024). Shen et al. (2023b) also recorded the effects of pretreatments on printability. In laser sintering 3D printing, lasers amalgamate powdered food substances. The procedure entails applying tiny layers of powder that the laser selectively sinters to produce the final 3D food item (Waseem et al. 2024). A recent study investigated the integration of microwave and ultrasound technologies with 4D printing to improve the conversion of ergosterol to vitamin D2 in mushroom waste prints. The findings revealed that vitamin D2 concentrations in items subjected to physical field pretreatment were 2.2–3.8 times more than in the untreated counterparts. The irradiation zone exhibited the most pronounced effect, whereas ultrasound treatment demonstrated the least impact, as indicated by the partial least squares regression analysis. The elevation in vitamin D2 levels was attributed to have resulted from pretreatments that render ergosterol more vulnerable to UV light, with an expanded irradiation region further amplifying the advantages (Chen et al. 2022).

7.2.4 Prospective Physical Stimuli for 4D Food Printing

To date, the reported physical stimuli were discussed above. This section is particularly dedicated to 4D food printing. Not all physical stimuli apply to food, as there are limitations in their application to food-related processes and conditions. This section summarizes the physical stimuli currently utilized in nonfood printing, focusing on futuristic applications for food materials.

7.2.4.1 Magnetic Field

Magneto-responsive materials utilized in 4D printing are increasingly documented in the biomedical sector, with their origins dating back around 2,600 years. These materials enable efficient and secure manipulation through noncontact techniques in diverse biomedical applications, such as magnetic resonance imaging, targeted drug delivery, and artificial muscles (Yarali et al. 2024). The swift reaction of these materials is ascribed to the presence of iron particles, which can rapidly acquire or abandon substantial magnetism upon the application or removal of an external magnetic field. For instance, Miyashita and colleagues devised a novel origami robot that can move at a velocity of 3.8 body lengths per second when subjected to an external magnetic field. This compact, autonomous robotic system has enhanced benefits, including excellent bioresorbability and the capacity for self-folding in reaction to thermal stimuli, rendering it appropriate for drug delivery applications. Further, magnetic forces can facilitate self-folding (Miyashita et al. 2015). In addition, applying a magnetic field during the printing process can effectively regulate the orientation of magneto-sensitive particles to produce anisotropic heterogeneous composites. By altering the orientation and density of the particles, these composites exhibit varying mechanical properties across different regions, emulating muscle-like characteristics where shape deformation occurs rapidly and dissipates abruptly upon the application and removal of an external magnetic field (Cezar et al. 2016).

Notwithstanding the benefits of magneto-responsive materials in biomedical applications, apprehensions persist regarding their elevated reactivity and propensity to aggregate. The magnetic field frequency must remain within the clinically safe 50–100 kHz range to avert potential tissue injury due to increased temperatures (Pretsch 2010). Investigating intelligent materials with magnetic responsiveness for food printing remains predominantly unexplored. Subsequent research can also focus on creating safer magneto-induced materials for food printing applications, including for targeted nutraceutical distribution within the body, as well as inducing shape and color alterations in food upon exposure to magnetic fields.

7.2.4.2 Electric Field

Electric fields can position and orient cells in predetermined ways, a phenomenon called galvanotaxis and electro tropism. A multitude of studies have concentrated on the advancement of electroactive materials for tissue regeneration. These scaffolds have exhibited favorable cellular viability and enhanced the osteogenesis of mesenchymal stem cells under electrical stimulation (Hwangbo et al. 2021; Soleymani and Naghib 2023). Jakus et al. (2015) reported that scaffolds containing a greater graphene concentration displayed an electrical conductivity of 0.08 S/cm. These scaffolds provide a promising foundation for nerve regeneration, as they facilitate the neurogenic development of human mesenchymal stem cells, exhibit superior handling properties and may be adjusted intraoperatively (Jakus et al. 2015). Besides electroactive materials, an alternative approach is employing cell-laden structures that may be activated by electrical stimulation to facilitate movement (Kalogeropoulou et al. 2024). This progress brings us nearer to operational integrated cellular systems, which may have various applications in programmable muscle tissue engineering and biomimetic machine design. Further, muscle and nerve tissues can greatly benefit from electrical stimulation in 4D printing applications. Nonetheless, issues such as localized heating, membrane breakdown, or cellular death may occur if the applied electric current is very high (Kim et al. 2024). Consequently, particular safeguards and accurate actuation are important when employing electric fields as stimuli in food printing, a domain that remains underexplored considering existing literature.

7.2.4.3 Optical Field

Using light as a stimulus in 4D printing enables rapid and noninvasive triggering, for instance, allowing for targeted illumination of specific areas within a printed object. Light can penetrate the printed structure and generate heat in intelligent materials; thus, it can even change the color and shape of the materials. Light-stimulated shape memory polymers are commonly employed in self-assembly structures, complex folding techniques, and surface deformations during 4D printing (Yan et al. 2023). The advantages of using light as a stimulus, such as fast switching, precise focusing, and sustainable properties, have garnered significant research interest. Light exposure can also induce various transitions, other than shape and color, the formation of zwitterionic species, charge generation, and photodimerizations (Sadraei and Naghib 2025). In the automotive and polymer sectors, initiatives have been undertaken to utilize light as a catalyst for additive manufacturing. For instance, a research team demonstrated the "bud-to-bloom" phenomenon using polymer materials as the "petals" of a "sunflower." They reported that the heat generated from absorbed light facilitated shape recovery, allowing repeatable global shrinking and localized bending through thermoresponsive gels and light-actuated fibers. The engineered capsules for therapeutic drug release and shape change were also reported due to light illumination (Figure 7.3). This approach has considerable potential for cancer medicines and treating multidrug resistance (Chinnakorn et al. 2023).

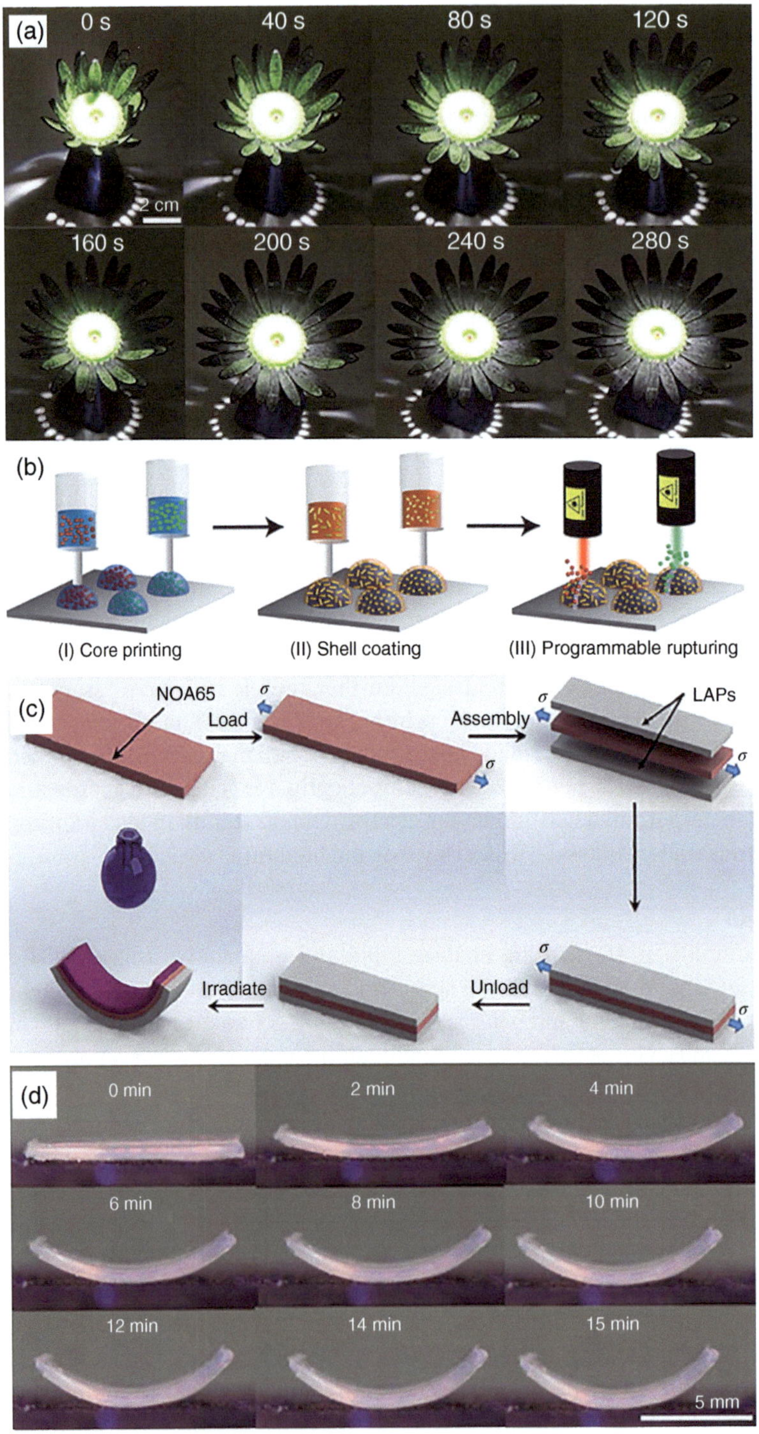

Figure 7.3 Light response in 4D-printed materials: (a) Printed "sunflower" from "bud to bloom" upon light illumination, (b) programmed rupture drug capsule, shape changes upon light stimuli (c) and (d). *Source:* Lui et al. (2019) / with permission of Elsevier.

While using light as a stimulus has opened new avenues for creating biomedical devices and food printing, further research is needed to address concerns about potential toxicity from photo-activated materials from biological sources, heat generation during photothermal conversion, and the stability of shapes. However, there is a limited knowledge base on 4D food printing due to light responses. In future research, their possibilities and technology-economic feasibilities for food applications must be investigated in detail.

7.3 Engineering Structures Through Chemical Stimuli

Alterations in physiological conditions, such as pH and ionic concentration changes, are critical indications of various diseases in biomedical applications. In food applications, numerous studies report pH stimuli to stimulate intelligent function in printed materials, and this aspect has been emphasized in this section.

7.3.1 pH

The impact of pH variations on 4D food-printed structures and their surrounding medium has been reported by many studies, especially in terms of their structural and functional characteristics (Chen et al. 2021b; Ghazal et al. 2019; Kang et al. 2023; Shanthamma et al. 2021). pH-induced color variations in plant-based food materials have attracted considerable attention. Typically, these pH stimuli can be applied internally or externally. In the case of internal stimuli, the material responds due to the internal reaction stimulating the pH variations with respective time changes. Conversely, in the case of external stimuli, the pH change of materials occurs due to the externally spraying solvents inducing the reaction of the printed materials with respect to time. Figures 7.4 and 7.5 illustrate the pH internal and external stimuli responses to the 4D food-printed samples. Further, a combination of external and internal stimuli is also reported. For instance, in 4D printing potato starch-based composites, color transformations were observed; herein, the pH solution acted as an external stimulus, and lemon juice jel acted as an internal stimulus (Ghazal et al. 2019). Natural pigments such as chlorophyll, carotenoids, and anthocyanins, prevalent in fruits and vegetables, are crucial for color appeal and advantageous for human health. However, chlorophyll, responsible for the green coloration in plants, is markedly sensitive to pH, with considerable color deterioration at reduced pH levels (Navaf et al. 2022). Likewise, carotenoid concentration diminishes in acidic environments. Anthocyanins exhibit a red color at lower pH and transition to yellow at elevated pH levels (Li et al. 2023a). Substances that react to pH variations, such as anthocyanins and certain proteins, can result in modified colors or textures according to the pH of their surroundings (Zhu et al. 2023). Subsequently, some studies report multiple responses due to the pH changes of the printed materials (Guo et al. 2021). This property facilitates the development of foods that exhibit visual colors regarding their condition or flavor profile, such as a dessert that alters color during consumption. Additionally, the pH changes are majorly influenced by bioactive compounds in many studies. For instance, curcumin was used in several 4D food printing works on pH-induced stimulation of color, texture and flavor changes in food materials (Cen et al. 2022; Li et al. 2023c; Shen et al. 2023a; Zhu et al. 2023).

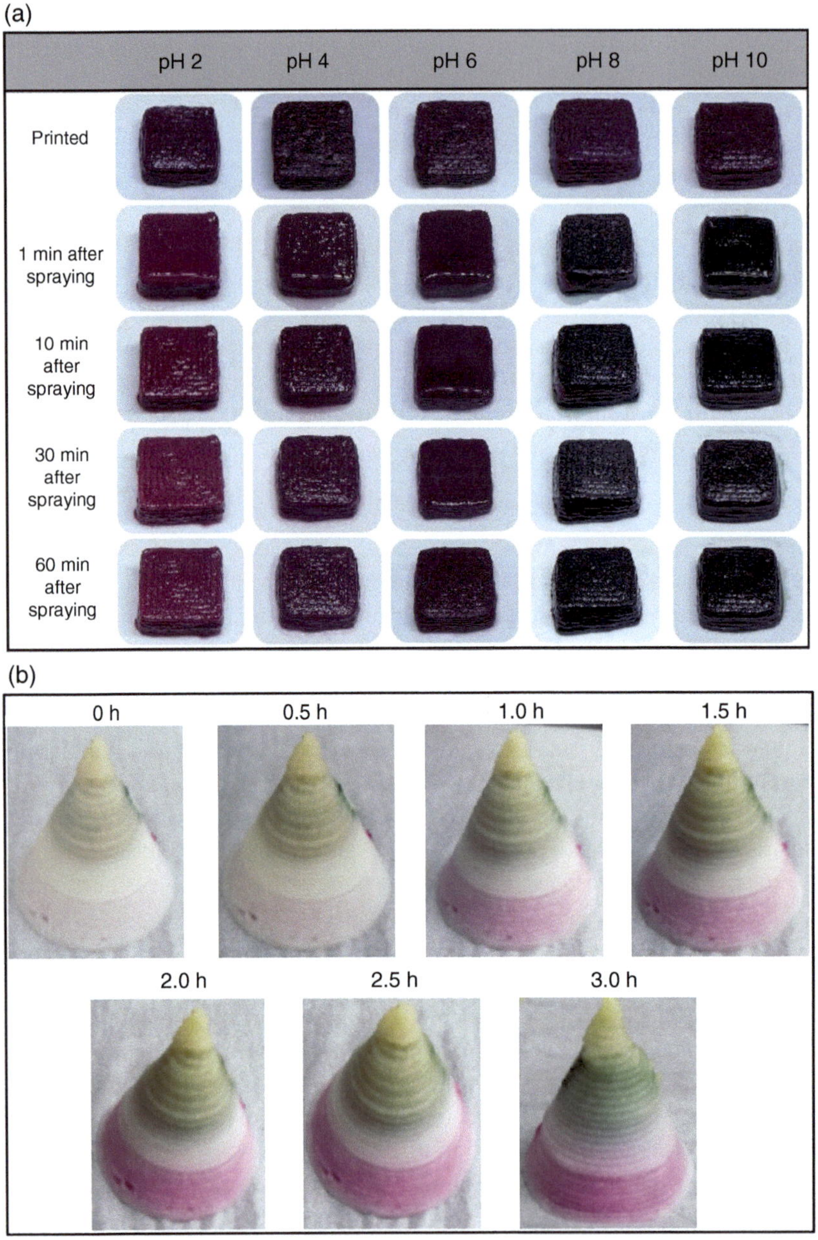

Figure 7.4 (a) 4D food-printed edible insect snacks' color changes due to external pH. *Source:* Kang et al. (2023) / Elsevier / CC BY-NC-ND 4.0, (b) 4D food-printed mashed potato spontaneous color changes due to internal pH. *Source:* He et al. (2020) / with permission of Elsevier.

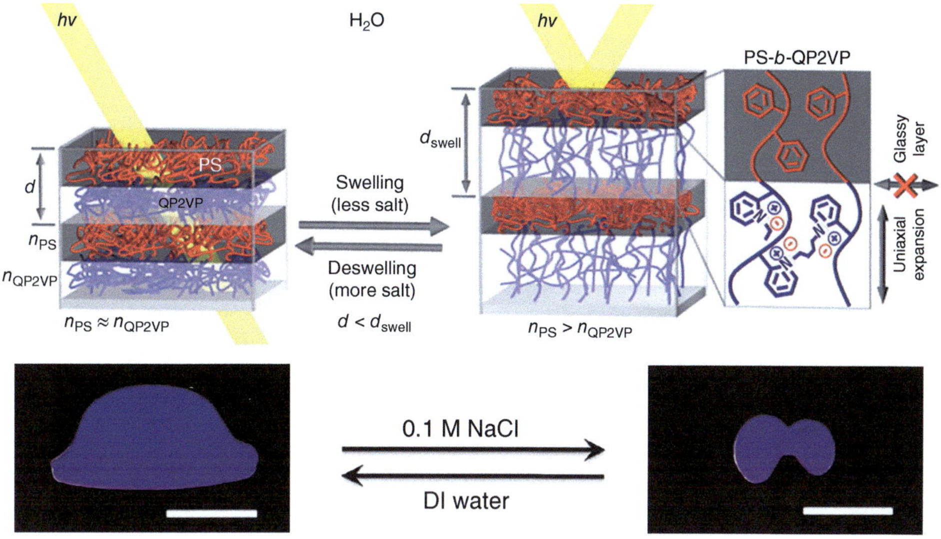

Figure 7.5 4D printing of hydrogels – ion concentration stimulation of shape-morphing behavior. *Source:* Lui et al. (2019) / with permission of Elsevier.

Kokane et al. (2024) and Singh et al. (2024) extensively documented the pH-induced changes in printed materials. Interestingly, multi-intelligent functions are also observed by the synergistic responses of pH and other stimuli (Wang et al. 2024). Various natural proteins, such as collagen, gelatin, exhibit pH responsiveness, thus extending this application into drug delivery, particularly for the human gastrointestinal tract.

7.3.2 Ionic Concentration

Ionic strength denotes the concentration of dissociated ions in aqueous solutions and can markedly affect the characteristics of 3D-printed materials. The influence of ionic concentration on hydrogel swelling depends on the chemical composition of the hydrogels. Ions enable physical interactions, including ionic bonding, within charged hydrogels. Divalent or trivalent cations are frequently utilized to facilitate ionic cross-linking in anionic polymers, exemplified by alginate cross-linked with divalent metal ions (Champeau et al. 2020). Nonetheless, investigations into the application of ionic strength as a stimulus in 4D printing technologies are still infancy.

Current research indicates that form morphing can be accomplished by modifying salt concentrations, especially within biomedical and food applications. For instance, in 4D food printing, the potato starch samples were printed, and the addition of salt and butter resulted in changed ion concentrations, significantly influencing the deformation behavior of the printed samples (He et al., 2021). The ionic concentration prominently changes the dielectric properties of the food samples. Shape transformation follows internal pressures induced by the swelling of diverse gels due to concentration variations, rendering this phenomenon advantageous for applications in soft robotics, actuation, and sensing (Li et al. 2024). Figure 7.6 presents shape changes in hydrogels due to ionic

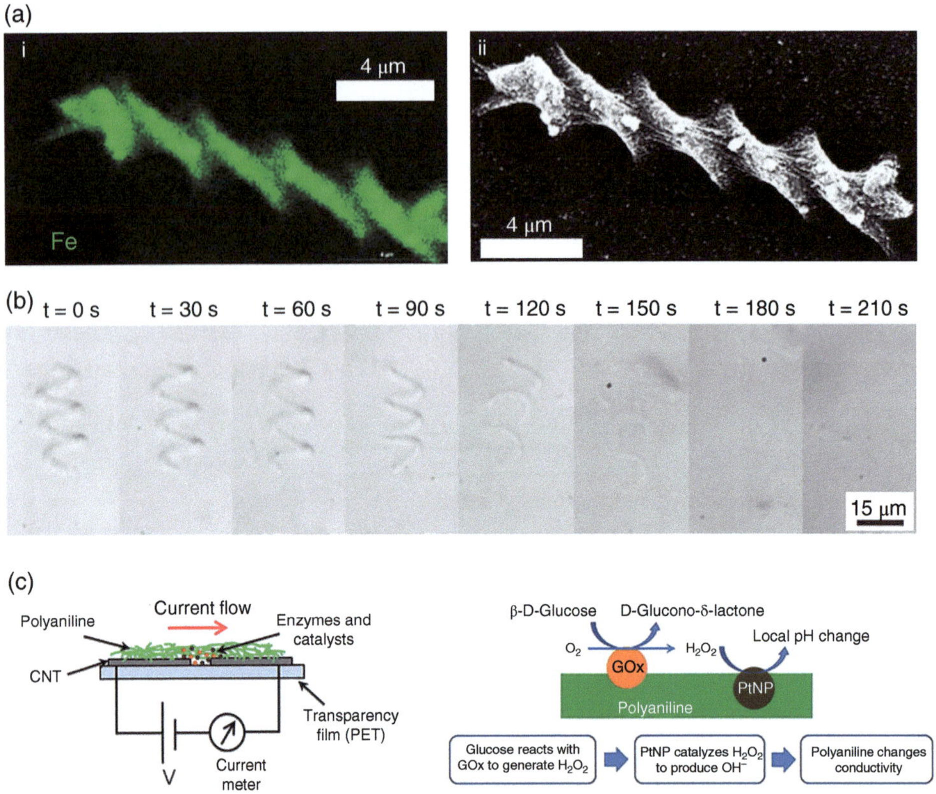

Figure 7.6 4D-printed microrobots' response to enzymes. (a) Microrobot, (b) series of images showing enzymatic degradation respective of time. *Source:* Lui et al. (2019) / with permission of Elsevier; (c) working of a 4D-printed glucose-responsive sensor. *Source:* Lui et al. (2019) / with permission of Elsevier.

concentration variations with time. However, few studies have explored this perspective in food additive manufacturing applications. In the future, the chemical characteristics of food materials, with their improvements in food science may yield a more efficacious understanding, emphasizing the behavior of food materials in reaction to fluctuating ionic concentrations.

7.4 Engineering Structures Through Biological Stimuli

In 4D food printing, intelligent materials can also react to biological and microbiological stimuli. Biological processes in the human body frequently rely on feedback-controlled communication that involves small biological molecules and biomacromolecules, including glucose, enzymes, nucleic acids, polypeptides, and proteins. An exemplary instance of this is the regulation of blood glucose levels, wherein insulin is released by islet cells in reaction to variations in blood sugar. Recent research has concentrated on endowing materials with certain characteristics that allow them to react to various stimuli. Likewise, with food, nutritional alterations may arise from microbial proliferation. Currently, this area remains underexplored, with a limited knowledge base specific to food applications. However, it is evident that there is ample scope to adapt these stimuli in 4D food printing. Accordingly, in this section, stimuli 4D printing for biomaterials is also discussed, in addition to the diverse biological triggers.

7.4.1 Microbes

4D food printing is a cutting-edge technology that can support the growth of plant or animal cells within a defined structure, resulting in alterations to food texture, flavor, and nutritional composition. This method aims to direct edible cells to proliferate within an edible matrix under regulated environmental conditions, employing temperature, humidity, and light stimuli. The approach can be extended to innovative tissue architectures, such as cultured meat, that replicate the nutritional profiles of authentic food (Dong et al. 2023). Researchers have exhibited elevated cell viability and extended proliferation within a sodium alginate hydrogel matrix through the successful integration of carrot callus tissue. During a 35-day cultivation period, scaffolds with differing quantities of callus tissue, incubated at an ideal temperature of 37°C, demonstrated a rich texture and nutritional value, resembling the attributes of actual food (Park et al. 2020). This approach demonstrates that cells can proliferate under suitable environmental conditions and experience substantial alterations in nutritional composition and texture. Furthermore, Rutzerveld (2014) presented the notion of "edible growth foods," demonstrating the creation of ready-to-eat snacks by incorporating seeds from edible plants, fungal spores, or yeast and bacterial cells into an edible nutrient medium. As these foods age, they alter nutritional composition and taste intensity, offering consumers a more varied gustatory experience. Recently, probiotic growth in printed starch-based gel inducing color and nutritional changes was observed by researchers (Chen et al. 2023b). Similarly, due to the probiotic growth, changes in the functional properties of printed food were observed (Chen et al. 2024). This sector is presently flourishing in food applications, especially in the advancement of plant-based meats, cellular agriculture, and bioprinting for both food and biomedical applications.

7.4.2 Enzymes

Enzymes are extremely specific and selective molecules that are essential in numerous biological processes, such as protein expression, cellular adhesion development, and muscle contraction. The increasing fascination with enzyme-responsive materials arises from the diverse array of enzymes inherently found in the body. Enzyme activation can be inherently initiated by the biological environment, with the degree of stimulation regulated by the body's condition (Faber et al. 2024). The technique entails the enzymatic breakdown of the drug-encapsulated hydrogels, facilitating on-demand administration to specific sites within the body (Chinnakorn et al. 2023). Recently, researchers have employed the fabricated microrobots for precise medication delivery. These magnetically actuated microrobots were capable of undergoing full degradation in the presence of collagenase and enzymes, resulting in no visible hazardous residues (Wang et al. 2018). This enzymatic degradability resolves prevalent concerns associated with additive manufacturing printed microrobots, including nondegradability and cytotoxic byproducts. Besides material degradation, enzymes can elicit numerous responses, such as modifications in supramolecular architecture, structural swelling or collapse, and changes in surface characteristics (Faber et al. 2024). Currently, the application of printed enzyme-responsive materials predominantly emphasizes enzymatically induced degradability. A major problem in utilizing these materials is managing their response behavior, as there exists a possibility of undesirable cross-reactivity with other enzymes in the body and a particular concern especially when DNA sequences or longer peptides are used (Lui et al. 2019). Nevertheless, these stimuli are mostly unexplored in food printing applications, indicating substantial potential for future progress in food additive manufacturing.

7.4.3 Glucose

With the global increase in diabetes mellitus, considerable emphasis has been placed on developing glucose-responsive materials for glucose monitoring and insulin delivery. Effective management of both type I and type II diabetes generally necessitates systematic blood glucose monitoring. In this regard, glucose-responsive materials are being incorporated into functional glucose-detecting devices. A glucose sensor employing graphene polylactic acid filament preconditioned with glucose dehydrogenase flavin adenine dinucleotide was reported based on an amperometric principle (Adams et al., 2018). Diabetes therapy necessitates not just blood glucose monitoring but also multiple daily subcutaneous insulin injections. Nonetheless, sustaining persistent patient monitoring and adherence can be difficult, hence elevating the risk of problems. This has increased interest in self-regulating glucose-responsive pharmaceuticals for regulated insulin administration. Contemporary methodologies for creating glucose-responsive insulin delivery systems predominantly depend on two mechanisms for the regulation of blood glucose levels: (i) a directly triggered model, wherein insulin is released via competitive binding of glucose with glycopolymer–lectin complexes, and (ii) a progressively activated model, in which insulin release occurs due to swelling induced by localized acidification resulting from glucose oxidase-based enzymatic reactions (Lui et al. 2019). Notwithstanding the known nature of these pathways, there are presently not many studies focusing on the fabrication of glucose-responsive devices utilizing these methods for insulin administration. Alternatively, researchers have proposed an innovative method whereby insulin administration is enabled by the secretion of β-cells in a hyperglycemic milieu. This entails the fabrication of a graft utilizing polylactic acid and the integration of stem cell-derived β-cell clusters to impart glucose responsiveness to the constructs. Their findings demonstrate that this cell-laden printed graft can markedly improve insulin production in reaction to glucose (Song and Millman 2016). A significant obstacle in the creation of glucose-responsive materials is the incorporation of enzymes in the manufacturing process. Consequently, sustaining favorable processing conditions is essential to retain enzyme bioactivity post fabrication. Moreover, glucose-responsive polymers designed for insulin delivery must exhibit biocompatibility to guarantee practical deployment. In the future, biosensor fabrications using the 4D printing approach may extend to a range of applications in food safety and quality.

7.5 Stimuli Effects on Smart/Intelligent Materials

The stimuli prominently affect the printed material in positive and negative ways. Depending on the material characteristics and stimuli concentration, the function of the printed materials varies. On the positive side, it can stimulate the function of the printed material with respect to time changes (and therefore can be referred to as 4D); conversely, on the negative side, higher stimuli concentration or unsuitable stimuli may destroy the nature of printed materials. Accordingly, numerous studies have demonstrated that stimuli considerably influence the concentration of bioactive compounds and their material composite behavior, underscoring their significance. 4D-printed foods containing bioactive components may demonstrate improved storage stability and extended shelf life (Kokane et al. 2024). Sometimes, the composition possesses negative effects if mixed with the 'wrong' materials. For instance, in compositions with bioactive molecules such as anthocyanin in food printing applications, color changes were observed by changing the pH levels from 2 to 10 at room temperature (Ghazal et al., 2023). Similarly, researchers investigated the stability of curcumin in printed material in response to thermal stimuli (drying techniques). Their findings indicated that curcumin retention reached a maximum of 62.17% with catalytic infrared drying, surpassing hot air drying by 34.86%, and microwave drying by 23.8%. Catalytic infrared drying efficiently eliminates moisture rapidly and uniformly while maintaining nutritional integrity, whereas hot air drying may result in the loss of essential nutrients due to extended convection durations. Conversely, microwave drying may result in the breakdown of curcumin with localized heating effects (Shen et al. 2023a).

All stimuli that influence printed materials, for instance, in terms of food texture, color, shape, flavor/aroma, and nutrition, are considered functional or intelligent due to their specific effects and have been well documented by researchers worldwide. Similarly, 4D multi-material food printing can enable multifunctionality in the printed constructs. Nevertheless, a synergistic approach involving multiple stimuli or frequent phase exposure and concentrations that exceed the threshold limits of the materials may compromise the functionality of 4D-printed foods. Additionally, greater emphasis should be placed on innovations in 4D printing materials. Comprehensive research is required to investigate novel materials and transformation mechanisms, determine optimal designs for controllable deformation components, and enhance printing processes. Furthermore, creating highly sensitive materials is essential to improve dynamic performance while reducing adverse effects on components during transformation.

7.6 Stimuli Effects on Sensory Properties of 4D-Printed Food

The comprehension of sensory characteristics and acceptance of 4D-printed food in everyday life is improved by sensory evaluation. Also, sensory attributes of printed food are affected by the conditions employed during its processing. The influence of internal and environmental factors, specifically pH levels, can significantly alter the color of foods. For example, at extremely low pH, foods may lose their palatability due to bitterness/acidity. Conversely, under such conditions, the color and texture of 4D-printed foods may be preferred more (Ghazal et al. 2019). In this line, several recent studies have documented the impact of pH on color changes in 4D-printed foods (Chen et al. 2024; He et al. 2020; Shanthamma et al. 2021). Ghazal et al. (2023) conducted a sensory evaluation of microwave-heated 3D-printed apple/potato starch gel incorporating varying quantities of microparticles (freeze-dried red cabbage juice-loaded whey protein isolate/gum arabic). Samples containing 60% microparticles scored the highest color ratings, and all 4D samples with different concentrations remained within acceptable limits for overall acceptability. Recently, to explain the multifunctions of printed foods, pH and thermal treatment were utilized; synergistic effects resulted in multi-intelligent functions (color, shape, flavor) in single printed food constructs (Chen et al. 2024).

In 4D food printing, sensory evaluation has been carried out by instrumental techniques (including E-nose and E-tongue). For example, an E-tongue assessment reported the sensory attributes of soybean protein isolates flavored with k-carrageenan and vanilla, and the sample displayed high levels of bitterness, astringency, umami, richness, and saltiness since the vanilla flavor was more responsive to temperature than the intrinsic flavor of soybean protein isolate (Phuhongsung et al. 2020). Similarly, for different fruit juice substitutes in 4D food printing, sensory results through E-tongue were reported by Ghazal et al. (2021). They elucidated the reactions of printed red cabbage juice and potato starch samples to pH solutions, in contrast to orange juice, apple juice, and lemon juice. The findings indicated that orange juice gel and vanilla samples achieved the highest aroma ratings, and the orange juice and lemon juice gels at pH 6 were the most palatable gels. An E-tongue study reported that a pH of 2 resulted in the highest level of sourness with negligible umami flavor. The dried 4D samples scored better acceptability ratings.

For 4D-printed meat products subjected to thermal stimuli, it was reported that consumers favored a 30% infill density because of its form, design, tactile quality, and familiarity. Nonetheless, several buyers preferred the 100% infill density owing to its pleasing taste and texture over "real meat." However, samples with reduced infill density were more palatable, indicating that both the infill ratio and thickness substantially influence sensory attributes (Dick et al. 2021).

Overall, the above cases highlight the substantial influence of stimuli on the sensory attributes of food, which is essential for assessing its appropriateness for consumption. Similarly, microbiological growth was observed by assessing the bioavailability of the printed products. Chen et al. (2023a) reported that the higher infill (100%) observed higher probiotic growth than the lower infill, thus changing the color and flavor of the printed objects. However, research in terms of sensory aspects needs to be carried out to support the regulations and safety requirements for 4D-printed foods.

7.7 Challenges Associated with Stimuli in 4D Food Printing

The significant efforts dedicated to additive manufacturing have propelled the emergence of 3D food printing applications, with 4D food printing set as the subsequent evolutionary advancement. Despite considerable advancements in 4D printing research, its practical application remains nascent. A significant difficulty is the low efficiency of stimuli procedures, which currently lack precision and often lead to excessive energy use. Suboptimal postprocessing settings or stimuli can negatively impact product quality, resulting in problems such as structural failure, oxidative discoloration, dehydration, and shrinkage. Mitigating the effects of excessive stimulus concentration on the quality of 4D-printed products will be essential in the future. Innovative methodologies, such as electric and magnetic fields or hybrid techniques like nonthermal, can improve stimulus responses of the printed materials.

Furthermore, advancing integrated 4D printing apparatuses incorporating printing and stimulation processes would facilitate more efficient, multiresponse behaviors by converting singular inputs into multiple programmed reactions. Although research has predominantly concentrated on microwave heating and pH alterations as stimuli, investigating additional stimuli such as light or variations in ionic concentration could broaden the potential in 4D food printing. Specifically, multimaterial 4D printing will facilitate the incorporation of various stimuli-responsive materials into a singular construction, creating new opportunities for innovation. Incorporating sensors and electronics into 4D-printed structures may result in intelligent materials that can detect and react to environmental alterations. The advancement of sophisticated simulation techniques will enhance the design and production processes. Developing precise computer models that forecast the behavior of smart materials under diverse stimuli will markedly enhance the efficiency and accuracy of 4D printing, facilitating more effective and practical applications.

A primary objective for the future is to establish monitoring systems that assess the magnitude of 4D transformations, facilitating accurate regulation of these alterations. Emerging technologies, including artificial intelligence, will augment predictability by accurately mimicking stimuli and material responses. Collaboration among scientists and engineers across disciplines such as food science, materials science, modeling, and engineering will be crucial for realizing the complete potential of 4D printing in food applications. Progress in multimaterial printing has enabled the fabrication of intricate structures with varied characteristics. Future advancements may provide enhanced control over mechanical and chemical characteristics.

7.8 Conclusion

The three primary domains of contemporary research in 4D food printing are induced color alteration, morphological transformation, and cellular cultivation. These temporal activities are stimulated by various stimuli, which enhance the sensory experience of consuming food. Whilst there are several stimuli explored for 4D printing applications, many approaches cannot work for food printing. Overall, it is essential to conduct further research to accurately regulate and enhance these dynamic features, as 4D food printing is still in its infancy. Future endeavors should focus on identifying novel stimuli that can induce favorable changes in food attributes, as well as the selection or synthesis of smart materials specifically designed for specific transformations. These developments will be essential for fully realizing the potential of 4D food manufacturing.

References

Adams, A., Malkoc, A., & La Belle, J. T. (2018). The development of a glucose dehydrogenase 3D-printed glucose sensor: A proof-of-concept study. *Journal of Diabetes Science and Technology*, *12*(1), 176–182. doi:10.1177/1932296817715272

Anandharamakrishnan, C., Moses, J. A., & Anukiruthika, T. (2022). *3D printing of foods*. John Wiley & Sons Ltd.

Billiet, T., Vandenhaute, M., Schelfhout, J., Van Vlierberghe, S., & Dubruel, P. (2012). A review of trends and limitations in hydrogel-rapid prototyping for tissue engineering. *Biomaterials*, *33*(26), 6020–6041. doi:10.1016/j.biomaterials.2012.04.050

Boyle, B. M., French, T. A., Pearson, R. M., McCarthy, B. G., & Miyake, G. M. (2017). Structural color for additive manufacturing: 3D-printed photonic crystals from block copolymers. *ACS Nano*, *11*(3), 3052–3058. doi:10.1021/acsnano.7b00032

Burke, R. M. (2024). New dimensions in dining: A review of the gastronomic evolution of 3D printed foods from 2013 to 2024 and beyond. *International Journal of Gastronomy and Food Science*, *38*, 101027. doi:10.1016/j.ijgfs.2024.101027

Cen, S., Li, Z., Guo, Z., Li, H., Shi, J., Huang, X., Zou, X., & Holmes, M. (2022). 4D printing of a citrus pectin/β-CD Pickering emulsion: A study on temperature induced color transformation. *Additive Manufacturing*, *56*, 102925. doi:10.1016/j.addma.2022.102925

Cezar, C. A., Roche, E. T., Vandenburgh, H. H., Duda, G. N., Walsh, C. J., & Mooney, D. J. (2016). Biologic-free mechanically induced muscle regeneration. *Proceedings of the National Academy of Sciences*, *113*(6), 1534–1539. doi:10.1073/pnas.1517517113

Champeau, M., Heinze, D. A., Viana, T. N., de Souza, E. R., Chinellato, A. C., & Titotto, S. (2020). 4D printing of hydrogels: A review. *Advanced Functional Materials*, *30*(31), 1910606. doi:10.1002/adfm.201910606

Chen, C., Zhang, M., Guo, C., & Chen, H. (2021a). 4D printing of lotus root powder gel: Color change induced by microwave. *Innovative Food Science & Emerging Technologies*, *68*, 102605. doi:10.1016/j.ifset.2021.102605

Chen, C., Zhang, M., Mujumdar, A. S., & Phuhongsung, P. (2021b). Investigation of 4D printing of lotus root-compound pigment gel: Effect of pH on rapid colour change. *Food Research International*, *148*, 110630. doi:10.1016/j.foodres.2021.110630

Chen, J., Teng, X., Zhang, M., Bhandari, B., Adhikari, B., & Yu, D. (2023a). 5D food printing with color change induced by probiotic growth in a starch-protein-based gel system. *Food and Bioprocess Technology*, *16*(10), 2304–2314. doi:10.1007/s11947-023-03064-7

Chen, J., Zhang, M., & Devahastin, S. (2021c). UV-C irradiation-triggered nutritional change of 4D printed ergosterol-incorporated purple sweet potato pastes: Conversion of ergosterol into vitamin D2. *LWT*, *150*, 111944. doi:10.1016/j.lwt.2021.111944

Chen, J., Zhang, M., Mujumdar, A. S., & Phuhongsunge, P. (2022). 4D printing induced by microwave and ultrasound for mushroom mixtures: efficient conversion of ergosterol into vitamin D2. *Food Chemistry*, *387*, 132840. doi:10.1016/j.foodchem.2022.132840

Chen, X., Teng, X., Zhang, M., Mujumdar, A. S., & Li, J. (2024). Functional changes in 5D printed starch-based gel systems caused by spontaneous growth of probiotics. *Food and Bioprocess Technology*, *17*(8), 2470–2482. doi:10.1007/s11947-023-03267-y

Chen, X., Zhang, M., & Tang, T. (2023b). Microwave-induced rapid shape change of 4D printed vegetable-based food. *Food*, *12*(11), 2158. doi:10.3390/foods12112158

Chen, X., Zhang, M., Tang, T., & Yu, D. (2024). The color/shape/flavor of yam gel with double emulsified microcapsules changed synchronously in 4D printing induced by microwave. *International Journal of Biological Macromolecules*, *260*, 129631. doi:10.1016/j.ijbiomac.2024.129631

Chinnakorn, A., Nuansing, W., Bodaghi, M., Rolfe, B., & Zolfagharian, A. (2023). Recent progress of 4D printing in cancer therapeutics studies. *SLAS Technology*, *28*(3), 127–141. doi:10.1016/j.slast.2023.02.002

Cui, C., Kim, D.-O., Pack, M. Y., Han, B., Han, L., Sun, Y., & Han, L.-H. (2020). 4D printing of self-folding and cell-encapsulating 3D microstructures as scaffolds for tissue-engineering applications. *Biofabrication*, *12*(4), 045018. doi:10.1088/1758-5090/aba502

Dai, S., Ravi, P., & Tam, K. C. (2009). Thermo- and photo-responsive polymeric systems. *Soft Matter*. doi:10.1039/b820044k

Derossi, A., Spence, C., Corradini, M. G., Jekle, M., Fahmy, A. R., Caporizzi, R., Devahastin, S., Moses, J. A., Le-Bail, A., Zhou, W., Zhang, M., Bhandari, B., & Severini, C. (2024). Personalized, digitally designed 3D printed food towards the reshaping of food manufacturing and consumption. *npj Science of Food*, *8*(1), 54. doi:10.1038/s41538-024-00296-5

Dick, A., Gao, Y., Bhandari, B., & Prakash, S. (2021). Influence of drying method and 3D design on the 4D morphing of beef products. *Applied Food Research*, *1*(2), 100017. doi:10.1016/j.afres.2021.100017

Dong, H., Wang, P., Yang, Z., & Xu, X. (2023). 3D printing based on meat materials: challenges and opportunities. *Current Research in Food Science*, *6*, 100423. doi:10.1016/j.crfs.2022.100423

Faber, L., Yau, A., & Chen, Y. (2024). Translational biomaterials of four-dimensional bioprinting for tissue regeneration. *Biofabrication*, *16*(1), 012001. doi:10.1088/1758-5090/acfdd0

Fahmy, A. R., Derossi, A., & Jekle, M. (2023). Four-dimensional (4D) printing of dynamic foodsefinitions, considerations, and current scientific status. *Foods*, *12*(18), 3410. doi:10.3390/foods12183410

Gao, R., Song, R., Shen, L., Zhao, X., Xue, L., Li, J., & Zheng, X. (2023). Evaluation of 3D printability of blueberry powder gel system under ultrasound pretreatment. *LWT*, *190*, 115577. doi:10.1016/j.lwt.2023.115577

Gauss, C., Pickering, K. L., & Muthe, L. P. (2021). The use of cellulose in bio-derived formulations for 3D/4D printing: A review. *Composites Part C: Open Access*, *4*, 100113. doi:10.1016/j.jcomc.2021.100113

Ghazal, A. F., Zhang, M., Bhandari, B., & Chen, H. (2021). Investigation on spontaneous 4D changes in color and flavor of healthy 3D printed food materials over time in response to external or internal pH stimulus. *Food Research International*, *142*, 110215. doi:10.1016/j.foodres.2021.110215

Ghazal, A. F., Zhang, M., & Guo, Z. (2023). Microwave-induced rapid 4D change in color of 3D printed apple/potato starch gel with red cabbage juice-loaded WPI/GA mixture. *Food Research International*, *172*, 113138. doi:10.1016/j.foodres.2023.113138

Ghazal, A. F., Zhang, M., & Liu, Z. (2019). Spontaneous color change of 3D printed healthy food product over time after printing as a novel application for 4D food printing. *Food and Bioprocess Technology*, *12*(10), 1627–1645. doi:10.1007/s11947-019-02327-6

Ghazal, A. F., Zhang, M., Mujumdar, A. S., & Ghamry, M. (2023). Progress in 4D/5D/6D printing of foods: Applications and R&D opportunities. *Critical Reviews in Food Science and Nutrition*, *63*(25), 7399–7422. doi:10.1080/10408398.2022.2045896

Guo, C., Zhang, M., Bhandari, B., & Devahastin, S. (2022). Investigation on simultaneous change of deformation, color and aroma of 4D printed starch-based pastes from fruit and vegetable as induced by microwave. *Food Research International*, *157*, 111214. doi:10.1016/j.foodres.2022.111214

Guo, C., Zhang, M., & Devahastin, S. (2021). Color/aroma changes of 3D-printed buckwheat dough with yellow flesh peach as triggered by microwave heating of gelatin-gum Arabic complex coacervates. *Food Hydrocolloids*, *112*, 106358. doi:10.1016/j.foodhyd.2020.106358

Guo, J., Zhang, M., Li, J., & Fang, Z. (2023). Using soy protein isolate to improve the deformation properties of 4D-printed oat flour butterfly. *Food and Bioprocess Technology*, *16*(5), 1165–1176. doi:10.1007/s11947-023-02994-6

He, C., Zhang, M., & Devahastin, S. (2020). Investigation on spontaneous shape change of 4D printed starch-based purees from purple sweet potatoes as induced by microwave dehydration. *ACS Applied Materials & Interfaces*, *12*(34), 37896–37905. doi:10.1021/acsami.0c10899

He, C., Zhang, M., & Devahastin, S. (2021). Microwave-induced deformation behaviors of 4D printed starch-based food products as affected by edible salt and butter content. *Innovative Food Science & Emerging Technologies*, *70*, 102699. doi:10.1016/j.ifset.2021.102699

He, C., Zhang, M., & Guo, C. (2020). 4D printing of mashed potato/purple sweet potato puree with spontaneous color change. *Innovative Food Science & Emerging Technologies*, *59*, 102250. doi:10.1016/j.ifset.2019.102250

Huang, Y., Zhang, M., Pattarapon, P., & Mujumdar, A. S. (2022). 4D printing of mixed vegetable gel based on deformation and discoloration induced by acidification and dehydration. *Journal of Food Process Engineering*, *45*(12). doi:10.1111/jfpe.14172

Hwangbo, H., Lee, H., Roh, E. J., Kim, W., Joshi, H. P., Kwon, S. Y., Choi, U. Y., Han, I.-B., & Kim, G. H. (2021). Bone tissue engineering via application of a collagen/hydroxyapatite 4D-printed biomimetic scaffold for spinal fusion. Applied. *Physics Reviews*, *8*(2). doi:10.1063/5.0035601

Ionov, L. (2014). Hydrogel-based actuators: Possibilities and limitations. *Materials Today*, *17*(10), 494–503. doi:10.1016/j.mattod.2014.07.002

Jakus, A. E., Secor, E. B., Rutz, A. L., Jordan, S. W., Hersam, M. C., & Shah, R. N. (2015). Three-dimensional printing of high-content graphene scaffolds for electronic and biomedical applications. *ACS Nano*, *9*(4), 4636–4648. doi:10.1021/acsnano.5b01179

Jeong, H. E., Lee, C. J., & Min, S. C. (2024). Improvement of jelly 3D printing using ultrasound treatment and calcium chloride. *Food Science and Biotechnology*, *33*(11), 2551–2557. doi:10.1007/s10068-024-01517-z

Joardder, M. U. H., Bosunia, M. H., Hasan, M. M., Ananno, A. A., & Karim, A. (2024). Significance of glass transition temperature of food material in selecting drying condition: An in-depth analysis. *Food Reviews International*, *40*(3), 952–973. doi:10.1080/87559129.2023.2204131

Kalogeropoulou, M., Díaz-Payno, P. J., Mirzaali, M. J., van Osch, G. J. V. M., Fratila-Apachitei, L. E., & Zadpoor, A. A. (2024). 4D printed shape-shifting biomaterials for tissue engineering and regenerative medicine applications. *Biofabrication*, *16*(2), 022002. doi:10.1088/1758-5090/ad1e6f

Kang, Z., Wang, Z., Zhang, W., & Fang, Z. (2023). 4D printing of edible insect snacks: Color changes with external pH stimulation. *Food Bioscience*, *55*, 102971. doi:10.1016/j.fbio.2023.102971

Kashiwakura, Y., Sogabe, T., Hiyama, Y., Arakawa, N., Fujii, T., Tochio, T., & Kawai, K. (2022). Prediction and control of glass transition temperature for hydrogenated starch hydrolysates and its impact on the texture modification of gummy. *Food Hydrocolloids*, *126*, 107467. doi:10.1016/j.foodhyd.2021.107467

Kim, J., DA, G., Debnath, P., & Saha, P. (2024). Smart multi-responsive biomaterials and their applications for 4D bioprinting. *Biomimetics*, *9*(8), 484. doi:10.3390/biomimetics9080484

Koirala, S., Prakash, S., Karim, A., & Bhandari, B. (2023). Shape morphing of foods: Mechanism, strategies, and applications. *Trends in Food Science & Technology*, *139*, 104135. doi:10.1016/j.tifs.2023.104135

Kokane, S. B., Anjaly, P., Thangalakshmi, S., & Arora, V. K. (2024). Current trends in additive manufacturing based 4D food printing technology: A review. *Future Foods*, *10*, 100450. doi:10.1016/j.fufo.2024.100450

Kokane, S. B., Arora, V. K., & Thangalakshmi, S. (2024). Potential of bioactive ingredients for healthy 4D food printing: Applications and challenges. *Food Reviews International*, 1–24. doi:10.1080/87559129.2024.2374808

Li, G., Zhan, J., Huang, J., Xu, E., Yuan, C., Chen, J., Yao, Q., & Hu, Y. (2023a). Enhanced fresh-keeping capacity of printed surimi by Ca2+-nano starch-lutein and its nondestructive freshness monitoring based on 4D printed anthocyanin. *International Journal of Biological Macromolecules*, *252*, 126543. doi:10.1016/j.ijbiomac.2023.126543

Li, S., Yang, H., Chen, G., Zheng, J., Wang, W., Ren, J., Zhu, C., Yang, Y., Cong, Y., & Fu, J. (2023b). 4D printing of biomimetic anisotropic self-sensing hydrogel actuators. *Chemical Engineering Journal*, *473*, 145444. doi:10.1016/j.cej.2023.145444

Li, T., Huang, Z., Tsui, G. C.-P., Tang, C.-Y., & Deng, Y. (2024). Recent advances in 4D printing of hydrogels. *Reviews on Advanced Materials Science*, *63*(1). doi:10.1515/rams-2024-0028

Li, X., Fan, L., Li, R., Han, Y., & Li, J. (2023c). 3D/4d printing of β-cyclodextrin-based high internal phase emulsions. *Journal of Food Engineering*, *348*, 111455. doi:10.1016/j.jfoodeng.2023.111455

Liu, Z., He, C., Guo, C., Chen, F., Bhandari, B., & Zhang, M. (2021). Dehydration-triggered shape transformation of 4D printed edible gel structure affected by material property and heating mechanism. *Food Hydrocolloids*, *115*, 106608. doi:10.1016/j.foodhyd.2021.106608

Lui, Y. S., Sow, W. T., Tan, L. P., Wu, Y., Lai, Y., & Li, H. (2019). 4D printing and stimuli-responsive materials in biomedical aspects. *Acta Biomaterialia*, *92*, 19–36. doi:10.1016/j.actbio.2019.05.005

Lv, Y., Wang, B., Cheng, Y., Lv, W., Zeng, S., & Xiao, H. (2024). Printing characteristics and microwave infrared-induced 4D printing of chestnut powder composite paste. *Journal of Food Engineering*, *382*, 112197. doi:10.1016/j.jfoodeng.2024.112197

Ma, C., Yan, J., Li, W., Wang, Y., McClements, D. J., Liu, X., & Liu, F. (2024). Enhanced printability of food-grade edible inks: Emulsions formulated with modified pea protein and sodium alginate. *Food Hydrocolloids*, *152*, 109946. doi:10.1016/j.foodhyd.2024.109946

Mahmood, A., Akram, T., Shenggui, C., & Chen, H. (2023). Revolutionizing manufacturing: A review of 4D printing materials, stimuli, and cutting-edge applications. *Composites Part B: Engineering*, *266*, 110952. doi:10.1016/j.compositesb.2023.110952

Maleki, S., Aarabi, A., Far, F. A., & Dizaji, H. Z. (2024). Heat moisture treatment and ultrasound-induced hydrothermal wheat starch modification: Techno-functional, microstructural and quality 3D printed characteristics. *International Journal of Biological Macromolecules, 276*, 133992. doi:10.1016/j.ijbiomac.2024.133992

Miyashita, S., Guitron, S., Ludersdorfer, M., Sung, C. R., & Rus, D. (2015). An untethered miniature origami robot that self-folds, walks, swims, and degrades. *2015 IEEE International Conference on Robotics and Automation (ICRA)*, 1490–1496. 10.1109/ICRA.2015.7139386

Navaf, M., Sunooj, K. V., Aaliya, B., Akhila, P. P., Sudheesh, C., Mir, S. A., & George, J. (2022). 4D printing: a new approach for food printing; effect of various stimuli on 4D printed food properties. A comprehensive review. *Applied Food Research, 2*(2), 100150. doi:10.1016/j.afres.2022.100150

Niu, D., Zhang, M., Mujumdar, A. S., & Li, J. (2023). Research on microwave-induced bidirectional deformation of coix seed compound materials in 4D printing. *Food and Bioprocess Technology, 16*(11), 2400–2415. doi:10.1007/s11947-023-03078-1

Noree, S., Pinyakit, Y., Tungkijanansin, N., Kulsing, C., & Hoven, V. P. (2023). Shape transformation of 4D printed edible insects triggered by thermal dehydration. *Journal of Food Engineering, 358*, 111666. doi:10.1016/j.jfoodeng.2023.111666

Park, S. M., Kim, H. W., & Park, H. J. (2020). Callus-based 3D printing for food exemplified with carrot tissues and its potential for innovative food production. *Journal of Food Engineering, 271*, 109781. doi:10.1016/j.jfoodeng.2019.109781

Phuhongsung, P., Zhang, M., & Bhandari, B. (2020). 4D printing of products based on soy protein isolate via microwave heating for flavor development. *Food Research International, 137*, 109605. doi:10.1016/j.foodres.2020.109605

Pretsch, T. (2010). Review on the functional determinants and durability of shape memory polymers. *Polymers, 2*(3), 120–158. doi:10.3390/polym2030120

Pulatsu, E., Su, J.-W., Lin, J., & Lin, M. (2022). Utilization of ethyl cellulose in the osmotically-driven and anisotropically-actuated 4D printing concept of edible food composites. *Carbohydrate Polymer Technologies and Applications, 3*, 100183. doi:10.1016/j.carpta.2022.100183

Rabaux, O., & Jérôme, C. (2025). Bioinspired morphing mechanisms for soft systems: A review. *Advanced Intelligent Systems, 7*(2), 2400331. doi:10.1002/aisy.202400331

Roppolo, I., Chiappone, A., Angelini, A., Stassi, S., Frascella, F., Pirri, C. F., Ricciardi, C., & Descrovi, E. (2017). 3D printable light-responsive polymers. *Materials Horizons, 4*(3), 396–401. doi:10.1039/C7MH00072C

Rutzerveld, C. (2014). Edible Growth-The use of additive manufacturing technologies to create an edible ecosystem. *Food Futurist*. https://www.chloerutzerveld.com/edible-growth

Sadraei, A., & Naghib, S. M. (2025). 4D printing of physical stimuli-responsive hydrogels for localized drug delivery and tissue engineering. *Polymer Reviews, 65*(1), 104–168. doi:10.1080/15583724.2024.2427184

Sajjad, R., Chauhdary, S. T., Anwar, M. T., Zahid, A., Khosa, A. A., Imran, M., & Sajjad, M. H. (2024). A review of 4D printing – technologies, shape shifting, smart polymer based materials, and biomedical applications. *Advanced Industrial and Engineering Polymer Research, 7*(1), 20–36. doi:10.1016/j.aiepr.2023.08.002

Shanthamma, S., Preethi, R., Moses, J. A., & Anandharamakrishnan, C. (2021). 4D printing of sago starch with turmeric blends: A study on pH-triggered spontaneous color transformation. *ACS Food Science & Technology, 1*(4), 669–679. doi:10.1021/acsfoodscitech.0c00151

Shen, C., Chen, W., Li, C., Chen, X., Cui, H., & Lin, L. (2023a). 4D printing system stimulated by curcumin/whey protein isolate nanoparticles: A comparative study of sensitive color change and post-processing. *Journal of Food Engineering, 342*, 111357. doi:10.1016/j.jfoodeng.2022.111357

Shen, D., Zhang, M., Mujumdar, A. S., & Li, J. (2023b). Advances and application of efficient physical fields in extrusion based 3D food printing technology. *Trends in Food Science & Technology, 131*, 104–117. doi:10.1016/j.tifs.2022.11.017

Shi, Y., Zhang, M., & Phuhongsung, P. (2022). Microwave-induced spontaneous deformation of purple potato puree and oleogel in 4D printing. *Journal of Food Engineering, 313*, 110757. doi:10.1016/j.jfoodeng.2021.110757

Singh, P., Pandey, V. K., Tripathi, A., Singh, R., Net, T. S. S., Ramniwas, S., & Pandiselvam, R. (2024). 4D food printing technology: Structural changes to culinary art and beyond. *Journal of Food Process Engineering, 47*(1). doi:10.1111/jfpe.14535

Soleymani, S., & Naghib, S. M. (2023). 3D and 4D printing hydroxyapatite-based scaffolds for bone tissue engineering and regeneration. *Heliyon, 9*(9), e19363. doi:10.1016/j.heliyon.2023.e19363

Song, J., & Millman, J. R. (2016). Economic 3D-printing approach for transplantation of human stem cell-derived β -like cells. *Biofabrication, 9*(1), 015002. doi:10.1088/1758-5090/9/1/015002

Sydney Gladman, A., Matsumoto, E. A., Nuzzo, R. G., Mahadevan, L., & Lewis, J. A. (2016). Biomimetic 4D printing. *Nature Materials, 15*(4), 413–418. doi:10.1038/nmat4544

Tang, T., Zhang, M., Mujumdar, A. S., & Teng, X. (2022). 3D printed white radish/potato gel with microcapsules: Color/flavor change induced by microwave-infrared heating. *Food Research International, 158*, 111496. doi:10.1016/j.foodres.2022.111496

Teng, X., Zhang, M., & Mujumdar, A. S. (2021). 4D printing: Recent advances and proposals in the food sector. *Trends in Food Science & Technology, 110*, 349–363. doi:10.1016/j.tifs.2021.01.076

Wang, J., Wang, Z., Song, Z., Ren, L., Liu, Q., & Ren, L. (2019). Biomimetic shape–color double-responsive 4D printing. *Advanced Materials Technologies, 4*(9). doi:10.1002/admt.201900293

Wang, N., Li, R., Wang, X., & Yang, X. (2024). 4D food printing: Key factors and optimization strategies. *Trends in Food Science & Technology, 145*, 104380. doi:10.1016/j.tifs.2024.104380

Wang, W., Yao, L., Zhang, T., Cheng, C.-Y., Levine, D., & Ishii, H. (2017). Transformative appetite: Shape-changing food transforms from 2D to 3D by water interaction through cooking. *Proceedings of the 2017 CHI Conference on Human Factors in Computing Systems*, 6123–6132. https://doi.org/10.1145/3025453.3026019

Wang, X., Qin, X., Hu, C., Terzopoulou, A., Chen, X., Huang, T., Maniura-Weber, K., Pané, S., & Nelson, B. J. (2018). 3D printed enzymatically biodegradable soft helical microswimmers. *Advanced Functional Materials, 28*(45). doi:10.1002/adfm.201804107

Waseem, M., Tahir, A. U., & Majeed, Y. (2024). Printing the future of food: The physics perspective on 3D food printing. *Food Physics, 1*, 100003. doi:10.1016/j.foodp.2023.100003

Xia, B., Abidin, M. R. Z., & Ab Karim, S. (2024). From tradition to technology: A comprehensive review of contemporary food design. *International Journal of Gastronomy and Food Science, 37*, 100977. doi:10.1016/j.ijgfs.2024.100977

Xiao, K., Zhang, Y., Pan, L., & Tu, K. (2024). Study on color and flavor changes of 4D printed white mushroom gel with microcapsules containing gelatin/β-cyclodextrin induced by microwave heating. *International Journal of Biological Macromolecules*, *279*, 135365. doi:10.1016/j.ijbiomac.2024.135365

Xu, D., Meng, X., Liu, S., Poisson, J., Vana, P., & Zhang, K. (2024). Dehydration regulates structural reorganization of dynamic hydrogels. *Nature Communications*, *15*(1), 6886. doi:10.1038/s41467-024-51219-7

Yan, S., Zhang, F., Luo, L., Wang, L., Liu, Y., & Leng, J. (2023). Shape memory polymer composites: 4D printing, smart structures, and applications. *Research*, *6*. doi:10.34133/research.0234

Yang, F., Guo, C., Zhang, M., Bhandari, B., & Liu, Y. (2019). Improving 3D printing process of lemon juice gel based on fluid flow numerical simulation. *Lwt*, *102*, 89–99. doi:10.1016/j.lwt.2018.12.031

Yarali, E., Mirzaali, M. J., Ghalayaniesfahani, A., Accardo, A., Diaz-Payno, P. J., & Zadpoor, A. A. (2024). 4D printing for biomedical applications. *Advanced Materials*, *36*(31). doi:10.1002/adma.202402301

Zhu, Y., Liu, Z., Zhang, X., He, C., & Zhang, X. (2023). Microwave-triggered 4D automatic color change in 3D-printed food materials incorporating natural pigments. *Food*, *12*(10), 2055. doi:10.3390/foods12102055

8

5D, 6D, 7D Printing and Beyond

CHAPTER MENU

8.1 Introduction

Emerging technologies such as 3D and 4D food printing are making waves in the food manufacturing sector. In the simplest terms, 3D printing involves the deposition of food products, often layer by layer, to fabricate food with a three-dimensional structure. In recent years, 3D printing has gained many applications in the food industry, with applications not limited to the personalization of nutrients (Keerthana et al., 2020; Wilson et al., 2020; Yoha et al., 2021), development of foods to meet specific health/age requirements (Derossi et al., 2020; Zhu et al., 2023), development of innovative food packaging designs and solutions (Nida et al., 2021, 2022a), strategies for food waste utilization (Muthurajan et al., 2021; Yoha & Moses, 2023; Keerthy et al., 2025), attractive food constructs (Santhoshkumar and Moses, 2025; Santhoshkumar et al., 2025), food fortification (Raja et al., 2023), meal customization (Escalante-Aburto et al., 2021; Subhasri et al., 2022; Ukkunda et al., 2025), nutraceutical delivery (Kavimughil et al., 2022, 2023), and the development of special foods for specific job roles (Santhoshkumar et al., 2023). In food

4D Printing of Foods: Fundamentals, Applications and Prospects, First Edition.
C. Anandharamakrishnan, Jeyan A. Moses, and P. Santhoshkumar.

printing, the rheological properties of the printing material supply significantly affect the deposition of subsequent layers. These include the viscosity, elasticity, and consistency of the materials (Anukiruthika et al., 2020; Santhoshkumar et al., 2024a). In addition, these characteristics significantly influence printing accuracy and the sensory quality of the resulting 3D-printed structures (Santhoshkumar et al., 2024b; Sun et al., 2018; Theagarajan et al., 2020). The printability of the inks can be enhanced by incorporating various additives such as hydrocolloids or natively printable materials, with the printing ink. These improve the rheological properties and postprinting stability of printed foods (Anandharamakrishnan et al., 2022).

4D printing represents an advancement of the 3D printing process, enabling the physical and/or chemical properties of a 3D-printed object to adapt in response to various environmental/external factors such as light, water, temperature, and pH. This adaptability results in alterations in color, shape, texture, nutrients, and flavor/aroma of 3D-printed food. These are expressed as changes in the fourth dimension, and thus the concept is termed 4D food printing (Chen et al., 2021a,b,c; He et al., 2020; Shanthamma et al., 2021; Shi et al., 2022). A recent report by Wang et al. (2024) exclusively documented the range of approaches in 4D food printing applications. The technology has been effectively deployed across multiple industries, including biomedical, electronics, biological sciences, and engineering (Ahmed et al., 2021; Santhoshkumar et al., 2024c). Currently, among all fields, the most notable applications are in the field of biomedical engineering. Whilst 3D and 4D printing demonstrate distinct advantageous applications within the food industry, specific limitations must be addressed. 3D printing enables the fabrication of complex structures, such as those featuring curved surfaces, most often by incorporating supplementary support material. 3D-printed structures are produced in a planar configuration, resulting in a potential vulnerability within the material integrity. Post-processing techniques and a step to remove supporting materials and any additional components are necessary for obtaining the "actual" shape. The extended processing time required for constructing structures increases printing costs and resources.

On the contrary, as an advancement, the 5D printing process facilitates the construction of buildings and other structures characterized by complex geometries and enhanced mechanical strength, thereby addressing existing challenges. This is achieved by eliminating the requirement for fillers or supplementary supporting components (Nida et al., 2022b). Such benefits also support a process involving minimal or zero waste generation. The surge in research and commercialization of food printing technologies has been fueled by the increasing demand for innovative food options and higher demands for customization and/or personalization. Figure 8.1a details an evolving landscape highlighting the transition from 2D to 6D printing; a detailed understanding of the material's composition and behavior is underscored. Using 5D and 6D printing techniques can facilitate the creation of intricate designs while requiring less building material than traditional 3D or 4D printing approaches (Vasiliadis et al., 2022). Further, the 5D printing process minimizes the necessity for additional support materials by incorporating three translational movements and two rotational axes (for instance, three axes in the head and two axes in the bed). These technological breakthroughs hold the potential to greatly increase its application range across numerous industries, including the food industry (Fahmy et al., 2023).

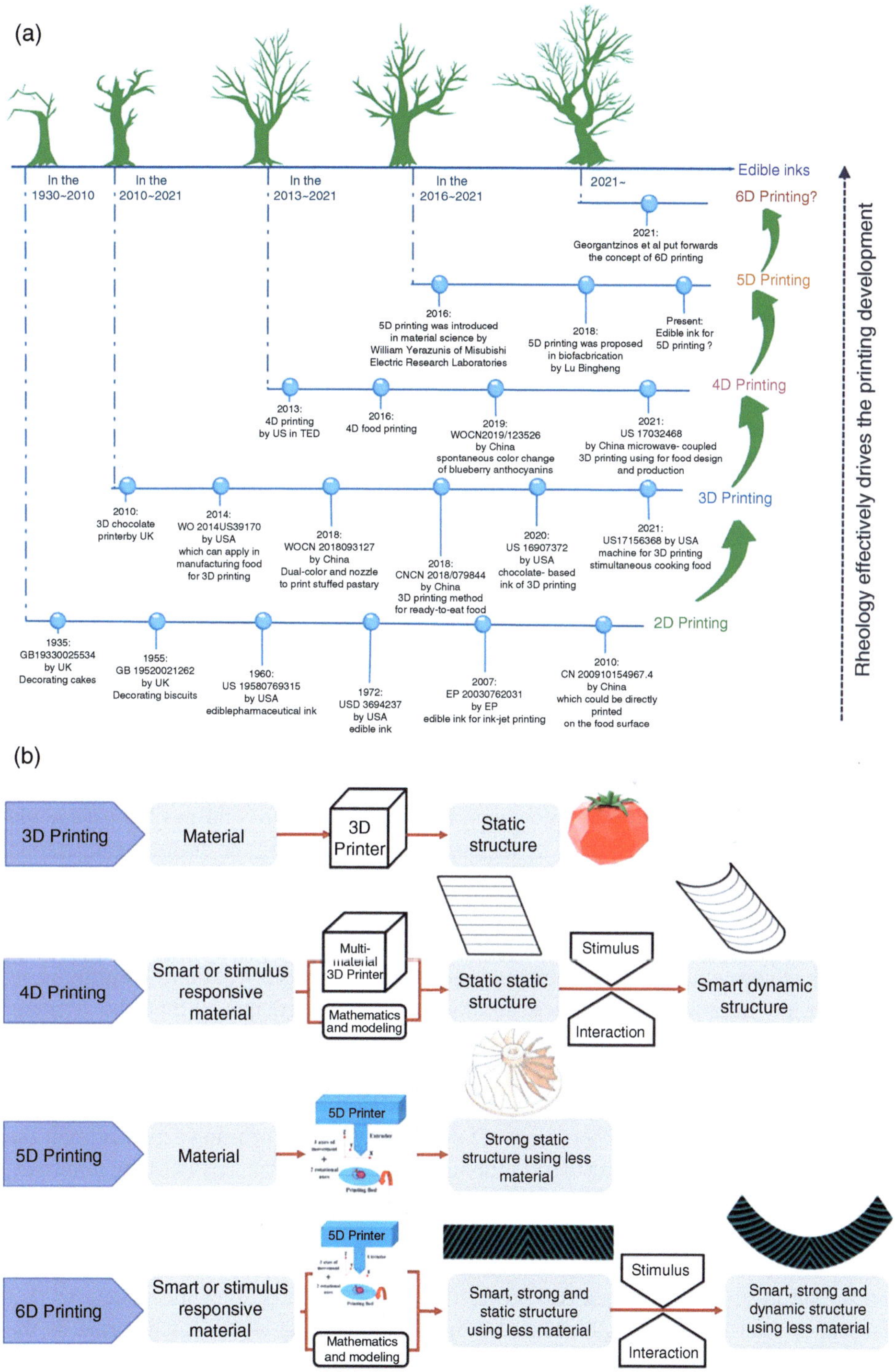

Figure 8.1 (a) Food printing and historical evolution from 2D to 6D. *Source*: Cheng et al. (2022) / with permission of Elsevier, (b) additive manufacturing perspective on dimensionality. *Source*: Yu et al. (2023) / with permission of Elsevier).

6D printing, which builds on 5D printing technology, makes it possible to create more durable materials with greater sensitivity to outside stimuli while increasing time and cost-effectiveness (Ghazal et al., 2023). Further, 7D printing can streamline manufacturing processes, cut prices, and enable the fabrication of more complicated goods than traditional manufacturing methods. This chapter delves into the fundamental definitions of 5D, 6D, and 7D food printing, their transformative effects, applications, research and development, and processing advantages. Although current applications are still evolving in the food industry, it is evident that several successful concepts/logic from other industry sectors will find food applications in the near future.

8.2 Evolving from 4D Printing – Need and Approaches

Additive manufacturing of food has experienced substantial innovations with evolution to different dimensions, from 2D to 3D to 4D to 5D to 6D to 7D. The technology is evolving with various applications in food, including culinary decoration, nutrition enhancement, and improved food intelligence function (Srivastava et al., 2023). Currently, the primary objective of 4D food printing is to modify the nutritional content, flavor, shape, and color of 3D-printed materials (Ghazal et al., 2023). In addition, it is rapidly growing and extending applications in the field of nutraceutical delivery, for instance, in achieving targeted or controlled or sustained release behaviors. 5D printing, a recent development in additive manufacturing, can achieve the production of intricate and complex food structures with minimal material consumption, demonstrating its significant application potential. In food production, 3D and 4D printing continue to be the most popular topics as they propel digital innovations and advance the intellectualization of food research and innovation. The discovery and integration of new smart materials, stimuli, and mechanisms could result in substantial advancements in 4D food printing, enabling the desired modification of various properties, including the nutraceutical delivery application. Nevertheless, 4D printing is often constrained by the necessity of supporting materials or the limited degree of freedom, which can increase processing time. These challenges have prompted the evolution of new additive manufacturing technologies, such as 5D, 6D, and 7D printing, as they provide improved capabilities. Additionally, it is anticipated that the ongoing development of additive manufacturing techniques will result in a new generation of smart materials with promising capabilities, which will substantially impact society, particularly as research and development further reduce the cost of these intricate processes.

In brief, 4D printing takes roots from 3D printing, while 5D printing expands upon the 3D framework by incorporating two additional axes (Haleem et al., 2019). In the same way, 6D printing is a development of 4D and 5D printing that includes a stimuli-responsive function (Nida et al., 2022b). The fundamental phenomena of various printing dimensions are illustrated in Figure 8.1b. Research on 4D, 5D, and 6D food printing has experienced a substantial increase in the past two years. The development of dynamic structures that respond to external signals is facilitated through 4D printing, which is supported by the design of 3D models. 3D printers, printing technology, pigments, building systems, and stimulants (such as water, heat, magnetic fields, light, and pH) are the fundamental components of 4D food printing (Cen & Meng, 2024). These stimulants facilitate controlled and predictable changes in the printed items over time. Numerous researchers have modified printers and printing programs to optimize 3D printing for 4D applications as technology has progressed (Ali et al., 2019). Nevertheless, the creation of flat-layered objects and the fabrication of integrated parts with curved surfaces that can enhance strength are restricted by the dependence of 4D printing technology on 3D printers (Wen et al., 2024). The technology's advancement is also hindered by the increased expenses associated with the printing and programming processes.

Regarding manufacturing constraints, 4D printing necessitates the use of supporting materials and post-processing procedures to eliminate these supports. The absence of additional printing degrees of freedom results in more intricate workflows and extended processing times. While 4D printing can generate intelligent material structures with complex configurations through sophisticated programming, it fails to provide the same degree of adaptability and adaptability in layer-by-layer construction as 6D printing technology (Georgantzinos et al., 2021; Ghazal et al., 2023). As a result, new additive manufacturing technologies, such as 5D and 6D printing, are evolving to overcome these constraints. Smart materials are an integral component of these technologies and are detailed in the following section.

8.2.1 Smart Materials

Ongoing progress in additive manufacturing is resulting in a new generation of sophisticated smart materials with unique structures and functionalities. For instance, advancements in multidimensional printing technologies are anticipated to substantially enhance the societal implications of additive manufacturing (Shi et al., 2024b). Owing to its benefits, multidimensional printing is expected to be rapidly embraced by firms engaged in producing printing apparatus, materials, and by consumers of additive manufacturing. It will similarly find uses in smart materials' technological and industrial domains such as 3D printing. The rheological qualities of food materials or inks are essential in food applications for the efficacy of 4D/5D/6D printing, as they affect printing results (Cheng et al., 2022). For instance, 5D printing can produce delicate, visually appealing food items with smooth, curved forms while ensuring high strength and complex designs. However, as in other cases, 5D-printed foods must also ensure stability during storage at fluctuating temperatures, transportation, and delivery obstacles that 3D and 4D printing cannot entirely address (Ghazal et al., 2023). Rheological parameters, including temperature ramping, yield stress, storage and loss modulus, creep recovery, and thixotropy, are essential for ensuring food products' requisite thermal endurance and structural integrity. Figure 8.2 shows the different rheological characteristics that explain the printability of edible materials. Moreover, it is important to carefully monitor and maintain the food's inherent flavor and nutritional in 5D printing (Anas et al., 2022). Better rheological behavior can preserve the flavor and nutrients of materials due to the high compactness of the matrix (Chen et al., 2022; Liu et al., 2017). The choice of food materials for 5D printing is a crucial consideration, although it presents challenges due to the intricacies of functioning over five dimensions and the substantial expenses linked to sophisticated equipment. Advancements in 5D printing necessitate interdisciplinary considerations among rheology, mathematics, engineering, and biology to support effectiveness in food applications (Cheng et al., 2022).

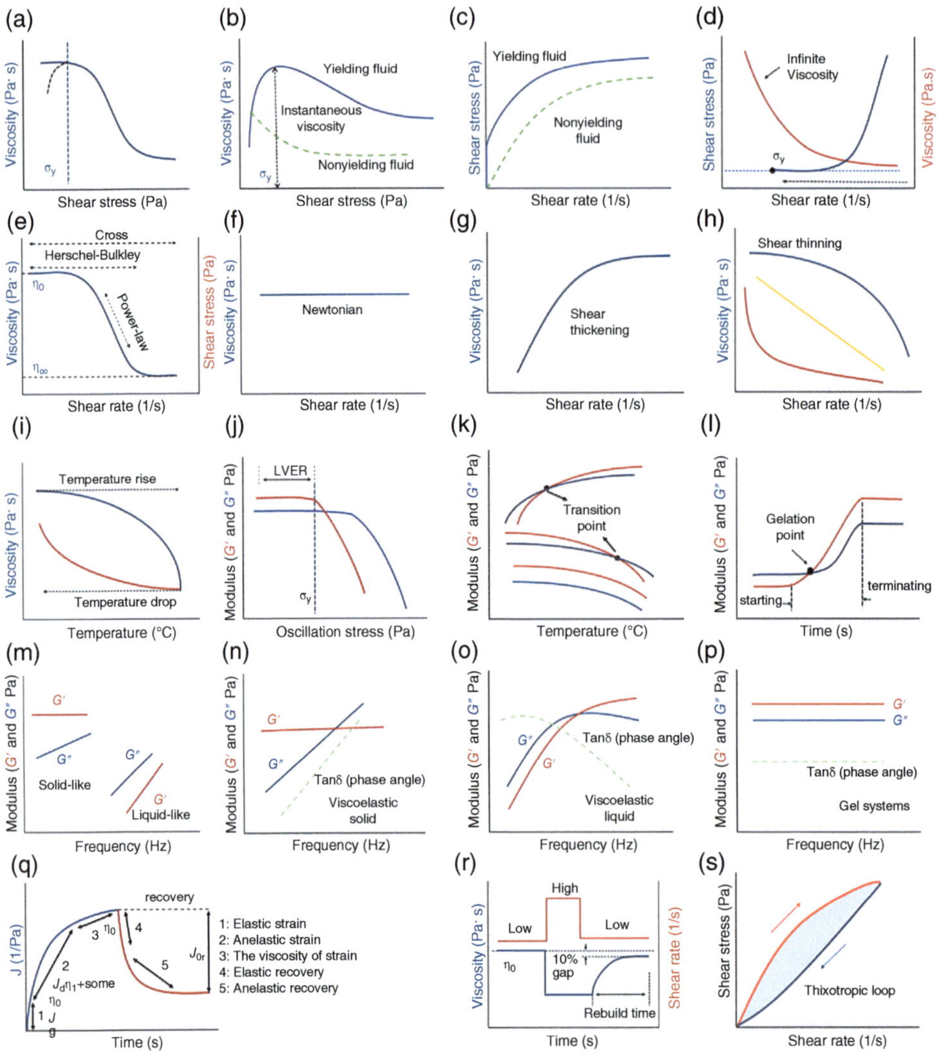

Figure 8.2 Rheological behavior of smart materials. (a–d) Yield stress, (e) σ_y is measured by transition point or low-end shear rate, fitting rheology models, (f–h) Newtonian and non-Newtonian behavior, (i) temperature ramp curve, (j) oscillatory sweep for the linear viscoelastic region, (k–l) crosslinking characterization by oscillatory temperature and time sweep; (m–p) frequency sweep, (q) creep recovery, (r) three interval thixotropy, (s) thixotropic loop test. *Source*: Cheng et al. (2022) / with permission of Elsevier.

In 6D printing, rheological analyses focus on ensuring structural integrity, especially emphasizing the thermal properties of smart materials. These occur because smart materials typically contain active groups that maintain their integrity only at specified temperatures. Moreover, 6D-printed materials composed of alternating polymer layers may undergo bending when subjected to heat, attributable to variations in their glass transition temperatures (Vasiliadis et al., 2022). Rheological testing is vital in determining temperature-dependent printing parameters, optimizing the process, and minimizing distortion in printed products. Through comprehensive rheological testing, producers can save expenses related to trial-and-error analyses of food material compositions and parameter modifications. Despite the elevated expenses associated with 6D printing in culinary applications, largely attributable to specialized materials and advanced machinery, the technology has considerable benefits. 6D printing integrates the structural integrity of 5D printing with the versatility in shape, color, and flavor of 4D printing, facilitating the creation of dynamic, high-performance products (Cheng et al., 2022). In terms of multidynamic function, through the complicated shape, multimaterial food intelligent material combination study needs to be carried out. Ultimately, 5D/6D/7D printing facilitates the production of intelligent food products that resonate with consumers, catering to diverse preferences across various market sectors and utilizing reduced material and lower infill density in accordance with the objectives of a contemporary, advanced society.

8.3 5D Printing

5D printing is a new subfield of additive manufacturing that is based on the technology of 3D printing. William Yerazunis, a senior principal research scientist at Mitsubishi Electric Research Laboratories (MERL), was the first to implement this concept, which was initially introduced by American institutions in 2016. In contrast to conventional 3D printing, which operates on three axes (*X*, *Y*, and *Z*), 5D printing incorporates two additional rotational movements, enabling printing from a total of five axes. They discovered that the cap produced with a 5D printer (3.7 MPa) exhibited greater mechanical strength compared to the structure made with a 3D printer (0.1 MPa). Similarly, the ethereal machines in Bengaluru, India, developed a hybrid "Halo 5D printer" to make jewelry and car parts (Nida et al., 2022b). The print head and the object being printed both have five degrees of freedom in 5D printing (Figure 8.3a). This method allows for the development of curved layers instead of planar ones. The printer head prints in a manner consistent with the object's curves, rather than the typical straight course of 3D printing, as the print section rotates. It leads to more durable parts due to the curved layers created during printing (Reddy & Devi, 2018).

The capacity to print from various angles and orientations enables 5D printing to generate intricate layers that are not feasible with conventional 3D printing. The curved layers of 5D printing offer supplementary strength to intricate designs, rendering it appropriate for producing objects that would necessitate substantial support and filler in 3D printing. A case study conducted at MERL revealed that 5D-printed objects are three to five times more durable than their 3D-printed counterparts. The horizontal layers in 3D printing, which generate vulnerable points, are responsible for the increased strength, whereas the multidirectional forces in 5D printing reinforce the structure of the printed objects. Furthermore, 5D printing is 25% more material efficient than 3D printing (Ghazal et al., 2023; Nida et al., 2022b). Typically, 5D printing has the potential to be used in a variety of applications, as it allows for the production of intricate shapes with improved strength and reduced material consumption. It is a substantial improvement over 3D and even 4D manufacturing technologies.

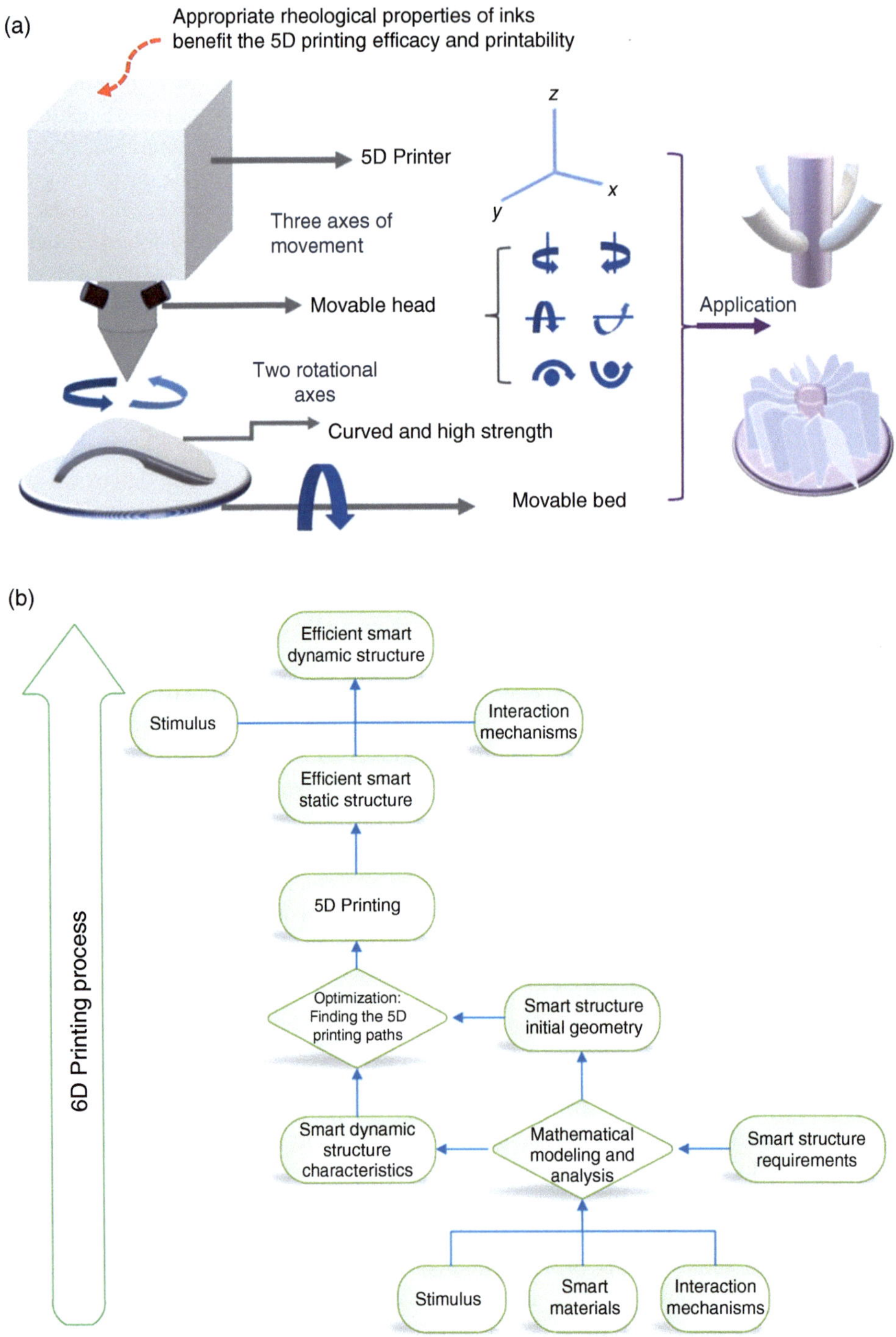

Figure 8.3 (a) Schematic explaining the concept of a 5D printer. *Source*: (Cheng et al. (2022)) / with permission of Elsevier and (b) A step-by-step approach to 6D printing. *Source*: Georgantzinos et al. (2021) / MDPI / CC BY 4.0.

8.3.1 5D Food Printing

Individuals express varying perspectives regarding 5D food printing, contingent upon their perspective on the dimensions of food apart from other perception-linked factors. Whilst 4D food printing has expanded the potential of 3D printing, it has not yet been able to improve multiple quality attributes postprinting to accommodate specific populations simultaneously. To resolve this, 5D food printing has been introduced (Chen et al., 2023a). In contrast to 4D printing, 5D printing can generate more complex and curved designs by employing fewer materials. Nevertheless, the food industry has not yet implemented this technology due to incompatibilities between the printer and the printing materials. 5D food printing can provide customization, similar to 3D printing, and allow for controllable changes in color, shape, and nutritional content over the course of the printed products (Hassoun et al., 2024). The reactions, synthesis, and transformations within the biological food matrix are the source of the food's postprinting modifications, enabling dynamic changes that can have a substantial positive impact on human health. Live cells or bacteria can be used to improve the properties of food by incorporating their metabolic processes. For instance, Chen et al. (2023a) have examined the multidimensional functional changes in 3D-printed food materials. They employed a starch-protein-based gel infused with probiotics as the raw material. The gel was subjected to fermentation following printing, resulting in improvements in product color and texture, and bioavailability of probiotics. It was reported that 3D-printed objects have the potential to demonstrate changes in biological activity over time; this has been explained as the fifth dimension. Importantly, the 5D printing approach described here showed capabilities to provide potential health benefits (Chen et al., 2023a). Similarly, another study on 5D food printing focused on the functional changes (titratable acidity, pH, pathogen quantity, texture properties, antioxidant properties, color, and flavor) during the growth of probiotics (Chen et al., 2024). Recently, through the 5D food printing approach, researchers developed food for infants, the color and texture modified by the growth of probiotics combined with anthocyanin microcapsules (Niu et al., 2025). Overall, research and reports in the segment of 5D printing of foods are currently limited.

8.4 6D Printing

6D printing is a cutting-edge additive manufacturing technology that integrates 4D and 5D printing. Using fused deposition modeling, researchers first introduced this technique, which can be viewed as an evolution of five-axis printing. Essentially, these smart materials respond to various stimuli (Georgantzinos et al., 2021). The key advantage of 6D printing is its ability to create complicated structures with greater strength while using less material. It blends the transformative capabilities of 5D printing, which enables objects to alter shape, color, or other attributes over time, with those of 4D printing, which is excellent at creating complex designs. Compared to 4D printing, this produces more lasting and effective products. Additionally, with the right setup, 6D printing's inherent flexibility can reduce production times and material consumption (Georgantzinos et al., 2021). The step-by-step process involved in a typical 6D printing process is illustrated in Figure 8.3b.

6D printing can produce comparable results while reducing manufacturing time and material consumption when optimal processing parameters are utilized (Georgantzinos et al., 2021). Samples may exhibit bending behavior upon exposure to heat, attributable to their composition of alternating layers of two distinct polymers (Kačergis et al., 2019). 4D-printed foods can undergo shape or color changes when exposed to stimuli (Wang et al., 2024). Such shape transformations can reduce the transportation and storage spaces required for different food products. Also, color improves the culinary appearance of the printed foods among consumers. However, the presence of these thin layers may adversely affect the product's structural integrity, leading to potential deformation and loss of shape during transit.

6D printing can enable the development of culinary products with complex geometries that are both dynamic and durable, thereby ensuring that they retain their unique shapes during handling and transport. Also, 6D printing can support the development of intelligent food products with varying infill types (e.g., V-type), reduced infill densities, and complex geometries. Such endeavors can expedite the potential for intelligent packaging for food products that are both lighter and stronger while simultaneously reducing material consumption (Ghazal et al., 2023). In the future, this may generate a more intelligent function with this dimension of printed food matrices. One significant disadvantage of 6D printing technology is that the initial setup costs are likely higher than those of other additive manufacturing methods. Nevertheless, the economic influence of 6D printing can be advantageous, particularly as production volumes and automation levels increase, resulting in decreased material waste (Georgantzinos et al., 2021; Ghazal et al., 2023).

In an interesting work on 6D printing, Swiss students from Zurich's University reported the novel six-axes printer with three axes in the print head and three in bed movement. These printers can move in six axes motion without material support. Although 5D printing could be achieved, challenges in adjusting the print head and bed simultaneously were reported and attributed to the limited time and resources available to develop the requisite software. With the reported configuration, the printer platform could only tilt to accommodate overhanging structures rather than tilting during extrusion to generate lighter curves. Nevertheless, the design enabled the efficient printing of hexagonal structures, as each side could be printed in a straight line along the same axis, thereby assuring uniform strength on all sides (Ghazal et al., 2023). Although the prototype has not yet been extensively implemented for other applications, it is being utilized for research and development purposes. Such 6D printer prototypes highlight requirements for additional developments, notably in the realm of software and control. When combined with 4D techniques, the technology has the potential to establish a new additive manufacturing method, and this is referred to as "smart 6D printing."

The primary benefit of 6D printing is its capacity to combine the complex structures of 5D printing with the ability to modify structure, color, and flavor over time in response to 4D printing inputs, thus resulting in more efficient and durable products (Srivastava et al., 2023). Currently, 6D printing is predominantly being explored in research settings, and there is a growing interest from researchers and businesses in various industries. As smart materials develop, the evolution of 6D-printed objects from 5D printing technology is possible. Although these regions are underdeveloped and offer substantial economic and research opportunities, the technique can potentially revolutionize food manufacturing and customization. The current high costs of 6D printing for culinary applications result from two primary factors: the use of unique edible inks with high entry barriers and the stringent standards associated with five axes operations, necessitating expensive and complex equipment.

8.5 7D Printing

7D printing provides scope for increased customization and flexibility, expediting prototyping and product development. It enables the production of intricate, detailed, and complex elements that are unattainable through conventional manufacturing methods. The technology has the potential to be implemented in a variety of sectors, such as aerospace and medical, to manufacture components that would be otherwise expensive and difficult to produce. Currently, 4D/5D/6D printing has a majority of applications in architecture, engineering, and the construction industry, and Rane et al. (2023a) have detailed their pros and cons. In the context of 7D printing, it still remains a theoretical definition, and there is no published report demonstrating the approach. Nevertheless, as an innovation, it is viewed to have huge potential for practical applications, such as reducing production costs by necessitating fewer resources and materials than conventional manufacturing processes. In other words, 7D printing is an environmentally responsible alternative, as it reduces the amount of waste produced during manufacturing (Srivastava et al., 2023). In terms of food application, 7D food printing has the potential to transform the production and consumption of food in attractive shapes and define intelligent functions through high-end edible smart materials. 7D printing can enhance this capability by facilitating the development of intricate edible designs and functional dynamics such as flavors, shape, color, and nutrition. Further, the approach has the potential to support food production in off-grid regions and generate novel prospects for food businesses and entrepreneurs. So far, no comprehensive study has provided a detailed analysis on food printing applications with 7D printing.

8.6 Current Research Status of 5D, 6D, and 7D Printing Technologies

Unlike 2D printing, which mainly serves surface decoration rather than improving food structure, 5D, 6D, and 7D food printing presents avenues for food innovations, encompassing customization, personalization of nutrients, and intelligent design features unattainable by conventional methods and 2D printing. The edible ink printing market is expanding, propelled by its ecofriendly characteristics. Currently, 3D and 4D food printing are prominent subjects, driving the digital manufacturing of food and furthering the intellectualization of food businesses. Although 5D and 6D printing are presently the only subjects of academic research, 7D printing remains in a much nascent and conceptual stage.

Rheological developments have improved edible ink print quality, enabling researchers to attain the best results for real-world uses. Since then, edible ink printing has progressed from 2D to 4D applications, covering anything from food intelligence to decorative aspects.

These state-of-the-art methods aim to produce goods with exceptional physical attributes. Complex shapes such as spherical candies and personalized packaging solutions are produced when these technologies are used in food and packaging. Additionally, sophisticated designs on dinnerware, such as glasses and bowls that need additional support, can be printed using advanced technology. Overall, the most popular uses of 5D printing are in tissue engineering and the production of biomedical devices. In the meantime, 6D printing is starting to become popular in the building industry. While 6D and 7D printing are still mostly in the research stage, 5D food printing has already begun to find uses in the food industry. Pizzas, cakes, and doughnuts can be automatically manufactured with the help of 5D food printing. The method makes it possible to precisely customize and produce esthetically pleasing food meals with intricate decorations. As advancements, for specific applications, 5D/6D/7D printing can improve efficiency, minimize waste, and decrease production times compared with conventional food manufacturing techniques.

8.7 5D and 6D Printer Configurations

A five-axis 3D printer fabricates components that exhibit greater durability compared to those produced by conventional 3D and 4D printing methods. This printer incorporates a movable extruder head and print bed, constituting two of the five axes in addition to the conventional x, y, and z axes. A printer utilizing fused deposition modeling (FDM) with five axes was designated as a 5D printer, and the operation it conducts was termed 5D printing. Similar to conventional 3D printers, these printers can use computer-aided design (CAD) data for production. Importantly, instead of adhering to a flat, layer-by-layer methodology, 5D printers navigate the contoured trajectories of the components being fabricated (Arikan, 2023). Although 3D and 5D printing utilizes identical CAD file inputs and materials, 4D printing incorporates smart materials, introducing a dimension in which structures can alter their shape over time. Conversely, 6D printing is more similar to 4D printing than to 3D or 5D printing, as it similarly employs smart materials. In 6D printing, the resultant component is a smart structure produced utilizing a five-axis printer and intelligent materials derived from the FDM approach (Arikan, 2023). The primary objectives of both 5D and 6D printing are to fabricate more robust components and reduce material waste. To enhance the productivity of these printers, recent technologies such as artificial intelligence (AI) and the Internet of Things (IoT) can be integrated into interesting applications (Rane et al., 2023b). In 6D printing approaches, for smart materials, multiintelligent functions can utilize multimaterial printing approaches. However, considering today's current technology readiness, using multiple materials in 6D printing may come with complexity in machinery and processes.

8.8 5D and 6D Printing: Non-food Applications

The possibilities of 5D and 6D printing are enormous in many fields, including engineering, dentistry, orthopedics, and medicine. Anas et al. (2022) reported how such developments outperform 3D printing, for instance, in terms of the physical properties of complex geometries, devices, and implants. As bio-ink encapsulation of living cells has the potential to compromise mechanical integrity, bioprinting relies on the rigidity and stability of bio-ink structures. The capacity to create parts and buildings with the necessary durability to meet industrial and safety standards makes 5D printing an important tool for the advancement of engineering and related fields in the future (Lai & Wang, 2023). 5D printing focuses on manufacturing curved, robust biostructures that preserve their form and mechanical qualities postprinting, a challenge for 3D printing (Foresti et al., 2020).

6D bioprinting allows for carefully manipulating biological components, such as cells and proteins, to produce complicated tissue structures with channels, holes, and microenvironments. With this level of precision, regenerative medicine can benefit from 3D-printed organs and tissues, opening the door to treatments for ailments such as cancer and heart disease (Khorsandi et al., 2024). Orthopedic surgery benefits considerably from 5D printing. For instance, human bones have curved, complicated structures, requiring implants to replicate their shapes with exceptional strength (Vasiliadis et al., 2022). Such implants cannot be made easily and precisely using traditional 3D printing methods. However, 5D printing makes short work of tailoring orthopedic implants with exact strength and curve, allowing for more precise surgical planning, instruction, and learning (Aruna et al., 2022; Haleem et al., 2019; Jeyaraman et al., 2022).

In dentistry, 5D printing presents new possibilities by enabling the fabrication of durable dental implants, crowns, braces, aligners, and bridges based on CAD data (Anas et al., 2022). Making artificial limbs, hands, or mouths that have complicated geometries and tremendous strength is also possible with this technology, making prosthetics a viable option. In particular, 6D printing may create unique medical devices that adapt closely to a patient's anatomy, giving lighter, stronger, and more comfortable implants. Accelerated gadget production enables patients to acquire care more promptly, increasing their outcomes (Girija & Sampath Kumar, 2024). Amaya-Rivas et al. (2024) note that 6D printing combines 3D printing with nanoprinting and microprinting, which enhances the potential for advanced diagnostics, therapy, and personalized prosthetics. The use of AI in multidimensional printing has the potential to be applied to drug delivery applications as well (Sharma, 2023).

5D printing has several potential applications outside of medicine, including the building and automobile industries, for the production of high-tech automobile parts and prototypes with complex designs. Homes with curving architectural details, which are challenging to accomplish using conventional building methods, are constructed using this printing approach (Kumaresan et al., 2026). However, the increased expense of incorporating two extra movement axes into the printer bed is a major obstacle to 5D printing. Further, 6D printing allows for integrating numerous components into one distinct structure, which allows for manufacturing lightweight, high-precision components useful in the aerospace and industrial industries. For instance, aircraft manufacturing relies on increased production speed, reduced prices, and enhanced design flexibility (Georgantzinos et al., 2021). 6D printing allows for more ecofriendly manufacturing processes, decreasing resource use and increasing accuracy. As 6D printing can manufacture complicated, personalized geometries, it will become an indispensable tool for industries such as electronics and automobile manufacturing that rely on accurate components (Ghazal et al., 2023). While 5D printing concentrates on enhancing material properties and enabling more intricate designs, 6D printing goes beyond these aspects by providing greater flexibility and control. These technologies make it possible to produce complex, personalized things in novel, more effective, precise, and strong ways. In medical and aerospace, 6D printing has demonstrated significant potential for creating customized implants and lightweight aircraft components. Additionally, it marks the beginning of an era of innovation in additive manufacturing. However, appropriate infrastructure and resources, as well as the expertise of highly qualified experts, are still needed to design and maintain these systems.

8.9 5D and 6D Printing: Food Applications

In the food industry, 5D printing enables the creation of highly intricate, curved food structures with minimal materials while enhancing the strength to maintain shape during transportation, a feat that conventional 3D printing cannot accomplish (Nida et al., 2022b). This technology enables the creation of diverse textures to accommodate varying consumer preferences by allowing food to be printed with flexible infill types and densities, ensuring that the final product retains its structure. The visual attractiveness of food products is improved by the smoother, more curved edges facilitated by the increased mobility of the print head and bed. Furthermore, 5D printing provides an optimal solution by creating such customized designs; food products frequently necessitate customized transportation equipment to accommodate their dimensions, thereby reducing the risk of injury during production and handling (Ghazal et al., 2023). For instance, laser-sintering techniques to produce intricate culinary items, including personalized chocolates and elaborate cakes, is a critical aspect of 5D printing and is also utilized to fabricate molds for confectionary products (Nida et al., 2022b).

These printing approaches can generate sophisticated food structures from edible materials, which generate novel flavors, textures, and tailored nutritional profiles, enabling personalized meal plans and customized supplements (Chen et al., 2023b; Chen et al., 2024; Hassoun et al., 2024). 5D printing also enhances food safety by guaranteeing uniform shape and size, facilitating contamination identification. The technology can also contribute to safer food production and reduce waste by extending the product's shelf life and reducing food spoilage (Shabir et al., 2024; Srivastava et al., 2023). These adaptabilities also enable multi-intelligent functions, for instance a broader range of colors, flavors, and textures in foods (Shi et al., 2024a). The precision of nutrient control improves the nutritional value of the food, and costs are reduced due to the reduced production time (Chen et al., 2024).

The potential for innovation is further increased by combining the advantages of 4D printing with the developments of 5D printing to create 6D printing. This technology can meet the specific needs of sectors such as the military, aerospace, and exploration missions by offering flexible, portable, and lightweight food options. Additionally, printing food packaging that is biologically active may enhance food preservation. 6D printing reduces resource consumption and is in line with sustainable practices by enabling the creation of intelligent foods with customized fillings and optimized infill densities (Teng et al., 2022). Furthermore, 6D food printing makes it possible to quickly and accurately create unique shapes and patterns for cookies, pastries, and desserts, doing anything with the need for conventional tools such as cookie cutters in food applications.

Additionally, it makes it possible to precisely customize sensory attributes such as flavor and texture, ensuring that meals are customized to each person's preferences. The technology ensures the creation of innovative solutions for extreme circumstances, such as astronaut food products, and customized gastronomic experiences (Enfield et al., 2022; Santhoshkumar et al., 2023). The potential for 6D printing to revolutionize food production and packaging is significant as it continues evolving, providing new sensory and functionality opportunities.

8.10 Prospects and Opportunities in 5D, 6D, and 7D Printing

Connecting rheology models for 5D and 6D printing materials is crucial to closing the gap between smart materials and food printing technologies and facilitating the mass production of innovative, reasonably priced food products. The multidisciplinary combination of rheology, material science, mathematics, chemistry, physics, biology, and engineering is necessary to understand the relationship between consumable ink structure and printability potential. Incorporating responsive or smart properties into food materials while maintaining their rheological behavior would facilitate the production of food products responsive to external stimuli, including heat, water, microwaves, salts, and magnetic fields. This would create new opportunities for the production of high-tech, flexible food. In addition, research into printing shape retention is crucial, as it directly affects the final product's quality. Specific additives could enhance the shape stability of printed foods. To preserve rheological stability, food materials must be optimized for resistance to environmental factors such as oxygen, light, storage time, and physical stress, as they frequently contain macro and micronutrients. Exploring the integration of diverse printing methods, such as the combination of 2D inkjet and 6D printing, could reduce production time and enable the development of sophisticated, multifunctional food products that delicately balance color, texture, shape, and flavor (Cheng et al., 2022). It would be essential to simplify the costly 5D/6D printing equipment for easier operation and to develop it for both industrial and research purposes in order to increase accessibility. In addition, the food industry could possibly be transformed by the expansion of the targeted applications of printed foods, including precision nutrition for various demographic groups, anti-counterfeit high-end food, bioactive foods with natural health benefits, and food that combines esthetic and functional properties. Institutions and companies could advance additive manufacturing by embracing these opportunities, which could drive significant innovations in fields such as engineering, medicine, and orthopedics, in addition to food manufacturing.

8.11 Conclusion

Most current 5D and 6D printing applications are in the field of biomedical engineering, and there exists considerable opportunity for further research and commercial advancement in the food sector, as the applications of 5D, 6D, and beyond remain relatively underexplored. With 5D and 6D printing, complex structures that are hard to make using traditional 3D printing methods can now be created. These advancements enable the creation of complex shapes and customizing meals without supplementary support materials. Furthermore, they offer the advantage of building structures with greater mechanical strength using less material than conventional printing techniques. For instance, 5D printing has the potential to create curved surfaces, which can be employed in food printing. Although 5D printing has already acquired considerable traction in food applications, even in nutrient enhancement applications, such advances remain underexplored in 6D and 7D printing.

References

Ahmed, A., Arya, S., Gupta, V., Furukawa, H., & Khosla, A. (2021). 4D printing: Fundamentals, materials, applications and challenges. *Polymer*, *228*, 123926. doi:10.1016/j.polymer.2021.123926

Ali, M. H., Abilgaziyev, A., & Adair, D. (2019). 4D printing: A critical review of current developments, and future prospects. *The International Journal of Advanced Manufacturing Technology*, *105*(1–4), 701–717. doi:10.1007/s00170-019-04258-0

Amaya-Rivas, J. L., Perero, B. S., Helguero, C. G., Hurel, J. L., Peralta, J. M., Flores, F. A., & Alvarado, J. D. (2024). Future trends of additive manufacturing in medical applications: An overview. *Heliyon*, *10*(5), e26641. doi:10.1016/j.heliyon.2024.e26641

Anandharamakrishnan, C., Moses, J. A., & Anukiruthika, T. (2022). *3D printing of foods*. Wiley.

Anas, S., Khan, M. Y., Rafey, M., & Faheem, K. (2022). Concept of 5D printing technology and its applicability in the healthcare industry. *Materials Today Proceedings*, *56*, 1726–1732. doi:10.1016/j.matpr.2021.10.391

Anukiruthika, T., Moses, J. A., & Anandharamakrishnan, C. (2020). 3D printing of egg yolk and white with rice flour blends. *Journal of Food Engineering*, *265*. doi:10.1016/j.jfoodeng.2019.109691

Arikan, N. A. (2023). *A short review of production with 5D and 6D printing*. 3rd Rumeli Engineering Education Symposium, 314–320.

Aruna, R., Mohamed Abbas, S., Vivek, S., Suresh, G., Meenakshi, C. M., Srinivasan, T., & Selva Ganapathy, K. (2022). *Evolution of 5D printing and its vast applications: a review* (pp. 691–714). https://doi.org/10.1007/978-981-19-0244-4_66

Cen, S., & Meng, Z. (2024). 4D printing: A novel application for structuring oils with fat-analog characteristics. *Trends in Food Science & Technology*, *153*, 104729. doi:10.1016/j.tifs.2024.104729

Chen, C., Zhang, M., Guo, C., & Chen, H. (2021a). 4D printing of lotus root powder gel: Color change induced by microwave. *Innovative Food Science & Emerging Technologies*, *68*, 102605. doi:10.1016/j.ifset.2021.102605

Chen, C., Zhang, M., Mujumdar, A. S., & Phuhongsung, P. (2021b). Investigation of 4D printing of lotus root-compound pigment gel: Effect of pH on rapid colour change. *Food Research International*, *148*, 110630. doi:10.1016/j.foodres.2021.110630

Chen, F., Zhang, M., Liu, Z., & Bhandari, B. (2021c). 4D deformation based on double-layer structure of the pumpkin/paper. *Food Structure*, *27*, 100168. doi:10.1016/j.foostr.2020.100168

Chen, J., Teng, X., Zhang, M., Bhandari, B., Adhikari, B., & Yu, D. (2023a). 5D food printing with color change induced by probiotic growth in a starch-protein-based gel system. *Food and Bioprocess Technology*, *16*(10), 2304–2314. doi:10.1007/s11947-023-03064-7

Chen, X., Teng, X., Zhang, M., Mujumdar, A. S., & Li, J. (2024). Functional changes in 5D printed starch-based gel systems caused by spontaneous growth of probiotics. *Food and Bioprocess Technology*, *17*(8), 2470–2482. doi:10.1007/s11947-023-03267-y

Chen, X., Zhang, M., & Tang, T. (2023b). Microwave-induced rapid shape change of 4D printed vegetable-based food. *Food, 12*(11), 2158. doi:10.3390/foods12112158

Chen, Y., Zhang, M., Sun, Y., & Phuhongsung, P. (2022). Improving 3D/4D printing characteristics of natural food gels by novel additives: A review. *Food Hydrocolloids, 123*, 107160. doi:10.1016/j.foodhyd.2021.107160

Cheng, Y., Fu, Y., Ma, L., Yap, P. L., Losic, D., Wang, H., & Zhang, Y. (2022). Rheology of edible food inks from 2D/3D/4D printing, and its role in future 5D/6D printing. *Food Hydrocolloids, 132*, 107855. doi:10.1016/j.foodhyd.2022.107855

Derossi, A., Husain, A., Caporizzi, R., & Severini, C. (2020). Manufacturing personalized food for people uniqueness. An overview from traditional to emerging technologies. *Critical Reviews in Food Science and Nutrition, 60*(7), 1141–1159. doi:10.1080/10408398.2018.1559796

Enfield, R. E., Pandya, J. K., Lu, J., McClements, D. J., & Kinchla, A. J. (2022). The future of 3D food printing: Opportunities for space applications. *Critical Reviews in Food Science and Nutrition, 63*. doi:10.1080/10408398.2022.2077299

Escalante-Aburto, A., Trujillo-de Santiago, G., Álvarez, M. M., & Chuck-Hernández, C. (2021). Advances and prospective applications of 3D food printing for health improvement and personalized nutrition. *Comprehensive Reviews in Food Science and Food Safety, 20*(6), 5722–5741. doi:10.1111/1541-4337.12849

Fahmy, A. R., Derossi, A., & Jekle, M. (2023). Four-dimensional (4D) printing of dynamic foods—definitions, considerations, and current scientific status. *Food, 12*(18), 3410. doi:10.3390/foods12183410

Foresti, R., Rossi, S., Pinelli, S., Alinovi, R., Sciancalepore, C., Delmonte, N., Selleri, S., Caffarra, C., Raposio, E., Macaluso, G., Macaluso, C., Freyrie, A., Miragoli, M., & Perini, P. (2020). In-vivo vascular application via ultra-fast bioprinting for future 5D personalised nanomedicine. *Scientific Reports, 10*(1), 3205. doi:10.1038/s41598-020-60196-y

Georgantzinos, S. K., Giannopoulos, G. I., & Bakalis, P. A. (2021). Additive manufacturing for effective smart structures: The idea of 6D printing. *Journal Of Composites Science, 5*(5), 119. doi:10.3390/jcs5050119

Ghazal, A. F., Zhang, M., Mujumdar, A. S., & Ghamry, M. (2023). Progress in 4D/5D/6D printing of foods: Applications and R&D opportunities. *Critical Reviews in Food Science and Nutrition, 63*(25), 7399–7422. doi:10.1080/10408398.2022.2045896

Girija, M., & Sampath Kumar, T. (2024). *State of art on evaluation of three- to six-dimensional novel additive manufacturing technology for biomedical applications.* Proceedings of the Institution of Mechanical Engineers, Part E: Journal of Process Mechanical Engineering. https://doi.org/10.1177/09544089241281985

Haleem, A., Javaid, M., & Vaishya, R. (2019). 5D printing and its expected applications in orthopaedics. *Journal of Clinical Orthopaedics and Trauma, 10*(4), 809–810. doi:10.1016/j.jcot.2018.11.014

Hassoun, A., Jagtap, S., Trollman, H., Garcia-Garcia, G., Duong, L. N. K., Saxena, P., Bouzembrak, Y., Treiblmaier, H., Para-López, C., Carmona-Torres, C., Dev, K., Mhlanga, D., & Aït-Kaddour, A. (2024). From Food Industry 4.0 to Food Industry 5.0: Identifying technological enablers and potential future applications in the food sector. *Comprehensive Reviews in Food Science and Food Safety, 23*(6). doi:10.1111/1541-4337.70040

He, C., Zhang, M., & Devahastin, S. (2020). Investigation on spontaneous shape change of 4D printed starch-based purees from purple sweet potatoes as induced by microwave dehydration. *ACS Applied Materials & Interfaces, 12*(34), 37896–37905. doi:10.1021/acsami.0c10899

Jeyaraman, M., Nallakumarasamy, A., & Jeyaraman, N. (2022). Industry 5.0 in orthopaedics. *Indian Journal of Orthopaedics, 56*(10), 1694–1702. doi:10.1007/s43465-022-00712-6

Kačergis, L., Mitkus, R., & Sinapius, M. (2019). Influence of fused deposition modeling process parameters on the transformation of 4D printed morphing structures. *Smart Materials and Structures, 28*(10), 105042. doi:10.1088/1361-665X/ab3d18

Kavimughil, M., Leena, M. M., Moses, J. A., & Anandharamakrishnan, C. (2022). 3D printed MCT oleogel as a co-delivery carrier for curcumin and resveratrol. *Biomaterials, 287*, 121616. doi:10.1016/j.biomaterials.2022.121616

Kavimughil, M., Mahalakshmi, L., Moses, J. A., & Anandharamakrishnan, C. (2023). Delivery of nutraceutical ingredients through three-dimensional food printing. In C. Anandharamakrishnan & P. Subramanian (Eds.), *Industrial application of functional foods, ingredients and nutraceuticals* (pp. 253–278). Elsevier. https://doi.org/10.1016/B978-0-12-824312-1.00008-X

Keerthana, K., Anukiruthika, T., Moses, J. A., & Anandharamakrishnan, C. (2020). Development of fiber-enriched 3D printed snacks from alternative foods: A study on button mushroom. *Journal of Food Engineering, 287*, 110116. doi:10.1016/j.jfoodeng.2020.110116

Keerthy, M. A., Dhanaselvam, K. R., Santhoshkumar, P., Ravikrishnan, V., & Moses, J. A. (2025). Improving the printability of watermelon rind using pectin: 3D printing optimization for the development of gummies. *Food Biomacromolecules, 2*(2), 143–156. doi:10.1002/fob2.70008

Khorsandi, D., Rezayat, D., Sezen, S., Ferrao, R., Khosravi, A., Zarepour, A., Khorsandi, M., Hashemian, M., Iravani, S., & Zarrabi, A. (2024). Application of 3D, 4D, 5D, and 6D bioprinting in cancer research: What does the future look like? *Journal of Materials Chemistry B, 12*(19), 4584–4612. doi:10.1039/D4TB00310A

Kumaresan, R., Kulandaivel, S., Samykano, M., Ngui, W. K., Harun, W. S. W., Noor, M. M., & Badadhe, A. M. (2026). Additive manufacturing: An overview of printing technologies. *Journal of Advanced Research in Applied Sciences and Engineering Technology, 59*(2), 245–260. 10.37934/araset.59.2.245260

Lai, J., & Wang, M. (2023). Developments of additive manufacturing and 5D printing in tissue engineering. *Journal of Materials Research, 38*(21), 4692–4725. doi:10.1557/s43578-023-01193-5

Liu, Z., Zhang, M., Bhandari, B., & Wang, Y. (2017). 3D printing: Printing precision and application in food sector. *Trends in Food Science and Technology, 69*. 10.1016/j.tifs.2017.08.018

Muthurajan, M., Veeramani, A., Rahul, T., Gupta, R. K., Anukiruthika, T., Moses, J. A., & Anandharamakrishnan, C. (2021). Valorization of food industry waste streams using 3D food printing: A study on noodles prepared from potato peel waste. *Food and Bioprocess Technology, 14*(10), 1817–1834. doi:10.1007/s11947-021-02675-2

Nida, S., Anukiruthika, T., Moses, J. A., & Anandharamakrishnan, C. (2021). 3D Printing of grinding and milling fractions of rice husk. *Waste and Biomass Valorization, 12*(1), 81–90. doi:10.1007/s12649-020-01000-w

Nida, S., Moses, J. A., & Anandharamakrishnan, C. (2022a). 3D extrusion printability of sugarcane bagasse blended with banana peel for prospective food packaging applications. *Sugar Tech, 24*(3), 764–778. doi:10.1007/s12355-021-01095-y

Nida, S., Moses, J. A., & Anandharamakrishnan, C. (2022b). Emerging applications of 5D and 6D printing in the food industry. *Journal of Agriculture and Food Research, 10*, 100392. doi:10.1016/j.jafr.2022.100392

Niu, D., Zhang, M., Mujumdar, A. S., & Yu, D. (2025). Investigation of 5D printing of children food based on anthocyanin microcapsules and spontaneous changes in probiotics. *Innovative Food Science & Emerging Technologies*, *100*, 103921. doi:10.1016/j.ifset.2025.103921

Raja, V., Moses, J. A., & Anandharamakrishnan, C. (2023). Effect of 3D printing conditions and post-printing fermentation on pearl millet fortified idli. *Journal of the Science of Food and Agriculture*, *103*(5), 2401–2412. doi:10.1002/jsfa.12410

Rane, N., Choudhary, S., & Rane, J. (2023a). 4D/5D/6D printing technology in the architecture, engineering, and construction (AEC) industry: Applications, challenges, and future advancements. *SSRN Electronic Journal*. doi:10.2139/ssrn.4624192

Rane, N., Choudhary, S., & Rane, J. (2023b). Enhanced product design and development using Artificial Intelligence (AI), Virtual Reality (VR), Augmented Reality (AR), 4D/5D/6D Printing, Internet of Things (IoT), and blockchain: A review. *SSRN Electronic Journal*. doi:10.2139/ssrn.4644059

Reddy, P. R., & Devi, P. A. (2018). Review on the advancements of additive manufacturing-4D and 5D printing. *International Journal of Mechanical and Production*, *8*(4), 397–402.

Santhoshkumar, P., & Moses, J. (2025). Impact of konjac glucomannan on hot extrusion 3D printability of cake gel. *International Journal of Biological Macromolecules*, *305*, 141162. doi:10.1016/j.ijbiomac.2025.141162

Santhoshkumar, P., Negi, A., & Moses, J. A. (2023). 3D printing for space food applications: Advancements, challenges, and prospects. *Life Sciences in Space Research*, *40*. doi:10.1016/j.lssr.2023.08.002

Santhoshkumar, P., Raja, V., Priyadarshini, S. R., & Moses, J. A. (2024a). Evaluating the 3D printability of pearl millet flour with banana pulp blends. *Journal of the Science of Food and Agriculture*, *104*(9). doi:10.1002/jsfa.13389

Santhoshkumar, P., Raja, V., Priyadarshini, S. R., & Moses, J. A. (2024b). Impact of baking and steaming of 3D printed Pearl millet-Banana pulp blends on physicochemical and sensory attributes. *International Journal of Gastronomy and Food Science*, *37*, 100962. doi:10.1016/j.ijgfs.2024.100962

Santhoshkumar, P., Ramu, D., Mahalakshmi, L., & Moses, J. A. (2024c). 3D printed edible electronics: Components, fabrication approaches and applications. *Biosensors and Bioelectronics*, *272*, 117059. doi:10.1016/j.bios.2024.117059

Santhoshkumar, P., Swathika, B. S., Negi, A., & Moses, J. A. (2025). Improving gel performance and hot extrusion three-dimensional printability of blueberry gel with guar gum. *Food Biophysics*, *20*(2), 1–15. doi:10.1007/s11483-025-09964-9

Shabir, I., Dar, A. H., Manzoor, S., Pandey, V. K., Srivastava, S., Shams, R., Dash, K. K., Bashir, I., Fayaz, U., Prithviraj, V., Singh, P., Rustagi, S., Ramniwas, S., & Pandiselvam, R. (2024). Advancements in food printing technologies and their potential culinary applications: A contemporary exploration. *Journal of Food Processing and Preservation*, *2024*, 1–17. doi:10.1155/2024/6621344

Shanthamma, S., Preethi, R., Moses, J. A., & Anandharamakrishnan, C. (2021). 4D printing of sago starch with turmeric blends: A study on pH-triggered spontaneous color transformation. *ACS Food Science & Technology*, *1*(4), 669–679. doi:10.1021/acsfoodscitech.0c00151

Sharma, K. S. (2023). Artificial intelligence assisted fabrication of 3D, 4D and 5D printed formulations or devices for drug delivery. *Current Drug Delivery*, *20*(6), 752–769. doi:10.2174/1567201820666221207140956

Shi, H., Zhang, M., & Mujumdar, A. S. (2024a). 3D/4D printed super reconstructed foods: Characteristics, research progress, and prospects. *Comprehensive Reviews in Food Science and Food Safety*, *23*(2). doi:10.1111/1541-4337.13310

Shi, Y., Yan, C., Song, B., Su, B., Wei, Q., Zhang, L., Wu, J., Wen, S., Liu, J., Cai, C., Yu, S., Li, C., Zhou, Y., Chen, A., Yang, L., Chen, P., Zou, Y., Tang, M., Chen, Y., ... Ouyang, Z. (2024b). Recent advances in additive manufacturing technology: Achievements of the rapid manufacturing center in huazhong university of science and technology. *Additive Manufacturing Frontiers*, *3*(2), 200144. doi:10.1016/j.amf.2024.200144

Shi, Y., Zhang, M., & Phuhongsung, P. (2022). Microwave-induced spontaneous deformation of purple potato puree and oleogel in 4D printing. *Journal of Food Engineering*, *313*, 110757. doi:10.1016/j.jfoodeng.2021.110757

Srivastava, S., Pandey, V. K., Singh, R., & Dar, A. H. (2023). Recent insights on advancements and substantial transformations in food printing technology from 3 to 7D. *Food Science and Biotechnology*, *32*(13), 1783–1804. doi:10.1007/s10068-023-01352-8

Subhasri, D., Dutta, S., Leena, M. M., Moses, J. A., & Anandharamakrishnan, C. (2022). Gastronomy: An extended platform for customized nutrition. *Future Foods*, *5*, 100147. doi:10.1016/j.fufo.2022.100147

Sun, J., Zhou, W., Yan, L., Huang, D., Lin, L., & ya. (2018). Extrusion-based food printing for digitalized food design and nutrition control. *Journal of Food Engineering*, *220*, 1–11. doi:10.1016/J.JFOODENG.2017.02.028

Teng, X., Li, C., Mujumdar, A. S., & Zhang, M. (2022). Progress in extrusion-based food printing technology for enhanced printability and printing efficiency of typical personalized foods: A review. *Food*, *11*(24), 4111. doi:10.3390/foods11244111

Theagarajan, R., Moses, J. A., & Anandharamakrishnan, C. (2020). 3D extrusion printability of rice starch and optimization of process variables. *Food and Bioprocess Technology*, *13*(6), 1048–1062. doi:10.1007/s11947-020-02453-6

Ukkunda, N. S., Santhoshkumar, P., & Moses, J. (2025). Impact of different natural sweeteners on 3D printability and post-printing quality of marzipan. *Future Foods*, *11*, 100557. doi:10.1016/j.fufo.2025.100557

Vasiliadis, A. V., Koukoulias, N., & Katakalos, K. (2022). From three-dimensional (3D)- to 6D-printing technology in orthopedics: Science fiction or scientific reality? *Journal of Functional Biomaterials*, *13*(3), 101. doi:10.3390/jfb13030101

Wang, N., Li, R., Wang, X., & Yang, X. (2024). 4D food printing: Key factors and optimization strategies. *Trends in Food Science & Technology*, *145*, 104380. doi:10.1016/j.tifs.2024.104380

Wen, Y., Che, Q. T., Wang, S., Park, H. J., & Kim, H. W. (2024). Elaboration of dimensional quality in 3D-printed food: Key factors in process steps. *Comprehensive Reviews in Food Science and Food Safety*, *23*(1). doi:10.1111/1541-4337.13267

Wilson, A., Anukiruthika, T., Moses, J. A., & Anandharamakrishnan, C. (2020). Customized shapes for chicken meat – based products: Feasibility study on 3D-printed nuggets. *Food and Bioprocess Technology*, *13*.

Yoha, K. S., Anukiruthika, T., Anila, W., Moses, J. A., & Anandharamakrishnan, C. (2021). 3D printing of encapsulated probiotics: Effect of different post-processing methods on the stability of Lactiplantibacillus plantarum (NCIM 2083) under static in vitro digestion conditions and during storage. *LWT*, *146*, 111461. doi:10.1016/j.lwt.2021.111461

Yoha, K. S., & Moses, J. A. (2023). 3D printing approach to valorization of agri-food processing waste streams. *Food*, *12*(1), 212. doi:10.3390/foods12010212

Yu, Q., Zhang, M., Bhandari, B., & Li, J. (2023). Future perspective of additive manufacturing of food for children. *Trends in Food Science & Technology*, *136*, 120–134. doi:10.1016/j.tifs.2023.04.009

Zhu, W., Iskandar, M. M., Baeghbali, V., & Kubow, S. (2023). Three-Dimensional printing of foods: A critical review of the present state in healthcare applications, and potential risks and benefits. *Food*, *12*(17), 3287. doi:10.3390/foods12173287

9

Consumer Perception Toward Food Printing Technologies

CHAPTER MENU

9.1 Introduction

The human body fundamentally needs food to sustain life and support well-being. Nutrients are essential although consumer perception greatly impacts food selection. In addition to its nutritious function, food offers sensory enjoyment, an aspect of increasing attention in food product development (van Eck & Stieger, 2020). With the growing global population and associated escalating food consumption, there are heightened fears of a potential Malthusian crisis. Nonetheless, progress in agricultural and food sciences has contributed to meeting this increasing need. With increasing affluence, urban and rural food systems have grown increasingly interlinked, influencing food accessibility, pricing, and consumer behavior beneficially and detrimentally (Ambikapathi et al., 2022). As fundamental food security becomes more guaranteed, the variety of food options increases. Food selections arise from intricate interactions among sensory inputs, perception, cognitive elements, and cultural acceptance, influencing food choices and can be categorized into multiple domains (Subhasri et al., 2022, 2023; Rodríguez-Parada et al., 2025). This encompasses internal food elements, including sensory and perceptual attributes, and external food aspects, such as accessible information, as well as the social and physical environments.

4D Printing of Foods: Fundamentals, Applications and Prospects, First Edition.
C. Anandharamakrishnan, Jeyan A. Moses, and P. Santhoshkumar.

In this context, personal-state aspects encompass biological characteristics, physiological requirements, psychological elements, habits, prior experiences, and cognitive aspects are those such as knowledge, abilities, attitudes, preferences, expected outcomes, and personal identity. Additionally, sociocultural factors, including cultural influences, economic situations, and political elements, additionally impact dietary selections (Chen & Antonelli, 2020). Furthermore, consumer meal selections are influenced by food accessibility, eating atmosphere, company, food coloration, and sensory attributes such as warmth, aroma, and timing of consumption (Rai et al., 2023), apart from other miscellaneous factors. Figure 9.1 delineates the diverse aspects influencing consumer food perceptions, highlighting a multisensory phenomenon.

More recently, advancements in food processing and preservation have also induced consumer apprehension. To overcome this, emerging technologies must maintain recognizable flavors, fostering consumer adoption and benefiting manufacturers (Bryant, 2019; Rai et al., 2023). Nonetheless, research investigating the determinants affecting consumer perceptions of food and their consequences for the food sector is still scarce. In reality, food businesses are presently experiencing a revolution. As people become more conscious of their food choices and seek personalized sensory experiences, new technologies are emerging to cater to these evolving consumption patterns. Numerous studies, as well as the market, provide alternatives for health-conscious consumers, including nutritionally fortified foods and products enhanced with essential nutrients to avoid or rectify deficiencies (Dary & Mora, 2002; Olson et al., 2021; Thakur et al., 2023; Tian et al., 2022; Ukkunda et al., 2025).

The food industry, renowned for its competitiveness and innovation, has a significant rate of product failures and market withdrawals. While the causes of this high failure rate vary, one of the most critical areas for improvement seems to be insufficient consumer research. Typically, food innovation is classified into different categories based on their degree and novelty type. Novel foods and their associated technologies represent the most disruptive category and are particularly susceptible to instigating mistrust and being rejected by consumers (Brunner et al., 2018). However, the early research about consumer perceptions, needs, and fears is particularly relevant when it comes to marketing these types of innovations. Adopting recent processing methods in food, such as advanced thermal and nonthermal technologies, and manufacturing processes, such as 3D printing, are considered emerging technologies in the food industry (Derossi et al., 2023). In-depth market analysis in terms of consumer perceptions is required for these types of technologies prior to the market launch of the product. In other words, consumer perception studies must go hand in hand with such emerging technologies.

9.2 Studies on Consumer Perception of Printed Foods

Printed foods are finding increasingly new commercial applications. Some interesting concepts and interventions are presented here. Nova Meat, a research organization, creates 3D-printed steaks utilizing green beans and seaweed protein. Similarly, Biozoon, a food technology enterprise, manufactures an array of 3D-printed dishes, including chicken and green bean meals, tailored for elderly consumers. The endeavor commenced in 2006 when

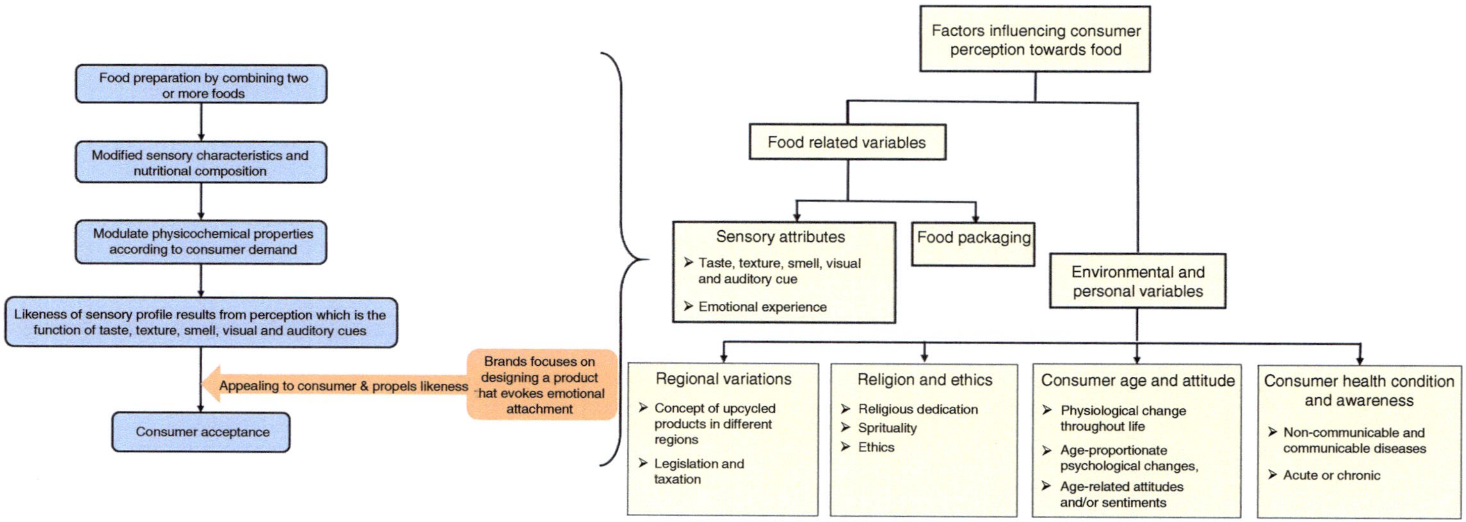

Figure 9.1 Factors and sensory attributes influencing consumer perception. *Source:* Rai et al. (2023) / Frontiers Media S.A. / CC BY 4.0.

Hod Lipson pioneered cookie fabrication using 3D food printers, leading to a diverse array of 3D-printed foods. Hershey's, the preeminent global chocolate producer, began producing diverse chocolates utilizing 3D printing technology. A 3D printing restaurant, "Food Ink," in London, garnered significant attention at its opening ceremony in April 2016 by offering a nine-course meal of 3D-printed food accompanied by 3D-printed dishes. Further, La Enoteca, a two-Michelin-star establishment in Barcelona, has employed 3D printing to craft coral-inspired dishes, and Wolvega (restaurant) in the Netherlands has produced 3D-printed meals featuring honeycomb-shaped beans and honey shaped like the iconic Eiffel Tower (Lee et al., 2021).

Researchers anticipate that 3D food printers will soon become prevalent household appliances, allowing individuals to print nutritionally customized meals suited to their requirements. This technology provides an engineering solution for tailored food design and individualized nutritional management. To support such endeavors, in recent times, many studies have reported aspects of consumer perception and marketing studies in additive manufacturing of foods. Researchers are utilizing different tools according to the focus of their studies; for instance, a questionnaire-based consumer survey was conducted to observe the attitude of consumers toward 3D-printed foods (Table 9.1). The authors compared the four possible application fields of 3D printers in food processing: fun, convenience, health, and nutrition (Brunner et al., 2018). The awareness about the 3D

Table 9.1 Questionnaire chart for attitudes toward 3D-printed food products by consumers.

Items	Scale
Previous knowledge	Measured on a six-point Likert scale (ranging from 1 = "do not agree at all" to 6 = "completely agree")
I've already read/heard a lot about 3D-printed food	
I already knew about these facts	
I know roughly how a 3D printer works	
I've already dealt with 3D food printing	
Benefit perception	Measured on a six-point Likert scale (ranging from 1 = "do not agree at all" to 6 = "completely agree")
I see an advantage in this application, also if I do not necessarily intend to make use of it (argument: fun)	
I see an advantage in this application ... (argument: convenience)	
I see an advantage in this application ... (argument: health)	
I see an advantage in this application ... (argument: nutrition)	
Willingness to consume	Measured on a six-point Likert scale (ranging from 1 = "do not agree at all" to 6 = "completely agree")
I would have no trouble consuming this product (argument: fun)	
I would have no trouble consuming this product (argument: convenience)	
I would have no trouble consuming this product (argument: health)	
I would have no trouble consuming this product (argument: nutrition)	

Table 9.1 (Continued)

Items	Scale
Fun to use	Measured on a six-point Likert scale (ranging from 1 = "do not agree at all" to 6 = "completely agree")
I would have fun producing such a food in my own kitchen (argument: fun)	
I would have fun ... (argument: convenience)	
I would have fun ... (argument: health)	
I would have fun ... (argument: nutrition)	
Cooking creativity	Measured on a six-point Likert scale (ranging from 1 = "do not agree at all" to 6 = "completely agree")
Cooking allows me to express my creativity	
When I cook, I like to try new recipes	
The best of cooking is to develop own recipes or optimise existing ones	
Food neophobia	Measured on a six-point Likert scale (ranging from 1 = "do not agree at all" to 6 = "completely agree")
I like to try new ethnic restaurants	
I am constantly sampling new and different foods	
I do not trust new foods	
If I do not know what is in a food, I won't try it	
I like foods from different countries	
Ethnic food looks too weird to eat	
At dinner parties, I will try a new food	
I am afraid to eat things I have never had before	
I am very particular about the foods I will eat	
I will eat almost anything	
Food technology neophobia	Measured on a six-point Likert scale (ranging from 1 = "do not agree at all" to 6 = "completely agree")
There are plenty of tasty foods around, so we don't need to use new food technologies to produce more	
The benefits of new food technologies are often grossly overstated	
New food technologies decrease the natural quality of food	
There is no sense trying out high-tech food products because the ones I eat are already good enough	
Convenience orientation	Measured on a six-point Likert scale (ranging from 1 = "do not agree at all" to 6 = "completely agree")
The less physical energy I need to prepare a meal, the better	
The ideal meal can be prepared with little effort	
Preferably, I spend as little time as possible on meal preparation	
I want to spend as little time as possible cooking	
At home, I preferably eat meals that can be prepared quickly	
Food involvement	Measured on a six-point Likert scale (ranging from 1 = "do not agree at all" to 6 = "completely agree")
I don't think much about food each day	
Cooking or barbequing is not much fun	
Talking about what I ate or am going to eat is something I like to do	
I enjoy cooking for others and myself	
When I eat, I do not wash dishes or clean the table	
I care whether or not a table is nicely set	

(Continued)

Table 9.1 (Continued)

Items	Scale
Health	Measured on a six-point Likert scale (ranging from 1 = "do not agree at all" to 6 = "completely agree")
It is important to me that the food I eat on a typical day:	
Contains a lot of vitamins and minerals	
Keeps me healthy	
Is nutritious	
Is high in protein	
Is good for my skin, teeth, hair, nails, etc.	
Is high in fibre and roughage	
Natural content	Measured on a six-point Likert scale (ranging from 1 = "do not agree at all" to 6 = "completely agree")
It is important to me that the food I eat on a typical day:	
Contains no additives	
Contains natural ingredients	
Contains no artificial ingredients	
Familiarity	Measured on a six-point Likert scale (ranging from 1 = "do not agree at all" to 6 = "completely agree")
It is important to me that the food I eat on a typical day:	
Is what I usually eat	
Is familiar	
Is like the food I ate when I was a child	
Digital native	Measured on a six-point Likert scale (ranging from 1 = "do not agree at all" to 6 = "completely agree")
I use computers for many things in my daily life	
I use the computer for leisure every day	
I am able to communicate with my friends and do my work at the same time	
I am able to use more than one application on the computer at the same time	
I use a lot of graphics and icons when I send messages	
I use pictures to express my feelings better	
I expect quick access to information when I need it	
I expect the websites that I visit regularly to be constantly updated	
Nutritional knowledge	Measured on a six-point Likert scale (ranging from 1 = "do not agree at all" to 6 = "completely agree")
Lentils contain only a few useful nutrients; therefore, their health benefit is not great	
The health benefit of fruit and vegetables lies only in the supply of vitamins and minerals	
A healthy diet means nothing else than eating vitamins	
Fat is always bad for your health; you should therefore avoid it as much as possible	
A healthy meal should consist of half meat, a quarter vegetables and a quarter side dishes	
If you have eaten high-fat foods, you can reverse the effects by eating apples	
Whole meal foods contain fiber, which is of no use for digestion	
To eat healthily, you should eat less; whatever foods you decrease does not matter	
If chips did not contain so much salt, you could eat more of them without any problem	
All dietary oils have the same ingredients; the oils differ only in taste	

Source: Brunner et al. (2018) / with permission of Elsevier.

printing approach successfully overcame food neophobia and persuaded consumers that this technology could aid in preparing healthy, personalized meals while adding a "fun" element to the food manufacturing process.

Lupton & Turner (2017) attempted to comprehend how consumers may react to food produced with printing technologies. The online study's group discussion underscored that public awareness of both the technology and the characteristics of the food produced with it was largely nonexistent and highly speculative. Several participants expressed concerns that the food produced with a printer may be inedible, unsafe, or nutritionally depleted. Interestingly, the term "printer" itself, which is frequently associated with the nonfood industry, appeared to have a detrimental effect on the participants' adoption. Perhaps "printers" are expected only to print synthetic/chemical-based dyes and pigments. Most participants' skepticism was not overcome by the presentation of images of 3D-printed food and meals, nor by the arguments that this technology could help reduce food waste and support meeting global dietary requirements. However, a limited number of individuals who identified themselves as adventurous eaters expressed an interest in consuming 3D-printed food.

Recently, the consumer willingness to pay premium prices for 3D-printed food in emerging economies was analyzed and reported by Yang et al. (2024). Surprisingly, the survey revealed that consumers are not willing to pay a higher price for 3D-printed foods despite holding more significant benefits to health. However, results may vary for different other study demographics and approaches. In this section, two case studies have been considered, each presenting interesting observations on how consumers perceive and accept/reject 3D-printed foods.

9.2.1 Case Study I: 3D-Printed Snack Bar for Soldiers

> Original study: An exploratory consumer study of 3D printed food perception in a real-life military setting published in the journal *Food Quality and Preference* by Caulier, S., Doets, E., & Noort, M.

3D food printing has many applications in the food industry, including personalization and customization of nutrients, shapes, and textures. Specifically, numerous studies have reported developing snack foods using the approach, utilizing different raw materials, including food materials and recovered fractions of agro-processing waste streams (Yoha & Moses, 2023). Caulier et al. (2020) conducted a 4-week study examining 12 soldiers' (age $= 32.1 \pm 7.6$ years) impressions of 3D-printed snack bars in actual military environments. In the initial week, soldiers received a standard conventionally produced snack bar. In the subsequent weeks, they received 3D-printed snack bars with customized features. During week 2, the emphasis was on texture customization, and in week 3, soldiers were allowed to select between two flavor profiles, sweet or savory. By the fourth week, they were presented with an assortment of ingredients comprising four varieties of dough (vanilla, oat, chocolate, and corn with parmesan) and thirteen distinct fillings (chocolate, blueberry, lemon, coconut, mango, kiwi, rhubarb, banana, raspberry, red pepper, eggplant, carrots, and green peas). The methodology is summarized in

Figure 9.2a. Evaluations were performed in weeks 1 and 4, prior to and after the recurrent consumption of the snack bars. Finally, these researchers observed that soldiers favored the 3D-printed snack bars as compared to the conventional counterpart, especially when bars were tailored with diverse textures and contents; these attributes markedly enhanced the overall preference and taste and texture ratings. In other words, customization in textures and fillings, as depicted in Figure 9.2b, significantly influenced soldiers' choices. The personalized snack bars from week 4 earned the highest preference ratings. The soldiers demonstrated enhanced interest in 3D-printed food products because of the ability to customize texture and flavor, reporting appetite suppression and enhanced diversity of available healthy choices. Overall, the substantial advantages of the technology are well recognized although it has certain technological limits, including material selection and printing constraints.

9.2.2 Case Study II: 3D-Printed Insect-Based Foods and 3D-Printed Meat

Original study: Contextual acceptance of novel and unfamiliar foods: Insects, cultured meat, plant-based meat alternatives, and 3D printed foods published in *Food Quality and Preference*, by Motoki, K., Park, J., Spence, C., & Velasco, C.

The materials used in 3D food printing have expanded to alternatives such as insect-based materials, protein isolates, and animal-based materials. In a recent survey conducted by Motoki et al. (2022), consumers' perception of unfamiliar foods such as insects, cultured meat, plant-based meat alternatives, and 3D-printed foods in three contexts: (i) Consumer acceptance of novel/unfamiliar foods, (ii) anticipated emotions, and (iii) actual product descriptions. The results demonstrated that dining with companions and participating in food festivals markedly affect the expected acceptability of insect-based foods, cultured meat, and 3D-printed foods in comparable manners (Figure 9.3). Anticipated emotions, whether favorable or unfavorable, significantly influence this process, and the increased expectation of happy emotions (or less negative emotions) enhances the probability of experimenting with unfamiliar meals. Also, environmental elements that facilitate the acceptance of plant-based meat alternatives are closely aligned with those that encourage the acceptance of conventional foods. These observations underscore the significance of situational elements in influencing individuals' receptiveness to novel foods and provide practical recommendations for efficiently introducing these items by fostering conducive conditions for their acceptance. In particular, the 3D food printing technologies make it possible to generate foods based on these novel raw materials that are similar to traditional foods; additionally, these foods are associated with less water consumption, giving them environmental benefits (González-Guzmán et al., 2024).

Lupton & Turner (2018) examined the implementation of innovative food technologies, including lab-grown meat and insect-derived ingredients, to enhance ethical consumption, food security, and environmental sustainability. The research collected perspectives from an online discussion forum comprising Australian participants, analyzing their views on food produced using 3D printing technology and unconventional ingredients. The questions asked to the participants are presented in Figure 9.4.

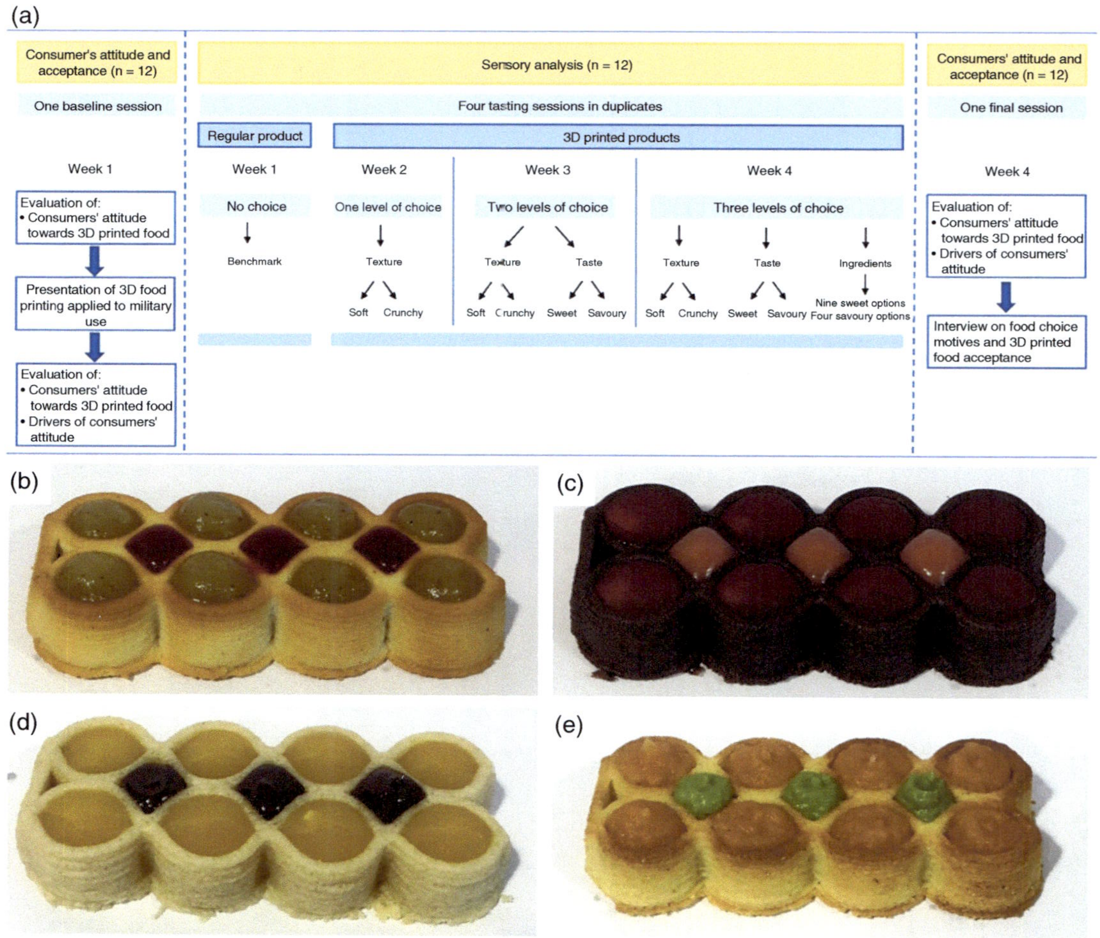

Figure 9.2 (a) Detailed research framework adapted for study to understand consumer preferences on 3D-printed snack bars for soldiers, modified infills and textures in 3D-printed snack bars: (b) vanilla biscuit – crunchy texture, kiwi and raspberry fillings, (c) chocolate biscuit – crunchy texture, raspberry and rhubarb fillings, (d) oat biscuit – soft texture, mango and chocolate fillings, (e) savory biscuit – crunchy texture, carrot and green peas fillings. *Source:* Caulier et al. (2020) / Elsevier / CC BY 4.0.

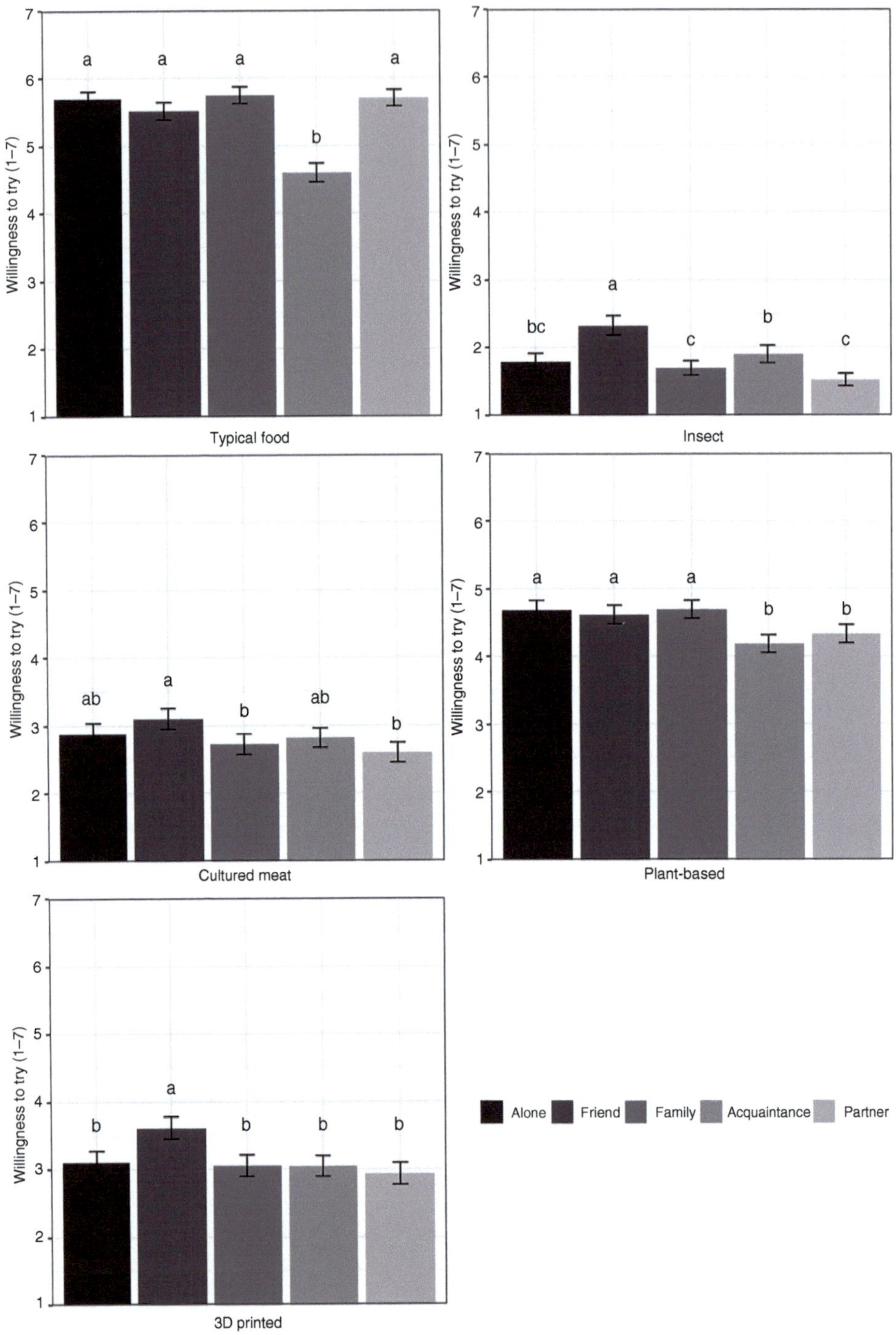

Figure 9.3 Influence of social situations on consumers' willingness to try novel/unfamiliar foods. *Source:* Motoki et al. (2022) / with permission of Elsevier.

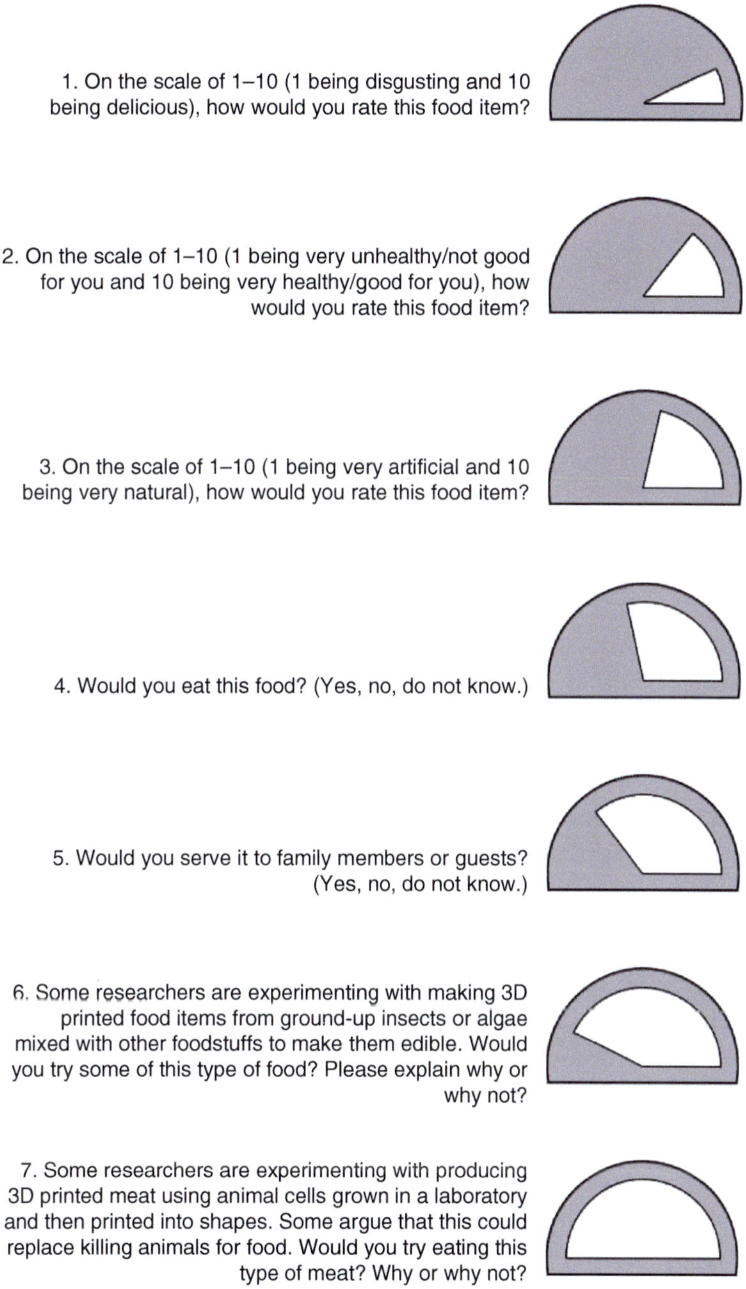

Figure 9.4 Questions asked to the participants in a survey to understand cultured meat and insect-based 3D-printed food preferences.

Participants recognized the societal advantages of 3D-printed foods derived from cultured meat and insects; however, their individual dietary preferences focused on health, flavor, and naturalness. Limited interest was shown in the consumption of such products. Cultured meat was met with skepticism, as numerous participants regarded it as excessively processed, “unnatural,” potentially detrimental, and deficient in flavor or nutrition. Insects, despite being perceived as more “natural” and nutritious than cultured meat, provoke intense feelings of disgust. These researchers concluded that substantial cultural obstacles must be overcome for 3D-printed foods, especially when such ingredients are used. They also reported that participants expressed various concerns about the future of food concerning the use of cultured meat in 3D printing technologies. A small number of individuals responded favorably. Despite their appreciation of meat, individuals already grappled with the ethical implications of consuming livestock and perceived this technology as a solution to their moral dilemmas. Others unwillingly recognized it as a lesser evil, considering the concerns of global starvation and environmental degradation. However, most participants rejected the idea of consuming such “unnatural” foods on a personal level. It was difficult for many individuals to overcome their initial aversion to ingesting insects although they recognized the nutritional value and “natural” classification of insect-based foods.

Insect-based cuisine was perceived as a daring alternative by those prepared to experiment with it. The potential impact of these innovations on agricultural practices and their implications for sustainability and food scarcity were not the subject of significant discourse. Although 3D-printed meat may offer guilt-free consumption, participants perceived it as unnatural and lacking in authenticity, flavor, and freshness. Insect-based 3D-printed food was perceived as synthetic, peculiar, and repulsive. Their research examined the cultural significance of consumer reactions to 3D-printed culinary technologies and unconventional ingredients. Cross-cultural research on food consumption suggests that the notions of what is considered “natural” and “food” evolve over time. Their results emphasize the disruptive nature of traditional norms and preferences by innovative food technologies. However, most participants did not perceive any personal benefit in consuming 3D-printed food, particularly those that contained ingredients they considered inedible. These concerns have not been adequately addressed in the prevailing narratives that commend the benefits of these food products, making it unlikely that consumers will adopt them. Nevertheless, this was an exploratory study that was conducted online. Participants responded to concepts and images rather than actual food products, which restricted the scope of the research.

9.3 Consumer Perception: 3D vs. 4D Food Printing

Consumers’ perception of 3D and 4D food printing is contingent upon their expectations, comprehension of the technologies, and perceived advantages. 3D food printing is generally perceived as a novel approach to the development of visually appealing, personalized, and intricate meals (Santhoshkumar et al., 2023). The customization options, including the ability to customize nutritional content, meet dietary requirements, and create distinctive culinary designs, are highly valued by consumers. It is frequently linked to modernity and creativity, making it popular among younger, tech-savvy demographics and those interested in unique dining experiences. The focus on sustainability, such as reducing food waste and using alternative ingredients, also resonates positively with environmentally conscious consumers (Reuben et al., 2025; Tyupova & Harasym, 2024). As an advancement, 4D food printing augments 3D printing by incorporating a transformative component that causes the food to fluctuate in color, texture, shape, or flavor over time, frequently in response to external factors such as temperature, moisture, or heat (Derossi et al., 2024). Curiosity and enthusiasm are sparked by this dynamic capability, particularly among adventurous eaters and those intrigued by experimental dining (Lupton & Turner, 2017). Nevertheless, food "changing" can also elicit skepticism or hesitancy, as some consumers may associate it with artificiality or complexity. Individuals who prioritize functionality and convenience in their food may perceive 4D printing as a means to improve their dining experiences, including the development of shape-changing snacks or other interesting innovations.

Recent documentations emphasize the advantages of 4D food printing and the advancements in programmable textural attributes (Oral et al., 2021). Although 3D food printing was initially designed to produce visually appealing and innovative shapes, researchers are now focusing on developing customized food products designed to meet consumers' nutritional and sensory requirements (Santhoshkumar et al., 2024; Ukkunda et al., 2025). Additionally, they strive to reduce food waste and increase the popularity of 3D food printing in the food industry, restaurants, and homes. An emerging application paradigm, 4D food printing, involves the response of specific food materials to external stimuli, resulting in timed changes in their properties. The enormous potential for producing food products with a broad range of quality attributes is suggested by the fact that shifts in color, aroma, and shape is achievable, as indicated by several studies. Researchers can generate a wide range of food textures by meticulously designing food formulations with specific properties, such as water mobility, shrinkage, and mechanical features, or by controlling the timing of stimuli (Shabir et al., 2024).

Programmable food texture is another recent advancement in 3D food printing. This approach entails the development of a 3D CAD model and the manufacturing of a food formula that provides specific mechanical properties. Such discoveries present thrilling opportunities to improve consumer perceptions of food texture, potentially leading to greater acceptability among consumers (Oral et al., 2021). Furthermore, it can potentially resolve concerns regarding chewing and swallowing in the elderly and assist in regulating satiety by regulating the duration of chewing (Raheem et al., 2021). These technologies can potentially revolutionize the relationship between the food industry and consumers, facilitating the transition to a more integrated manufacturing and consumption system influenced by human-computer interactions (Hassoun et al., 2024). Overall, both technologies are positively regarded for their innovation and potential. However, 3D printing is perceived as a practical and creative instrument, whereas 4D printing is regarded as a futuristic and experimental concept with the potential to redefine the food experience completely. The adoption of either technology by consumers will be contingent upon resolving concerns regarding cost, simplicity of use, and naturalness, while simultaneously emphasizing the distinctive advantages each technology can offer to their dining and culinary experiences.

9.4 Consumer Behavior and Food Choices Toward Food Printing

The food industry is experiencing remarkable innovations, and consumer behavior toward 3D-printed food is increasingly influenced by its numerous positive attributes. Its capacity to generate visually spectacular, intricate designs is a source of fascination for many, as it provides distinctive dining experiences that enhance presentation and creativity (Califano & Spence, 2024). It is highly appealing that the technology has the potential to personalize meals to meet the nutritional requirements, preferences, and dietary restrictions of each individual (Motoki et al., 2022). In addition, 3D food printing encourages sustainability by minimizing food waste, utilizing alternative constituents such as plant-based proteins, and facilitating the efficient utilization of resources (Muthurajan et al., 2021; Reuben et al., 2025; Yoha & Moses, 2023). The fusion of technology and food is particularly popular among younger and tech-savvy consumers, who value the convenience and innovation it brings to modern lifestyles. The potential of the technology to revolutionize food manufacturing and satisfy the needs of health-conscious, ecofriendly, and adventurous consumers is increasing as it becomes more accessible and affordable (Hassoun et al., 2022).

There is a widespread concern that food produced by a 3D printer may be hazardous, inedible, or nutritionally deficient. Consumer adoption is adversely affected by the term "printer," which is frequently associated with nonfood applications. The overcoming of skepticism was not significantly influenced by displaying images of 3D-printed food or emphasizing the benefits, such as reducing food waste and mitigating global famine (Brunner et al., 2018). Nevertheless, adventurous eaters are interested in experimenting with 3D-printed food. Typically, the negative preferences can be related to two important concerns: (i) Food neophobia and (ii) Technology unfamiliarity.

9.4.1 Food Neophobia

The reluctance or avoidance of taking new foods is referred to as "food neophobia." This mindset can negatively affect the willingness to experience and consume a range of cuisines. The direct influential imposes involved in food neophobia are illustrated in Figure 9.5. Studies demonstrate that those exhibiting food neophobia show more unfavorable perceptions of novel foods than those more receptive to new culinary experiences (Bareen et al., 2025). Furthermore, customers may possess unfavorable views not just toward the novel meals but also toward the technologies employed in their production, highlighting the necessity of evaluating consumer perceptions before developing such food products (Wang et al., 2023; Wendt & Weinrich, 2023). To overcome this, researchers have developed a Food Technology Neophobia Scale, a psychological measurement that assesses an individual's acceptance of foods created using innovative technologies (Cox & Evans, 2008; Siddiqui et al., 2022). This scale additionally aids in forecasting the prospective commercial performance of these products. Such happenings demonstrate critical considerations of the many elements influencing printed foods, including sensory characteristics, individual and contextual influences, consumer age, health, and awareness.

Consumer attitude toward 3D-printed food is adversely affected by neophobia, and the reluctance to ingest foods manufactured using new technologies. The rejection of technology

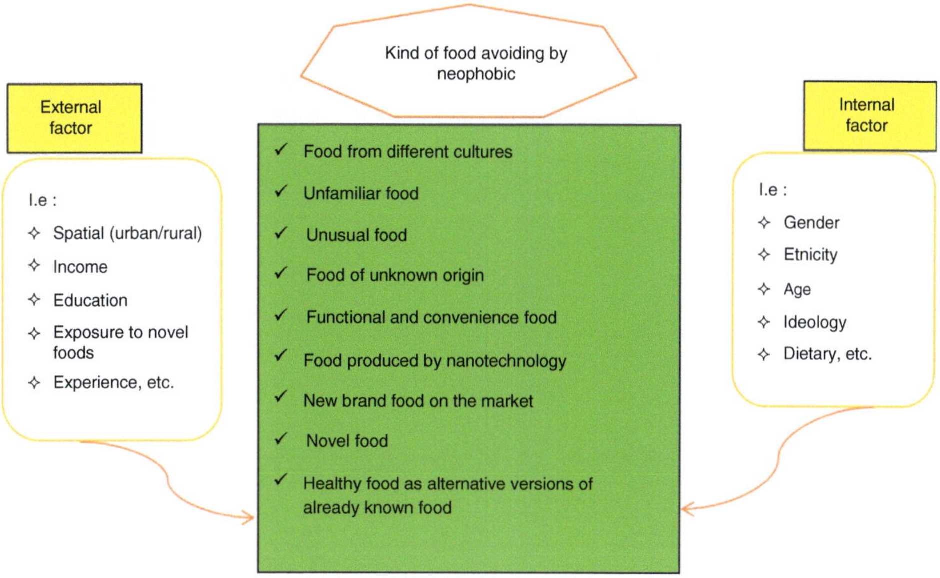

Figure 9.5 Food avoidance and food neophobia. *Source:* Siddiqui et al. (2022) / MDPI / CC BY 4.0.

is influenced by its perceived/assumed unnaturalness, unfamiliarity, and apprehension. Consumers' concerns regarding the safety and edibility of 3D-printed foods and their products are often exacerbated by their lack of understanding and experience (Wendt & Weinrich, 2023). For many, 3D food printing appears unnatural due to its high-tech, digitally driven, and highly processed nature, raising concerns about potential risks. For instance, some consumers feel that 3D-printed foods are heavily processed and therefore are unhealthy. It may also be challenging to reconcile the association between 3D printing and food production, as 3D printing has historically been associated with nonfood industries (Ng et al., 2022).

On the contrary, implementing 3D printing technology in the food industry has resulted in the creation of diverse alternative foods. Sustainable ingredients possessing high nutritional value yet low palatability, such as algae, insects, and leaves, can be processed into powders and used in printing material supplies. This facilitates the production of foods with modified shapes and flavors that are more esthetically pleasing and easier to swallow. In the context of food printing, individuals who experience food neophobia often fall into this category for three main reasons: (i) they perceive the food as inedible, (ii) they consider it potentially hazardous, or (iii) they believe it has lower nutritional value (Siddiqui et al., 2022). Additionally, the results of a survey conducted by Silva et al. (2024) on 3D-printed food and consumer behavior patterns are shown in Figure 9.6. The findings suggest that gender plays a role in the level of knowledge about 3D printing technology, and 51.6% of consumers were open to trying 3D-printed food products. Those seeking healthier, personalized food options were more inclined to accept 3D-printed foods. Finally they reported 54.1% of consumers stated that they would be willing to purchase 3D-printed food if it were nutritionally superior to comparable products already available on the market.

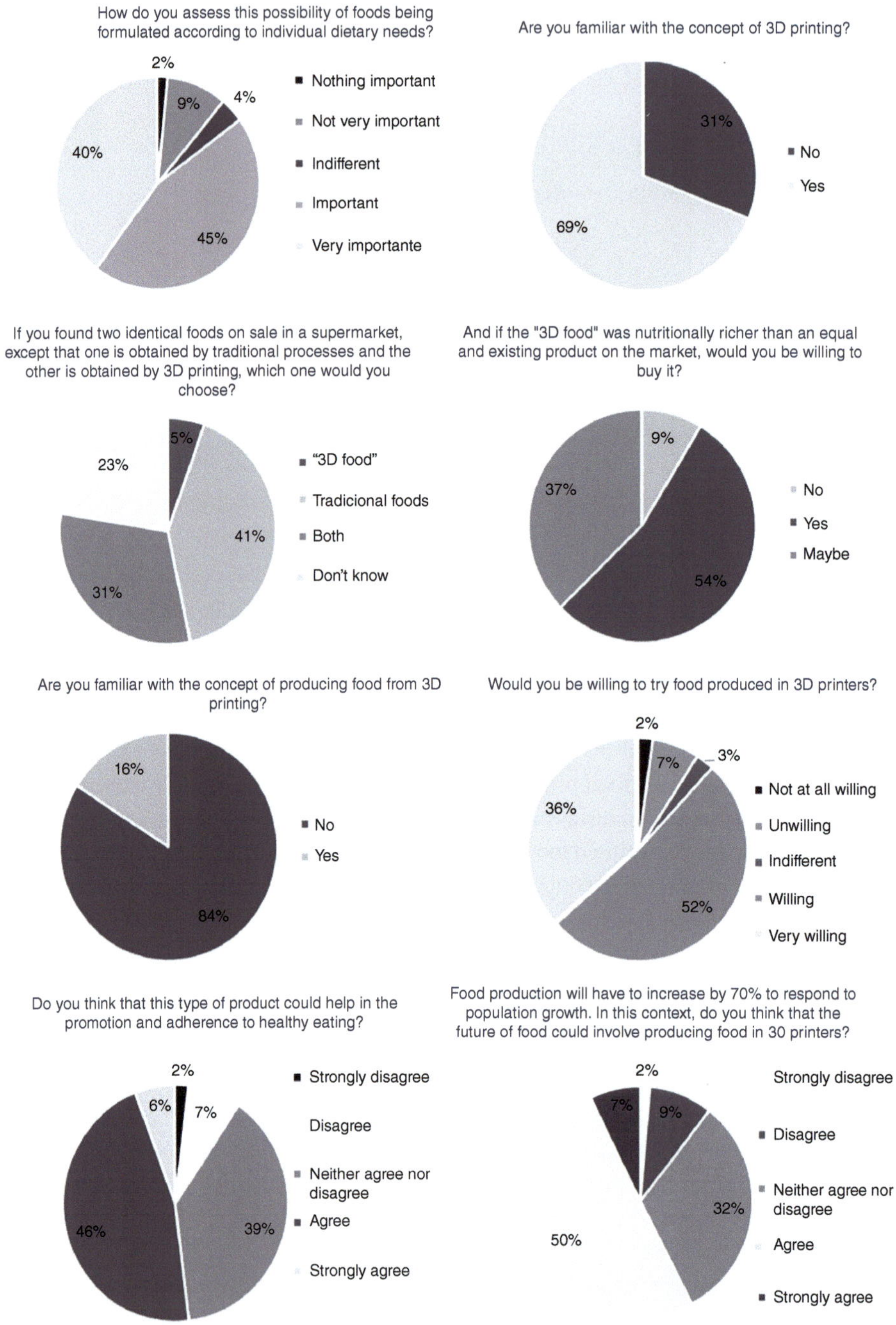

Figure 9.6 Results of the survey of consumer perception toward 3D-printed foods. *Source:* Silva et al. (2024) / with permission of Elsevier.

9.4.2 Technology Unfamiliarity

Consumer unfamiliarity with the technology contributes to negative opinions about food printing technologies. Individuals are significantly more likely to overlook processed foods or goods with which they are unfamiliar. The concept of printing foods is relatively novel, with few individuals possessing the requisite knowledge or experience to discuss it. Moreover, the method may utilize alternative components, such as insects or laboratory-cultivated animal protein, which are predominantly rejected by numerous persons (Lupton & Turner, 2018). Notwithstanding these ingredients' environmental and health benefits, there is often opposition from some who perceive insects or "alternative meat" as inedible. Conversely, familiarity with food provides reassurance and alleviates fear and distrust (Ng et al., 2022; Silva et al., 2024). Another example is the prospects of using food printing to valorize food processing waste streams such as potato peel or grape pomace or others and upcycling them into main food streams. Unfortunately, a low preference for foods prepared from "discards" or "waste" may be the actual reason why some consumers show distrust of food printing.

9.5 Changing Consumer Dynamics: Increasing Awareness and Popularity

In general, the food printing approach has its advantages and disadvantages from an application standpoint. In terms of sensory, nutrition, and appearance, printed products typically show increased popularity than conventional products, and this has been widely recognized. Importantly, consumers are becoming increasingly aware of the advantages and capabilities of 3D printing. However, while technologists and researchers are more cognizant of 3D printing, the general population is not similarly informed. Certainly, in the next couple of years, with increasing public awareness, consumers' willingness toward food printing technologies with improve phenomenally. Many studies reported that esthetic appeal, structural changes in culinary art, sustainable production in the food industry, and value addition to the food improve the consumer mentality positively on printed foods. For this, sufficient awareness is required. It is unlikely that consumers will be convinced to incorporate these innovative technologies and food products into their consumption practices unless the actors and agencies responsible for delineating and advocating these concepts address the concerns and establish the significances that influence localized daily food choices and preferences.

9.6 Tools and Approaches for Enhancing Consumer Knowledge

In most studies on food printing and consumer perception, attitudes toward the technology were determined using online or offline surveys conducted in particular/selected regions and population groups. Such studies have revealed conclusive thoughts about consumer perception of food printing technologies. It is certain that awareness among consumers must be improved, and therefore, technical tools must be used to create awareness among consumers about 3D-printed products. Suggested strategies to improve consumer perception of printed foods are as follows:

- Utilize interactive tools, practical experiences, and instructional resources for consumer understanding
- Digital platforms and applications to offer simulations, ingredient repositories, and recipe-sharing networks
- Educational tools such as video tutorials, webinars, and online courses to aid in technology knowledge acquisition
- Practical workshops and tasting events to provide a tactile experience for food printers and sample creations
- Guided experimentation kits to facilitate home experimentation
- Augmented reality (AR) and virtual reality (VR) enhances the experience by allowing the viewing of culinary designs or virtual printing instructions
- Educational publications to provide insights into trends and methodologies
- Social media platforms and online communities to facilitate the exchange of experiences, recipes, and troubleshooting advice

Overall, it is evident that with increasing technology familiarity, printed foods will certainly gain interesting application grounds in the food industry. This being said, it is important to recollect that many then-novel technologies in the food industry were always looked upon skeptically; today, they are concepts of mainstream food processing/manufacturing/value addition.

9.7 Conclusion

The initial impression customers form and the influence of the primary information source are vital in molding perceptions about food printing technologies. Carefully constructed communication can cultivate favorable attitudes, and the consumers' modest baseline knowledge enhances the liking behavior of printed products. This chapter provides essential factual information regarding the technology's perception and delineates several application scenarios that effectively enhance consumer preferences. Nevertheless, certain initial apprehensions and reservations prove challenging despite customized communication intended to alleviate these issues. Food technology neophobia and technology unfamiliarity are by far important considerations that require careful thought and planning as part of food printing exercises. With weighted benefits, food printing is certainly all set to revolutionize the food manufacturing sector.

References

Ambikapathi, R., Schneider, K. R., Davis, B., Herrero, M., Winters, P., & Fanzo, J. C. (2022). Global food systems transitions have enabled affordable diets but had less favourable outcomes for nutrition, environmental health, inclusion and equity. *Nature Food*, *3*(9), 764–779. doi:10.1038/s43016-022-00588-7

Bareen, M. A., Prakash, S., Sahu, J. K., Bhandari, B., & Naik, S. (2025). Understanding the intention of consumers towards 3D food printing: Exploratory study of psychological factors and sensory analysis. *Journal of Food Science and Technology*, 1–12. doi:10.1007/s13197-025-06261-8

Brunner, T. A., Delley, M., & Denkel, C. (2018). Consumers' attitudes and change of attitude toward 3D-printed food. *Food Quality and Preference*, *68*, 389–396. doi:10.1016/j.foodqual.2017.12.010

Bryant, C. J. (2019). We can't keep meating like this: Attitudes towards vegetarian and vegan diets in the United Kingdom. *Sustainability*, *11*(23), 6844. doi:10.3390/su11236844

Califano, G., & Spence, C. (2024). Consumer preference and willingness to pay for 3D-printed chocolates: A discrete choice experiment. *Future Foods*, *9*, 100378. doi:10.1016/j.fufo.2024.100378

Caulier, S., Doets, E., & Noort, M. (2020). An exploratory consumer study of 3D printed food perception in a real-life military setting. *Food Quality and Preference*, *86*, 104001. doi:10.1016/j.foodqual.2020.104001

Chen, P.-J., & Antonelli, M. (2020). Conceptual models of food choice: Influential factors related to foods, individual differences, and society. *Food*, *9*(12), 1898. doi:10.3390/foods9121898

Cox, D. N., & Evans, G. (2008). Construction and validation of a psychometric scale to measure consumers' fears of novel food technologies: The food technology neophobia scale. *Food Quality and Preference*, *19*(8), 704–710. doi:10.1016/j.foodqual.2008.04.005

Dary, O., & Mora, J. O. (2002). Food fortification to reduce vitamin A deficiency: International Vitamin A Consultative Group Recommendations. *The Journal of Nutrition*, *132*(9), 2927S–2933S. doi:10.1093/jn/132.9.2927S

Derossi, A., Di Palma, E., Moses, J. A., Santhoshkumar, P., Caporizzi, R., & Severini, C. (2023). Avenues for non-conventional robotics technology applications in the food industry. *Food Research International*, *173*, 113265. doi:10.1016/j.foodres.2023.113265

Derossi, A., Spence, C., Corradini, M. G., Jekle, M., Fahmy, A. R., Caporizzi, R., Devahastin, S., Moses, J. A., Le-Bail, A., Zhou, W., Zhang, M., Bhandari, B., & Severini, C. (2024). Personalized, digitally designed 3D printed food towards the reshaping of food manufacturing and consumption. *NPJ Science of Food*, *8*(1), 54. doi:10.1038/s41538-024-00296-5

González-Guzmán, M. E., Del-Aguila-Arcentales, S., Alvarez-Risco, A., Rojas-Osorio, M., & Yáñez, J. A. (2024). Intention to purchase foods based on insects, arachnids, and arthropods, processed by 3D printing in panama consumers. *International Journal of Food Science*, *2024*(1). doi:10.1155/2024/9094666

Hassoun, A., Cropotova, J., Trif, M., Rusu, A. V., Bobiş, O., Nayik, G. A., Jagdale, Y. D., Saeed, F., Afzaal, M., Mostashari, P., Khaneghah, A. M., & Regenstein, J. M. (2022). Consumer acceptance of new food trends resulting from the fourth industrial revolution technologies: A narrative review of literature and future perspectives. *Frontiers in Nutrition*, *9*. doi:10.3389/fnut.2022.972154

Hassoun, A., Jagtap, S., Trollman, H., Garcia-Garcia, G., Duong, L. N. K., Saxena, P., Bouzembrak, Y., Treiblmaier, H., Para-López, C., Carmona-Torres, C., Dev, K., Mhlanga, D., & Aït-Kaddour, A. (2024). From Food Industry 4.0 to Food Industry 5.0: Identifying technological enablers and potential future applications in the food sector. *Comprehensive Reviews in Food Science and Food Safety*, *23*(6). doi:10.1111/1541-4337.70040

Lee, K. H., Hwang, K. H., Kim, M., & Cho, M. (2021). 3D printed food attributes and their roles within the value-attitude-behavior model: Moderating effects of food neophobia and food technology neophobia. *Journal of Hospitality and Tourism Management*, *48*, 46–54. doi:10.1016/j.jhtm.2021.05.013

Lupton, D., & Turner, B. (2017). 'Both fascinating and disturbing': Consumer responses to 3D food printing and implications for food activism. In *Digital food activism* (pp. 151–167). Routledge.

Lupton, D., & Turner, B. (2018). Food of the future? Consumer responses to the idea of 3D-printed meat and insect-based foods. *Food and Foodways*, *26*(4), 269–289. doi:10.1080/07409710.2018.1531213

Motoki, K., Park, J., Spence, C., & Velasco, C. (2022). Contextual acceptance of novel and unfamiliar foods: Insects, cultured meat, plant-based meat alternatives, and 3D printed foods. *Food Quality and Preference*, *96*, 104368. doi:10.1016/j.foodqual.2021.104368

Muthurajan, M., Veeramani, A., Rahul, T., Gupta, R. K., Anukiruthika, T., Moses, J. A., & Anandharamakrishnan, C. (2021). Valorization of food industry waste streams using 3D food printing: A study on noodles prepared from potato peel waste. *Food and Bioprocess Technology*, *14*(10), 1817–1834. doi:10.1007/s11947-021-02675-2

Ng, W. E., Pindi, W., Rovina, K., & Mantihal, S. (2022). Awareness and attitude towards 3D food printing technology: The case of consumer responses from Klang Valley, Malaysia. *Food Research*, *6*(4), 364–372.

Olson, R., Gavin-Smith, B., Ferraboschi, C., & Kraemer, K. (2021). Food fortification: The advantages, disadvantages and lessons from sight and life programs. *Nutrients*, *13*(4), 1118. doi:10.3390/nu13041118

Oral, M. O., Derossi, A., Caporizzi, R., & Severini, C. (2021). Analyzing the most promising innovations in food printing. Programmable food texture and 4D foods. *Future Foods*, *4*, 100093. doi:10.1016/j.fufo.2021.100093

Raheem, D., Carrascosa, C., Ramos, F., Saraiva, A., & Raposo, A. (2021). Texture-modified food for dysphagic patients: A comprehensive review. *International Journal of Environmental Research and Public Health*, *18*(10), 5125. doi:10.3390/ijerph18105125

Rai, S., Wai, P. P., Koirala, P., Bromage, S., Nirmal, N. P., Pandiselvam, R., Nor-Khaizura, M. A. R., & Mehta, N. K. (2023). Food product quality, environmental and personal characteristics affecting consumer perception toward food. *Frontiers in Sustainable Food Systems*, *7*. doi:10.3389/fsufs.2023.1222760

Reuben, S. J., Chandra, S., Praveena, B., Santhoshkumar, P., & Moses, J. A. (2025). 3D printing of alternative proteins: Approaches, challenges and advances. *Sustainable Materials and Technologies*, *43*, e01285. doi:10.1016/j.susmat.2025.e01285

Rodríguez-Parada, L., De la Rosa, S., Sánchez Salado, J., Desmet, P., & Pardo-Vicente, M. (2025). Edible innovations: Testing the WOW impact of 3D printed chocolate packaging. *Food Quality and Preference*, *123*, 105337. doi:10.1016/j.foodqual.2024.105337

Santhoshkumar, P., Negi, A., & Moses, J. A. (2023). 3D printing for space food applications: Advancements, challenges, and prospects. *Life Sciences in Space Research*, *40*. doi:10.1016/j.lssr.2023.08.002

Santhoshkumar, P., Raja, V., Priyadarshini, S. R., & Moses, J. A. (2024). Impact of baking and steaming of 3D printed Pearl millet-banana pulp blends on physicochemical and sensory attributes. *International Journal of Gastronomy and Food Science*, *37*, 100962. doi:10.1016/j.ijgfs.2024.100962

Shabir, I., Dar, A. H., Manzoor, S., Pandey, V. K., Srivastava, S., Shams, R., Dash, K. K., Bashir, I., Fayaz, U., Prithviraj, V., Singh, P., Rustagi, S., Ramniwas, S., & Pandiselvam, R. (2024). Advancements in food printing technologies and their potential culinary applications: A contemporary exploration. *Journal of Food Processing and Preservation*, *2024*, 1–17. doi:10.1155/2024/6621344

Siddiqui, S. A., Zannou, O., Karim, I., Kasmiati, Awad, N. M. H., Gołaszewski, J., Heinz, V., & Smetana, S. (2022). Avoiding food neophobia and increasing consumer acceptance of new food trends—A decade of research. *Sustainability*, *14*(16), 10391. doi:10.3390/su141610391

Silva, F., Pereira, T., Mendes, S., Gordo, L., & Gil, M. M. (2024). Consumer's perceptions and motivations on the consumption of fortified foods and 3D food printing. *Future Foods*, *10*, 100423. doi:10.1016/j.fufo.2024.100423

Subhasri, D., Dutta, S., Leena, M. M., Moses, J. A., & Anandharamakrishnan, C. (2022). Gastronomy: An extended platform for customized nutrition. *Future Foods*, *5*, 100147. doi:10.1016/j.fufo.2022.100147

Subhasri, D., Moses, J. A., & Anandharamakrishnan, C. (2023). New food structures and their influence on nutrition, health and well-being. In *Food structure engineering and design for improved nutrition, health and well-being* (pp. 17–39). Elsevier. doi:10.1016/B978-0-323-85513-6.00014-1

Thakur, S., Singh, A., Insa, B., & Sharma, S. (2023). Food fortification in India as malnutrition concern: a global approach. *Sustainable Food Technology*, *1*(5), 681–695. doi:10.1039/D3FB00079F

Tian, Y., Zhu, H., Zhang, L., & Chen, H. (2022). Consumer preference for nutritionally fortified eggs and impact of health benefit information. *Food*, *11*(8), 1145. doi:10.3390/foods11081145

Tyupova, A., & Harasym, J. (2024). Valorization of fruit and vegetables industry by-streams for 3D printing—A review. *Food*, *13*(14), 2186. doi:10.3390/foods13142186

Ukkunda, N. S., Santhoshkumar, P., & Moses, J. A. (2025). Impact of different natural sweeteners on 3D printability and post-printing quality of marzipan. *Future Foods*, *11*, 100557. doi:10.1016/j.fufo.2025.100557

van Eck, A., & Stieger, M. (2020). Oral processing behavior, sensory perception and intake of composite foods. *Trends in Food Science & Technology*, *106*, 219–231. doi:10.1016/j.tifs.2020.10.008

Wang, K., Cong, L., Mirosa, M., Hou, Y., & Bremer, P. (2023). Food technology neophobia scales in cross-national context: Consumers' acceptance of food technologies between Chinese and New Zealand. *Journal of Food Science*, *88*(8), 3551–3561. doi:10.1111/1750-3841.16693

Wendt, M.-C., & Weinrich, R. (2023). A systematic review of consumer studies applying the food technology neophobia scale: Lessons and applications. *Food Quality and Preference*, *106*, 104811. doi:10.1016/j.foodqual.2023.104811

Yang, M., Gao, J., Yang, Q., Al Mamun, A., Masukujjaman, M., & Hoque, M. E. (2024). Modeling the intention to consume and willingness to pay premium price for 3D-printed food in an emerging economy. *Humanities and Social Sciences Communications*, *11*(1), 274. doi:10.1057/s41599-024-02776-1

Yoha, K. S., & Moses, J. A. (2023). 3D printing approach to valorization of agri-food processing waste streams. *Food*, *12*(1), 212. doi:10.3390/foods12010212

10

Research and Commercialization Status

Food Printing Technologies

CHAPTER MENU

10.1 Introduction

Additive manufacturing is a relatively new innovation that is revolutionizing culinary arts by bringing digital gastronomy into the domain of traditional cooking. This has been a game changer because it lets people alter, customize, and personalize the taste, color, and texture of food. It also lets people make unique shapes, create individualized diet plans, and make meals for people with specific dietary requirements (Ukkunda et al., 2025; Zhu et al., 2023). Further, making great meals with healthy ingredients encourages people to eat better. It promotes sustainability by decreasing food waste and using ingredients better in ways that enhance transportation, storage, and reusing (Baiano, 2022; da Silva & Sehnem, 2025). For instance, 3D food printing technology can be used to prepare/customize a variety of foodstuffs, including pasta, pizza, meat, chocolate, and even cake decorations (Subhasri et al., 2022; Santhoshkumar and Moses 2025; Santhoshkumar et al., 2025; Keerthy et al., 2025). The additive

4D Printing of Foods: Fundamentals, Applications and Prospects, First Edition.
C. Anandharamakrishnan, Jeyan A. Moses, and P. Santhoshkumar.

manufacturing technology (sometimes called food layered manufacturing or FLM) models, builds, and fabricates complex food structures layer by layer using computer-aided designs and software.

What follows is an incomplete list of what is edible: carbohydrates, proteins, seaweed, chocolate, animal purees, vegetable and fruit juices, and powders (Anandharamakrishnan et al., 2022; de Farias et al., 2024). Among the various types of 3D printing processes, extrusion-based 3D printers are the most widely recognized and frequently reported by scientists in the context of food printing (Hussain et al., 2022). This type of printing is especially favored for its ability to handle diverse ingredients and create intricate shapes, making it particularly suitable for food applications (Santhoshkumar et al., 2023). In recent days, many types of 3D food printers have become commercially available in the market, and Table 10.1 summarizes them. Nonetheless, 5D and 6D food printers and printing applications are still in the research phase.

It is certain that the printable ink's rheological and textural behavior can vary based on the ingredients used; 3D-printing food products may not always achieve the desired results. To successfully 3D print food via extrusion printing, the ink must possess a gel-like consistency and exhibit pseudoplastic behavior; that is, its viscosity must decrease with the shear rate (Santhoshkumar et al., 2024). This quality ensures smooth flow through the nozzles, which is essential for producing well-defined shapes and structures in the final product (Anukiruthika et al., 2020). The use of emulsions, which have desirable rheological qualities, has lately gained prominent interest in the field of food printing (Shahbazi et al., 2024). The structure and adhesion of the printed foods can be preserved by these formulations when they are made into colloidal emulsions. Small, surfactant-stabilized droplets of incompatible liquids, such as oil and water, are the building blocks of traditional emulsions. However, many traditional surfactants may be synthetic and potentially detrimental to health.

Additive manufacturing, particularly food printing, enhances consumer perception of printed products. Beyond consumer appeal, this technology contributes to the United Nations' sustainable development goals (SDGs) (Reuben et al., 2025). By promoting resource efficiency, reducing food waste, and enabling customized nutrition, 3D food printing supports goals such as responsible consumption and production, good health and well-being, and sustainable industry innovation (Derossi et al., 2024). The advancement to 4D and 5D food printing has introduced new possibilities; with 4D printing, foods can alter shape, color, or nutritional content over time (Teng et al., 2021), and 5D printing allows the creation more complicated structures with advanced printer configurations (Ghazal et al., 2023). Nevertheless, as dimensions expand, so do technological complexities.

The focus of this chapter is to present the current status of food printing technologies, considering two aspects: (i) research and (ii) commercial usage. A detailed bibliometric analysis was conducted, and results are presented in subsequent sections of this chapter. Overall, this chapter will provide a clear overview of the present status of food printing technologies, considering global-level advancements. To achieve this, historical and current trends have been considered, and the way how research and commercial interests are currently shifting and would therefore likely establish similarly in the near future is presented.

Table 10.1 Commercially available 3D food printers.

3D Printer	Organization	Products	End users	Product features	Product image	Technological impact
3D Dessert Decorator	Beehex	Decoration for cookies, cupcakes, cake, etc.	Mid- to small-sized bakeries	High precision, easy to use, and fast setup		Increases productivity and consistency in small bakeries
Deco-Pod	Beehex	Decoration for cookies, cupcakes, cake, etc.	Direct contact store (e.g., grocery stores, bakeries)	Compact design, user-friendly interface, and customizable designs		Enhances customer engagement through customization and reduces manual labor
Cake Writer™	Beehex	Decoration for cookies, cupcakes, cakes, etc.	In-store bakeries	Fast printing speed, versatile decoration options, and minimal training required		Reduces labor costs and increases product consistency and appeal
Chef 3D	NA	Pizza	Restaurants	Multilayer printing, customizable toppings, and robust build	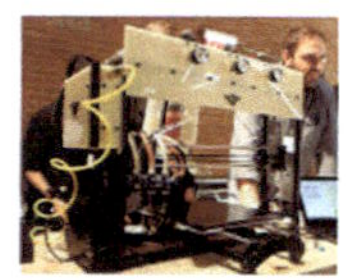	Revolutionizes restaurant food preparation and enables customized dining experiences

(*Continued*)

Table 10.1 (Continued)

3D Printer	Organization	Products	End users	Product features	Product image	Technological impact
ChefJet 3D Printer	3D Systems	Confection, cake decoration	Industrial kitchen, mixologist, and molecular gastronomists	High-resolution printing, large build volume, and supports multiple materials		Enables complex and high-quality food decorations and enhances culinary creativity
Brill 3D Culinary Studio				Precision printing, customizable recipes, and durable construction	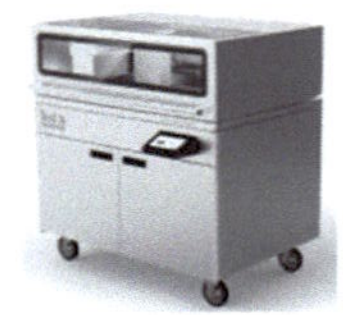	Supports innovative culinary techniques and improves efficiency in professional kitchens
Choc Creator V2.0 Plus	Choc Edge	Couverture chocolate	Chocolates	Fine detail printing, easy to clean, temperature control		Choc Creator V2.0 Plus
mycusini® 2.0	mycusini	Chocolate	Home users and small chocolates	Compact size, affordable, user-friendly software		mycusini® 2.0
Chocola3d	Chocola3d	Chocolate, cheese, dough, vegetables, and fruit pastes	Education institutions, homes, confectioneries, and cafes	Versatile material compatibility, educational applications, and reliable performance		Chocola3d

Procusini® 5.0 3D Food Printer	Procusini	Chocolate	Caterings, hotels, confectioneries, gastronomies, bakeries	High-quality output, multiple material support, and user-friendly interface	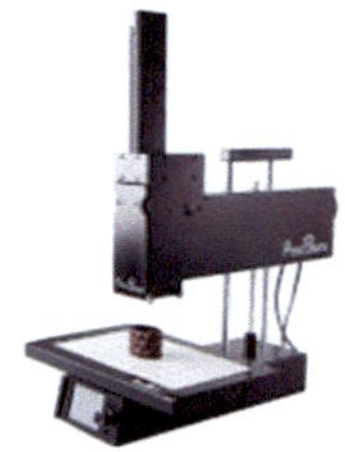	Ideal for professional environments; it enhances dessert and decoration offerings
Focus 3D Food Printer	byFlow	Chocolate, meringue, fruit, cookie, and meat	Pastries, chocolates	Portable, versatile, high-resolution printing	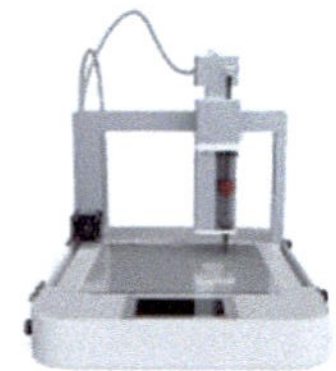	Enables on-site food customization, broadens product offerings
Commercial Art Pancakes Printer F5	ZBOT	Pancake	Pancake House	Detailed pancake designs, fast operation, easy maintenance		Attracts customers with unique pancake art and increases efficiency
PancakeBot 2.0	PancakeBot	Pancake	Pancake House	Customizable designs, user-friendly, compact size		Offers creative breakfast options and engages customers

(*Continued*)

Table 10.1 (Continued)

3D Printer	Organization	Products	End users	Product features	Product image	Technological impact
FoodBot-S2	Changxing Shiyin Technology Co., Ltd.	Chocolate, biscuits, milk candy, pastries, jams, and mashed potatoes	Vending machine, store, and educational institution	Multimaterial support, high precision, easy integration		Expands automated food offerings, suitable for various retail environments
Natural Machines Foodini	Natural Machines	Chocolate, cake, vegetables, cheese, cupcakes, meat, and cookies	Restaurants, hospitals, R&D companies, and educational institutions	Versatile material use, intuitive interface, robust design		Supports innovative food solutions and enhances nutritional offerings
WiibooxSweetin 3D Food Printer	Wiiboox	Chocolate, mashed potatoes, fruit jam, biscuits, cream candy, bean paste	Restaurants, home users	High precision, multiple recipe support, compact design	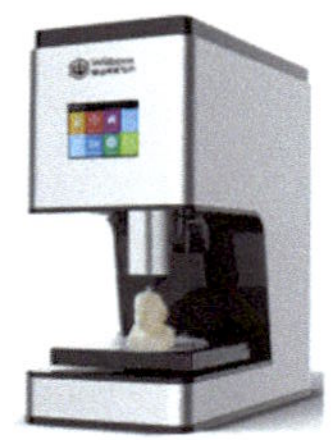	Facilitates creative food preparation at home and in restaurants and broadens culinary possibilities

Source: Neamah & Tandio (2024) / Elsevier / CC BY 4.0.

10.2 Bibliometric Analysis – Food Printing Technologies

A bibliometric analysis is an essential quantitative evaluation employed to measure the development and progression of certain subjects or research domains. It can facilitate the identification of emergent themes and the assessment of research output. This section presents the bibliometric methodologies used to assess research trends in the field of additive manufacturing and 3D/4D food printing, analyzing contributions from journals, patents, authors, institutions, and countries. The section intends to extensively map, quantify, and analyze the adoption of additive manufacturing as an innovative and sustainable trend in the field of food processing/manufacturing.

10.2.1 Scientific Publications

The body of knowledge surrounding food printing is undergoing rapid expansion, drawing in academic and research interest from various parts of the globe. This plays a vital role in scientific decision-making, offering researchers a holistic view of products and processes, enhancing service strategies for customers, and guiding investors in funding scientific research. The analysis employed herein utilizes bibliometric techniques to assess the current state of art in food printing, offering insights into prominent researchers/authors, significant journal publications, highly cited articles, and countries/groups engaged in collaborations. The goal is to motivate upcoming investigations in this area, especially among those who are working in this fascinating field of research.

10.2.1.1 Search Criteria and Database

An inquiry was performed in the Scopus database on July 5, 2025, focusing exclusively on research and review articles published between 2005 and 2026. Conference papers, book chapters, conference reviews, books, notes, editorials, brief surveys, correspondence, and errata were omitted from the search. Two sets of keywords were utilized.

In the initial phase, keywords focused on additive manufacturing of food, with terminology such as:

> *"3D OR three-dimension OR three-dimensional OR 4D OR four-dimension OR four-dimensional OR 5D OR five-dimension OR five-dimensional OR 6D OR six-dimension OR six-dimensional OR additive-manufacturing AND food AND printing"*

This search yielded 2683 peer-reviewed articles on additive manufacturing in food. The volume of publications has markedly increased in recent years (Figure 10.1a), signifying expansion in this domain, particularly among the scientific community. This can also explain the growing commercial interest in the field of food printing.

A second investigation focused on 4D food printing, utilizing certain keywords:

> *"4D OR four-dimension OR four-dimensional AND food AND printing"*

This inquiry revealed 156 articles indexed in Scopus pertaining to 4D food printing. Since its initial publication in 2019, the quantity of articles has consistently risen, exceeding 30 documents released yearly during the past three years, as illustrated in Figure 10.1b.

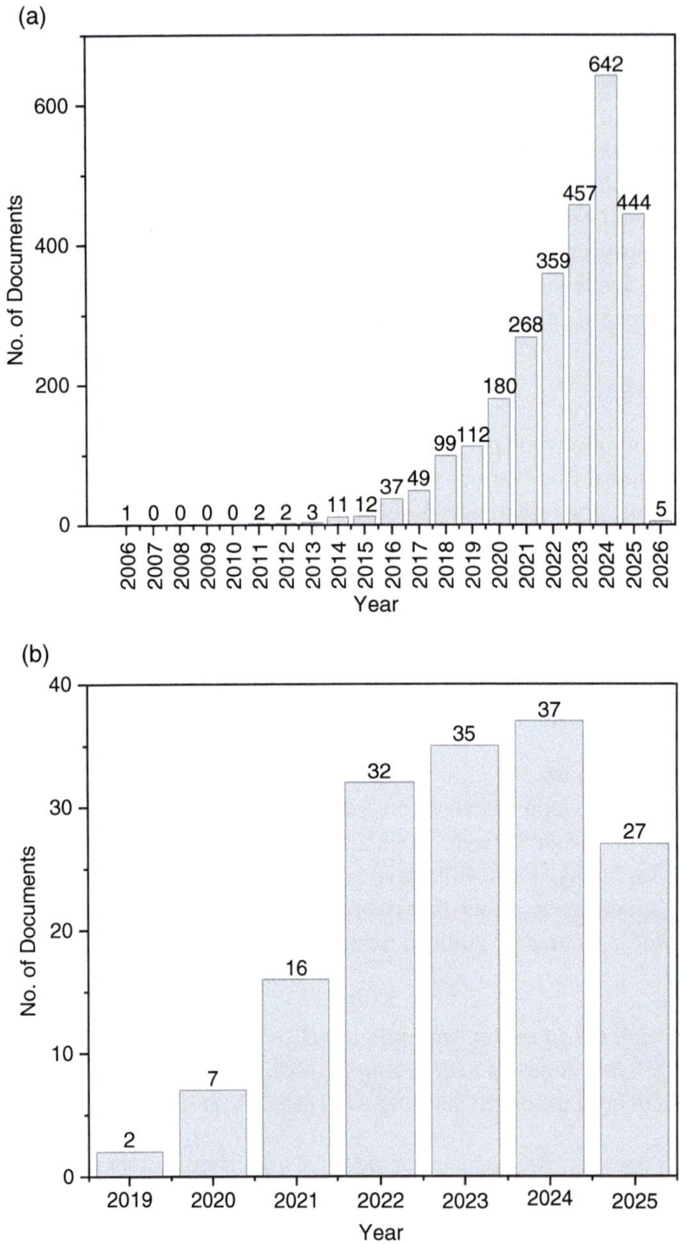

Figure 10.1 Scientific publication trends. (a) All additive manufacturing techniques for food application, (b) 4D printing of foods.

Then, bibliometric analysis was performed utilizing Biblioshiny module in RStudio (Version 4.4.1) to do analyze, generate graphs and visuals (Aria and Cuccurullo 2017). Furthermore, VOS viewer (Version 1.6.19) was employed to create word network maps.

10.2.1.2 Search Approach and Results

A three-plot diagram (Figure 10.2) illustrates the connections among leading authors, journals, and their designated research areas (keywords), emphasizing significant contributors in the domain of food printing. Similarly, Figure 10.3 illustrates the timeline of publications categorized by year and journal, emphasizing the advancement of studies in food

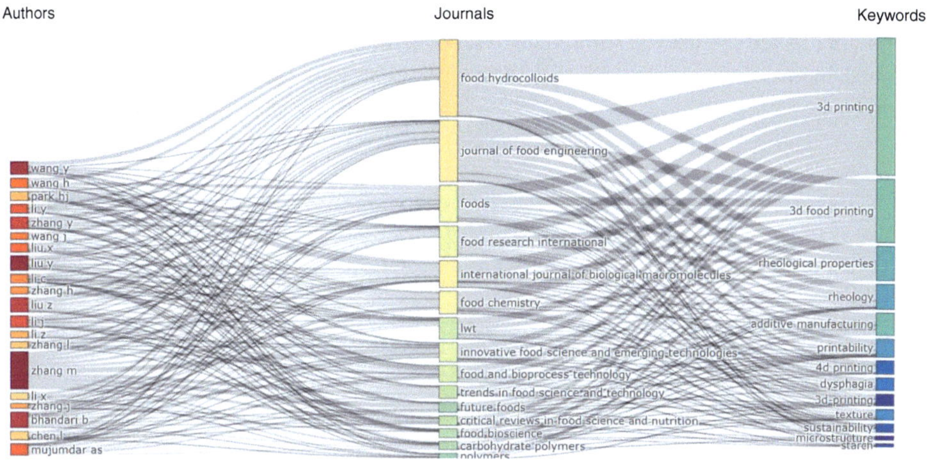

Figure 10.2 Three-field plot interaction between the leading authors, peer-reviewed journals, and the authors' keywords in the field of food printing technologies.

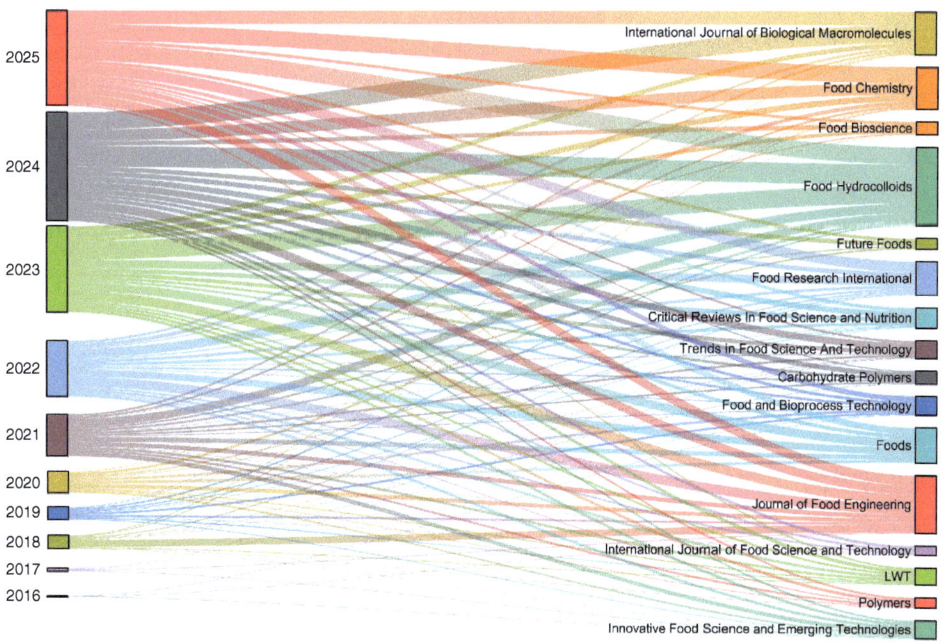

Figure 10.3 Sankey diagram – journals and their publication years in the field of food printing technologies.

printing. Several articles on this topic have recently been published in peer-reviewed journals such as Food Hydrocolloids and Journal of Food Engineering. This trend could be linked to the alignment of these journals' aims and scopes with the emerging field of food printing research and may reflect a preference among authors and readers for such studies. Nonetheless, various underlying factors could play a role in this trend, many of which are complex to establish.

In the fields of additive manufacturing and 4D printing of foods, Food Hydrocolloids (journal, published by Elsevier) holds the first place in terms of the number of documents published in this field. Following this, different other food-specific journals are listed (Figure 10.4).

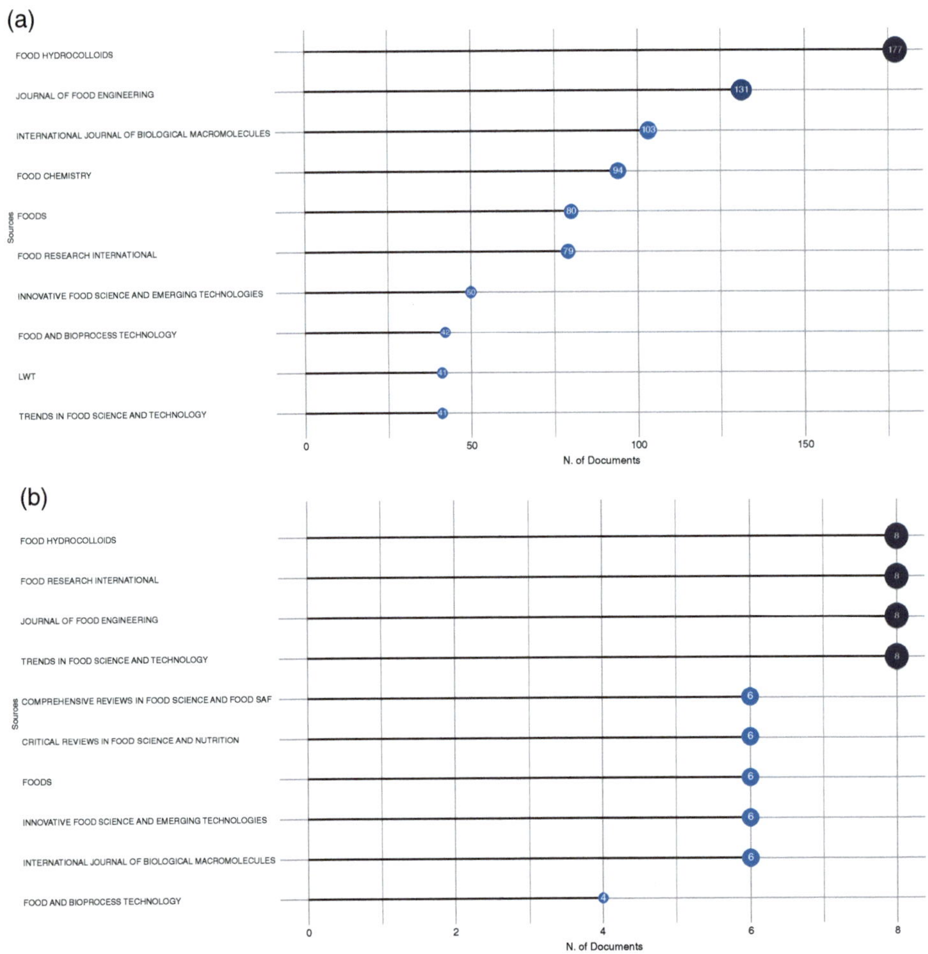

Figure 10.4 Most relevant journals based on their published documents. (a) All additive manufacturing techniques for food application, (b) 4D printing of foods.

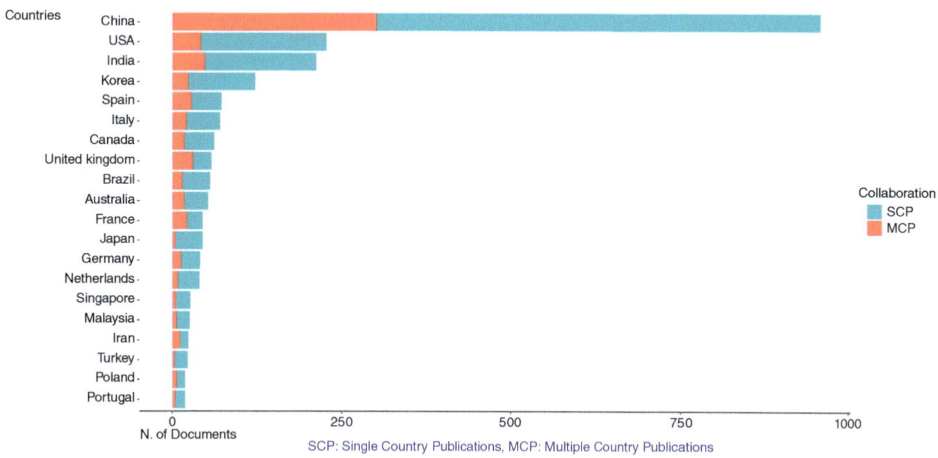

Figure 10.5 Country wise publication trends in the field of additive manufacturing with a high volume of publications from corresponding authors.

Many publications in food printing can be linked to specific corresponding authors and countries, indicating strong research activity in these regions. To present the observations clearly, Figure 10.5 explains the concentrated efforts and resources invested by leading countries in advancing food 3D/4D printing and additive manufacturing technologies. A careful analysis of their publications can provide a better understanding of the types of food printers used, the kinds of applications for which they have conducted specific research, and the most popular foods being considered for such applications. In short, it is evident that only a handful of food printers used have been fabricated in-house based on specific requirements. Although this is the current trend, with a widening application range in the food industry and the creation of customized regional diets/foods with food printing technologies to meet other specific requirements, it is expected that current trends will be significantly altered.

Similarly, Figure 10.6a shows a ranking of the top 10 authors in the field of additive manufacturing and 4D food printing, as determined by their publication count in the field. Figure 10.6b highlights the author's contributions to food printing. These visuals illustrate the author's influence on the food printing landscape and their ongoing contributions to the field. It is also evident that every year, several new research groups from different (new) countries have shown interest in research and development in the field of food printing. With gaining applications at the commercial scale and the increasing need for novel approaches and troubleshooting interventions, the upcoming decade is expected to witness a surge in scientific focus in the field of food printing.

It is evident that Jiangnan University, China, contributed to the maximum number of peer-reviewed articles in the field. In this regard, Figure 10.7 illustrates the top 20 institutions/organizations worldwide, considering their publication count in additive manufacturing and 4D printing of foods. These can be directly linked with the levels of research funding being allocated for R&D in the field of food printing in different countries and

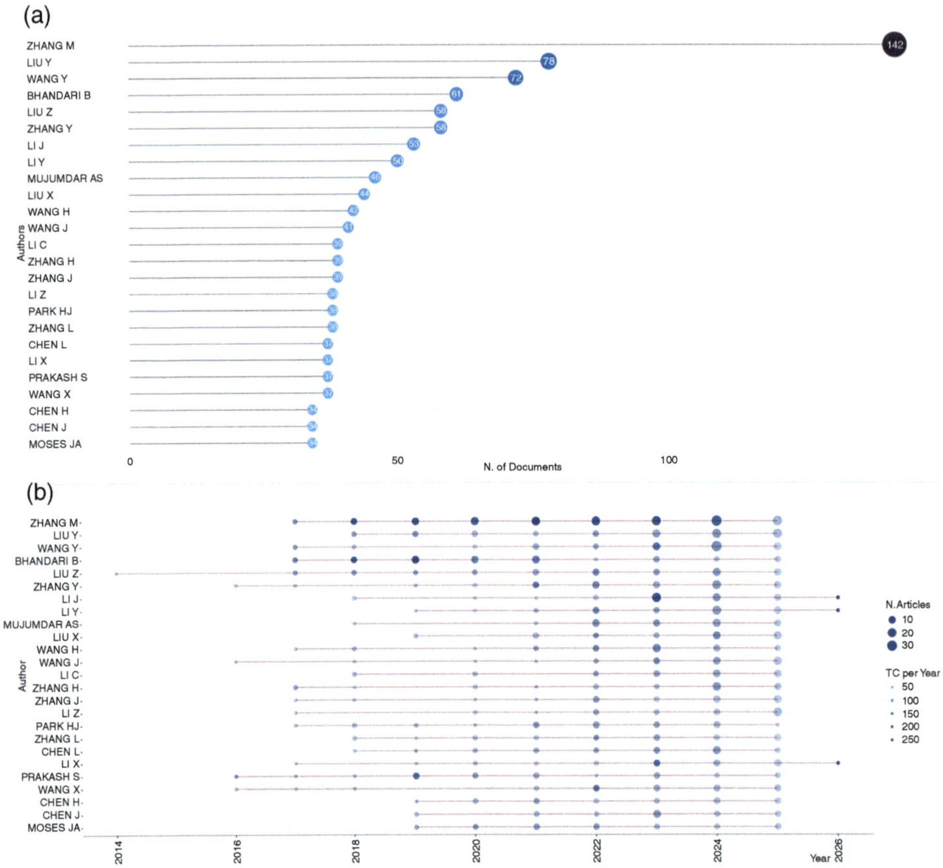

Figure 10.6 (a) Publication counts of the popular researchers/authors in the field of food printing technologies, (b) authors' publication production over the years in the field of additive manufacturing of food.

their institutions/organizations. Further, Figure 10.8 illustrates the most frequently cited publications in this domain, signifying the most substantial scientific contributions in additive manufacturing and 4D printing of foods.

Intriguing word map graphics depict the progression of additive manufacturing in the culinary sector. Figure 10.9a illustrates a treemap diagram that depicts the author's keywords, indicating the frequency of each term utilized in their works and the percentage of utilization of selected terms that constitute the treemap. This is a graphic depiction of keyword frequency and trends within the realm of food printing. The graphic emphasizes specific subjects, like material characterization via rheological studies and investigations into specific materials, such as chitosan biomaterials, beyond the general concepts of "3D food printing" and "additive manufacturing." It additionally encompasses novel methodologies, such as cultured meat and 4D printing.

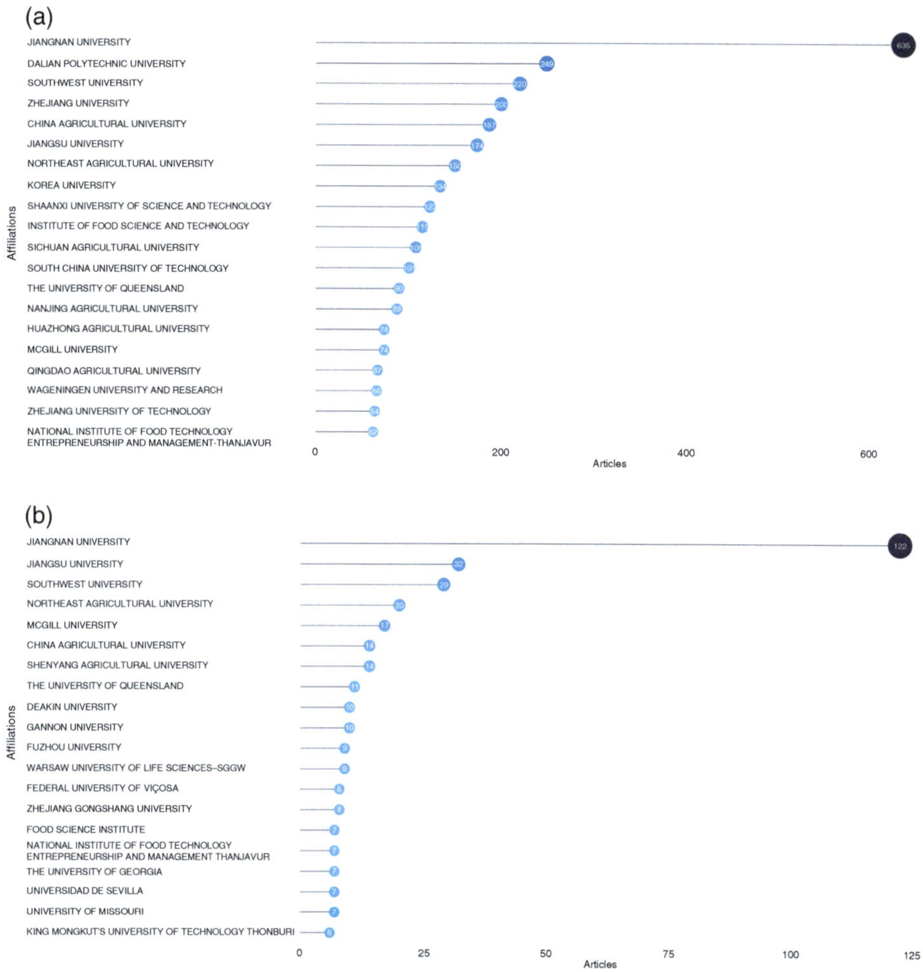

Figure 10.7 Most relevant affiliations contributing to peer-reviewed publications in the field. (a) All additive manufacturing techniques for food application, (b) 4D printing of foods.

Subsequently, Figure 10.9b displays a network map that associates numerous phrases pertinent to 3D food printing, elucidating the interrelations across diverse study topics and applications within this domain. Comparable representations, such as treemaps and word network maps, are presented for 4D printing in food, as illustrated in Figure 10.9c and d.

Moreover, the thematic map in Figure 10.10 illustrates an analysis performed using R Studio, focusing solely on 4D food printing. The map is segmented into four quadrants, each denoting distinct research themes. Quadrant 1, titled "*Niche Themes*," encompasses nascent topics such as tissue engineering, printing characteristics, 5D printing, and stimuli-responsive polymers, signifying highly innovative domains.

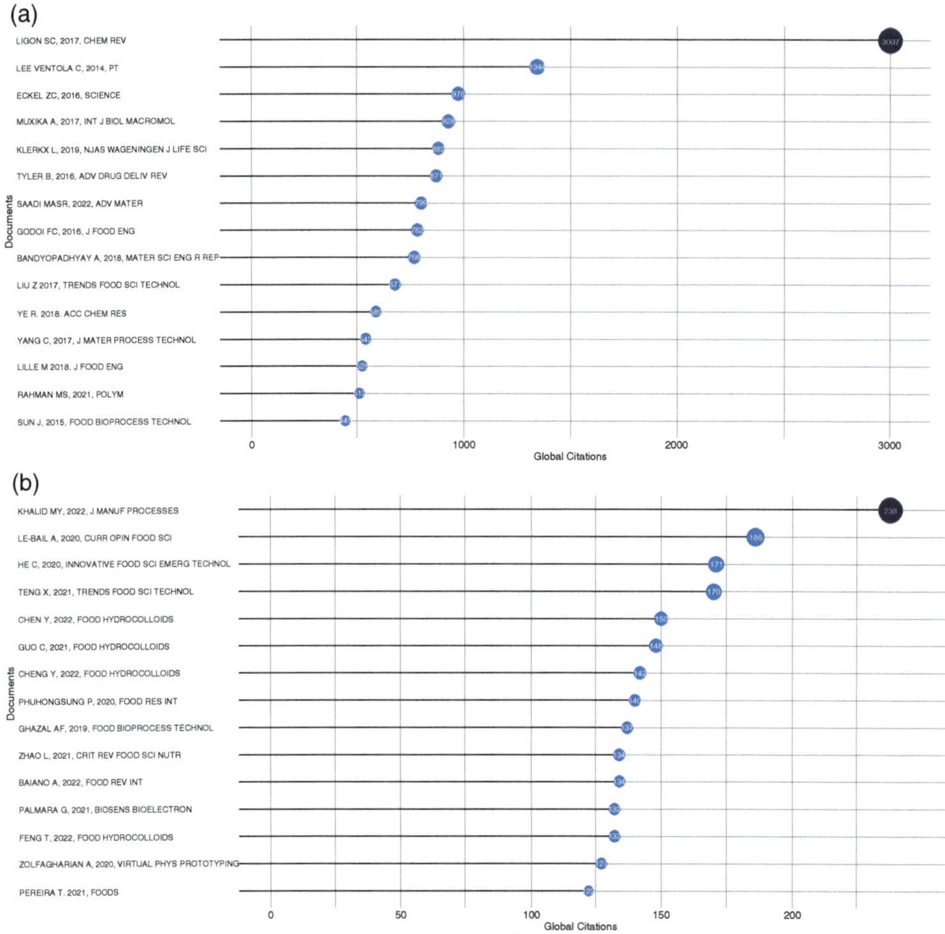

Figure 10.8 Highly cited documents and their journals. (a) All additive manufacturing techniques for food application, (b) 4D printing of foods.

Quadrant 2, "*Motor Themes*," highlights pivotal topics with growth potential, including 4D food printing, 5D, and 6D printing, as well as edible inks crucial areas for advancing additive manufacturing in food. Quadrant 3, "*Basic Themes*," comprises well-explored topics with further development prospects, such as dehydration in 3D food printing and printability and 4D printing (non-food). Lastly, Quadrant 4, "*Emerging/Declining Themes*," presents topics that may be saturated or present diversification opportunities for future interests.

It is important to note that any bibliometric analysis would have its own merits and constraints. For instance, search results may be heavily dependent on the search keywords or the search database. Therefore, to present a broader understanding of the field and current trends, peer-reviewed articles are listed in Table 10.2, presenting different other bibliometric analyses. Most of these reports focus on 3D food printing.

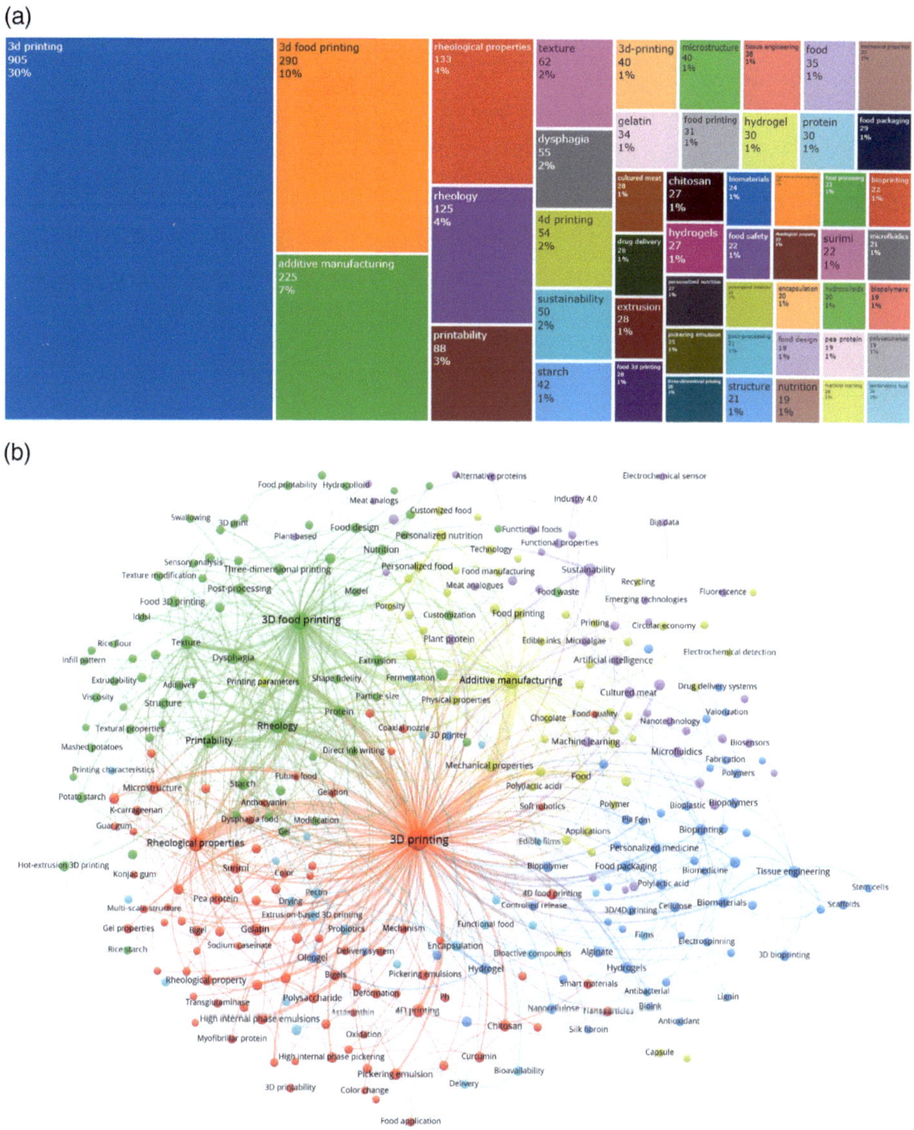

Figure 10.9 Authors' keyword map in additive manufacturing of food: (a) TreeMap (words and values in each box in the order of *author keyword, documents used this keyword, usage of words in percentage among mentioned keywords in treemap*), (b) network visualization of keywords; Authors' keyword map in 4D printing of foods, (c) TreeMap (words and values in each box in the order of *author keyword, documents used this keyword, usage of words in percentage among mentioned keywords in treemap*), (d) network visualization of keywords.

(c)

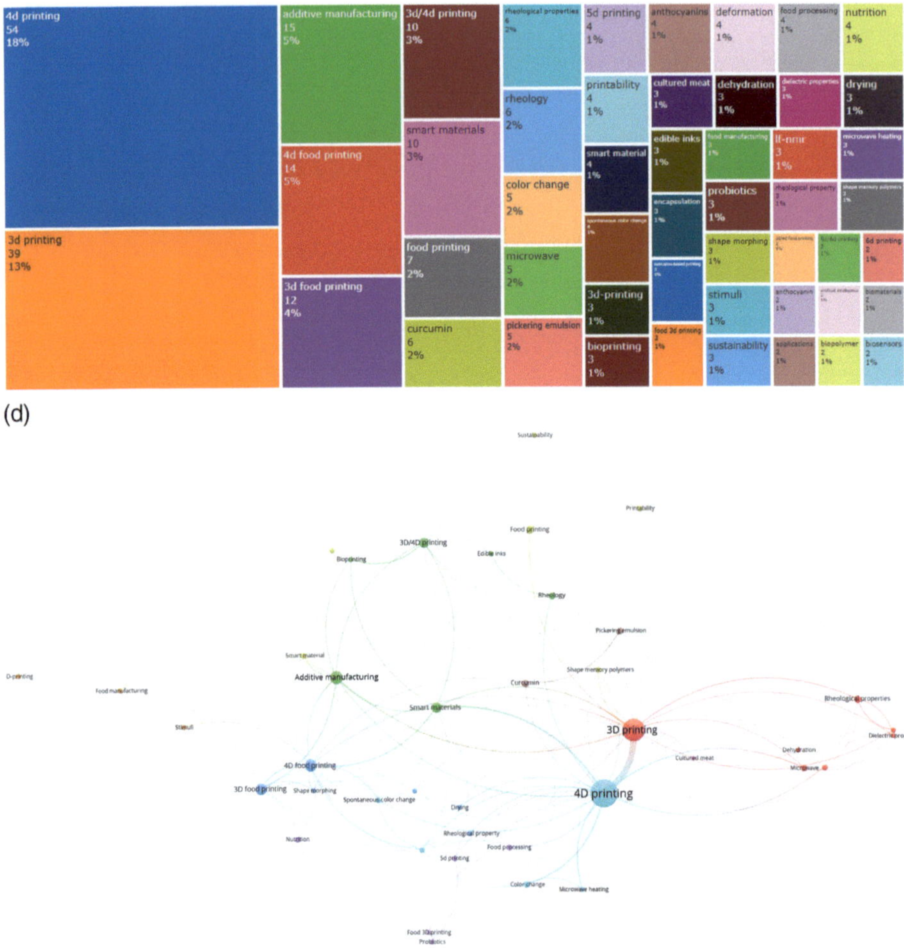

(d)

Figure 10.9 (Continued)

10.2.2 Patents

The application of 3D printing in food manufacturing has garnered the interest of inventors and prospective users, resulting in substantial progress since the initial patents were issued. Early applications primarily focused on its potential for cakes. Notably, patent USD 6,280,784 B1 specifies a method for leveraging 3D printing to manufacture delicacies with intricate structures, such as customized birthday cakes. The food formulation included a volatile component to preserve its liquid consistency during dispensing, which is swiftly eliminated post application. Once the volatile ingredient is eliminated, the formulation solidifies, creating a rigid structure capable of supporting additional layers. In recent years, there has been a proliferation of discoveries pertaining to novel food products, such as fortified rice, and breakthroughs in printing technology, with several developments recorded in patent offices. These patents demonstrate

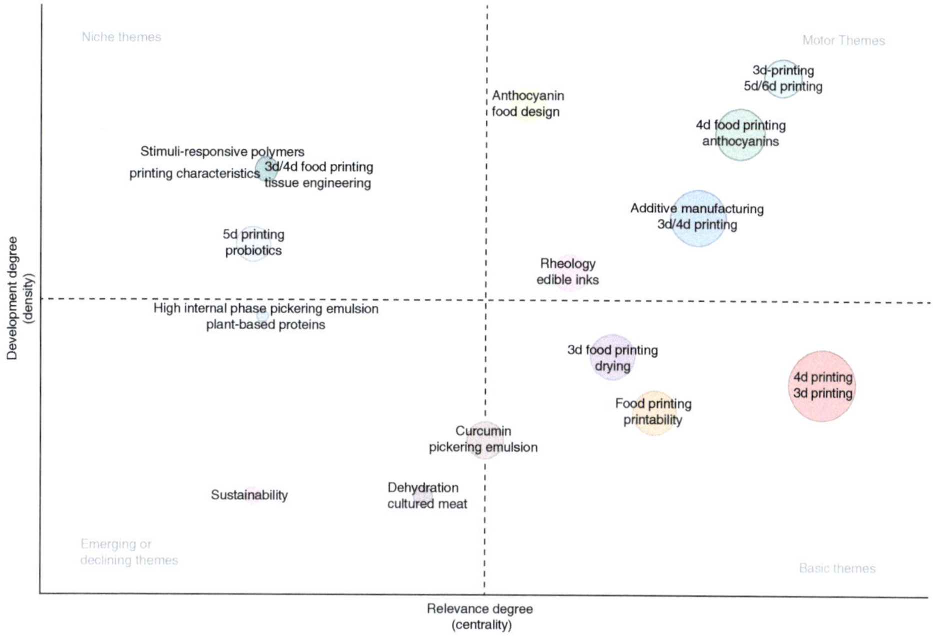

Figure 10.10 Thematic map – 4D printing of foods (*x*-axis and *y*-axis indicate the degree of impact and importance; the size of the circle indicates the frequency of the clusters).

Table 10.2 Further reading – bibliometric analyses on 3D food printing landscape.

Year of publishing	Title of the article	References
2024	3D-printed foods for dysphagia: A bibliometric review	Shao et al. (2024)
2024	Additive manufacturing in space food: A bibliometric review	Scatto et al. (2024)
2024	Scientometrics and visualized analysis of 3D food printing: A boost to future food customized development	Wang et al. (2024)
2024	Innovating gastronomy through information technology: A bibliometric analysis of 3D food printing for present and future research	Elshaer et al. (2024)
2024	Bibliometric mapping analysis of Pickering emulsion applied in 3D food printing	de Farias et al. (2024)
2024	Research landscape and trending topics on 3D food printing – a bibliometric review	Bi et al. (2024)
2023	3D Printing, insects and food: A bibliometric analysis	Singh (2023)
2022	A bibliometric analysis of 3D food printing research: A global and African perspective	Fasogbon & Adebo (2022)
2021	Drawing the scientific landscape of 3D food printing. Maps and interpretation of the global information in the first 13 years of detailed experiments, from 2007 to 2020	Derossi et al. (2021)

innovative techniques and technology for creating superior food products and customized 3D printers capable of processing various materials and compositions. These developments underscore the increasing convergence of food science and additive manufacturing, facilitating individualized nutrition and tailored food solutions. Recently, Baiano (2022) reported the most significant patents related to 3D food printing, published globally since 2013. These patents encompass several elements, including 3D food printers, printing techniques, and materials.

10.2.2.1 Search Criteria and Database

A patent search was conducted on the World Intellectual Property Organization (WIPO) website (www.wipo.int) using the keyword search [EN_ALLTXT: "3D food printing"] on July 5, 2025, considering all offices worldwide.

10.2.2.2 Search Results

Results are summarized based on the data listed on the WIPO website. Figure 10.11 summarizes the number of patents generated over the years; overall, 732 patents have been published related to the 3D printing of food from around the globe. However, in the field of 4D printing of foods, to date, 17 patents have been published, from China (12), USA (2), India (1), and Japan (1). In terms of trademarks, search results did not reveal any trademark in this field, but five entries have been published in the design database. In recent times, the production of new printers has been booming in this field, thus can be seen from the visuals in terms of food printing evolution.

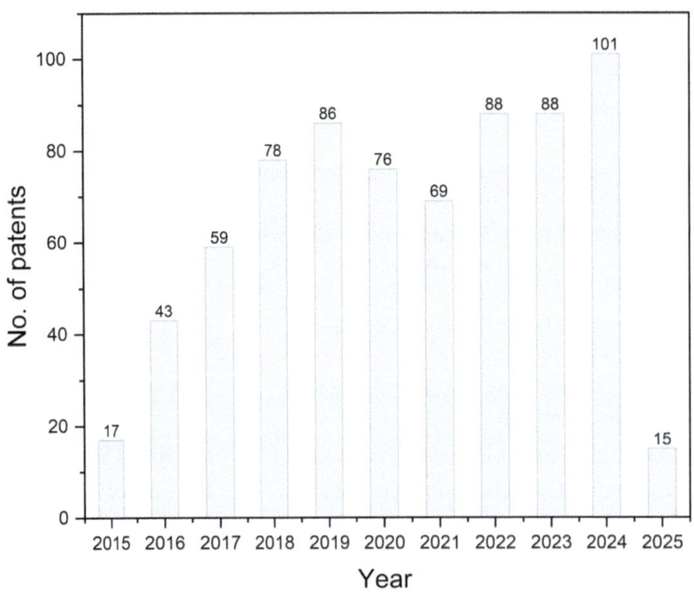

Figure 10.11 Number of patents published in the field of food printing technologies.

10.3 Research Funding for Food Printing Technologies

Significant economic prospects and substantial investment in food printing have advanced in recent years. The commercial value of 3D food printing is projected to reach $42.5 million by 2025, with a compound annual growth rate (CAGR) of around 54.75% in 2018. Reports have highlighted the CAGR for the periods 2016–2026, 2017–2024, 2020–2025, and 2021–2028, yielding values of around 20.21%, 50%, 16.1%, and 52.30%, respectively. The industry could attain $1 billion by 2027 in view of its potential benefits. Key entities contemplating utilizing food 3D printers include restaurants, aerospace firms, confectioneries, bakeries, and healthcare facilities (Neamah & Tandio, 2024).

Over 100 agencies have been recognized as financiers of 3D food printing research. The National Natural Science Foundation of China has the highest number of reported publications, followed by the National Key Research and Development Program of China. This trend corresponds with the data presented in Figure 10.12, indicating that China and Jiangnan University, in particular, are the foremost contributors to 3D food printing-related publications. Certainly, governments retain a vital responsibility in advancing local manufacturing, acceptance, and commercialization of 3D printing technology and products (Kolade et al., 2022), hence propelling the growth of the food printing sector. Substantial investments will be crucial for advancing this progress and resulting in transformative innovations.

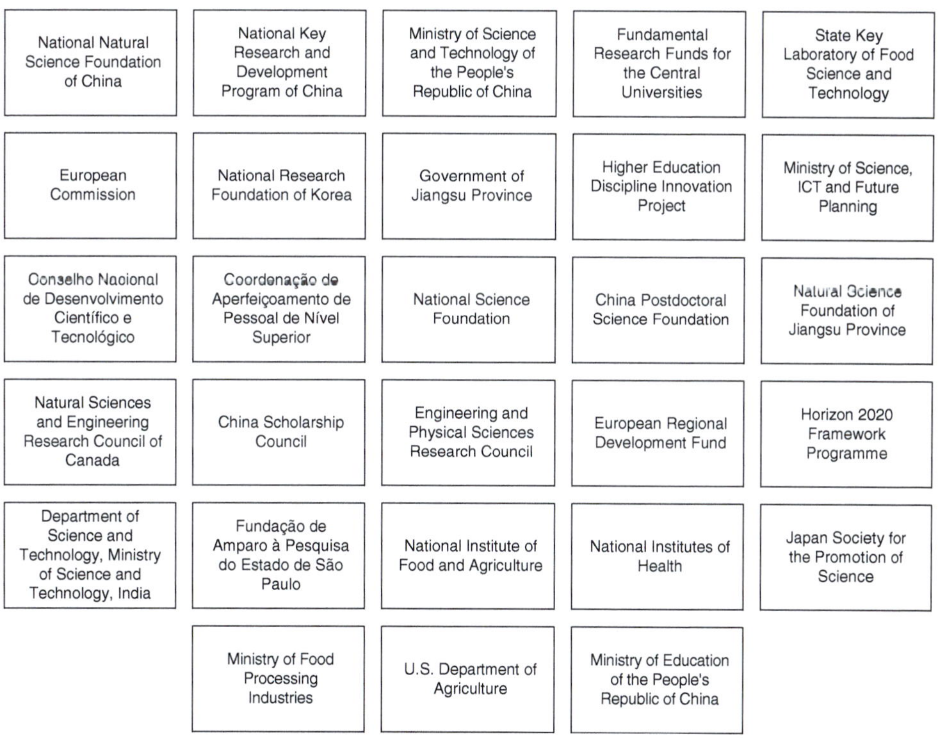

Figure 10.12 Top 20 funding agencies based on published documents acknowledging them.

Recently, the delivery of nutraceuticals and micronutrients has gained traction in additive food manufacturing. Prominent entities in the food sector, such as Barilla (Italy) and Nestlé (Switzerland), have collaborated with makers of 3D food printers to derive polymers from food products designated for printing, underscoring their enthusiasm for 3D food printing and its market viability. Global customers exhibit varied culinary preferences and utilize unique ingredients, challenging 3D food printers to meet regional demands, particularly in Asia and Africa. This necessitates the creation of printers that integrate local components and preferences while also guaranteeing the availability of ingredients for the production of varied cuisines (Market Research Database, 2022). A further problem is that not all food kinds are responsive to 3D printing; meals that lack adequate structural integrity may be incompatible with this procedure. Therefore, in addition to forming foods, these printers must advance to produce a variety of textures, emulating both conventional and innovative food consistencies.

10.4 Current Commercialization Status and Globalization of Food Printing Technologies

Rapid advancements are being made in the field of food printing, combining culinary science with cutting-edge technologies. The development of printing innovations, such as the invention of multi-material/head/nozzle combinations, has made it possible to produce complex, multisensorial meals with increased printing precision and speed (Fujiwara et al., 2025). This is reinforced by improvements in ingredient formulations, which include sustainable and nutritious materials. On the commercial front, 3D food printing is being utilized to create eating experiences that are both gourmet and individualized (Derossi et al., 2024; Sohel et al., 2025). This technology is being utilized by a multitude of businesses in order to personalize meals to meet the dietary requirements and tastes of certain individuals. Additionally, high-end culinary artists use it to create sophisticated and visually attractive food presentations. With these improvements, the food sector is experiencing a surge in both the creative potential and the practicality of food printing applications (Rinshana et al., 2025).

The advancement of digital technologies, electronics, and communication has rendered 3D model designs more accessible, facilitating the easy sharing of concepts among designers, food technologists, and culinary specialists. The rapid downloading of source files has rendered food printing according to individual tastes more viable. Advanced communication technologies enable the aggregation of culinary expertise and artistic abilities from diverse professions on integrated internet platforms, hence facilitating 3D and 4D food printing technologies (Hassoun et al., 2024). Numerous websites enable professionals to post 3D design models that users can download on demand for food printing (Anandharamakrishnan et al., 2022). These improvements indicate that 3D and 4D food printing may become accessible in small and large industrial enterprises and domestic kitchens, underscoring the significance of globalizing this printing technology. This globalization may diminish time, energy, transportation, and packaging expenses.

A centralized cloud infrastructure could provide global access to 3D models, allowing users to download designs as required. In addition to optimizing the food supply chain, additive manufacturing can mitigate product failure by conforming to consumer preferences (Varvara et al., 2021). Furthermore, it serves as a prototype instrument, evaluating product appeal before expanding production. Consumers can customize and acquire 3D designs through e-commerce platforms using PC or mobile applications. Standardizing 3D printing technology to guarantee interoperability among software platforms will mitigate quality discrepancies in produced objects, even when the same designs are utilized across several printers. Emerging market trends suggest that additive manufacturing could transform food trading by facilitating online exchanges of food-printing designs. This technology has the potential to revolutionize conventional food trade methods by providing flexibility, cost-effectiveness, product diversity, and personalization. It streamlines the supply and distribution chain, reducing expenses associated with manufacturing, shipping, distribution, packaging, and warehousing; also, this technological approach contributes to SDG (Hai Alami et al., 2024).

3D- and 4D-printed foods can transform global food commerce by facilitating the localized creation of tailored items. With the advancement of internet technologies, tracking and traceability in design, process planning, manufacturing, and customer service are more connected, notwithstanding geographical distribution. The 4.0 and 5.0 Industrial Revolutions have propelled smart manufacturing, resulting in the widespread adoption of efficient and economical procedures across practically all sectors (Hassoun et al., 2024). Innovative technology combined with traditional manufacturing techniques creates chances for intelligent factories, potentially leading to the globalization of production. Implementing computer-based additive manufacturing technologies offers vast design, engineering, and manufacturing opportunities, establishing a basis to satisfy bespoke worldwide market requirements.

Notwithstanding the benefits of additive manufacturing of food, the current market potential has undergone a significant transformation with the rise of 3D printing, benefiting from three decades of experience. It is shifting toward more advanced technologies, particularly in terms of dimension and printing methods. Particularly for 4D printing, market studies reported a current value of approximately $63 million in 2019, with a substantial increase expected to $537 million by 2025. In comparison, the 3D printing market is anticipated to expand significantly, surpassing $32 billion by 2023, primarily driven by the defense and aerospace sectors, with North America leading in market share. Regionally, North America is expected to dominate the global 4D printing landscape, followed by the Asia-Pacific region.

Regarding materials, the 4D printing market is forecasted to be led by carbon fiber, which is anticipated to hold around 62% of the market share due to its exceptional mechanical, thermal, and conductive properties. Smart fabrics and wood are also expected to play significant roles due to their weight and strength characteristics. Looking at end applications, the military sector is projected to account for 25% of the 4D printing market by 2025, with aerospace following closely behind (Rastogi & Kandasubramanian, 2019). The future of 4D printing is promising, with the potential to revolutionize a range of fields, including advanced ballistics, weaponry, therapeutic and surgical applications, manufacturing, underwater warfare equipment, and enhanced educational tools.

As noted earlier, 4D food printing technology is still in the early stages of commercialization, which means that costs may be currently high compared to production output.

However, this emerging technology promotes environmental sustainability by minimizing the waste of natural resources, ultimately lowering long-term expenses and providing sustainable alternatives to existing products (Thorakkattu et al., 2025). The 4D printing market is poised for a wave of eco-friendly product launches and collaborations across the value chain, enhancing resource efficiency and promoting sustainable consumption.

10.5 Techno-feasibility and Challenges

Food printing has certain challenges that must be resolved to harness its capabilities fully. Challenges associated can be broadly categorized as consumer acceptability, economic feasibility, and regulatory and technical considerations. Consumer acceptability can pop up as an obstacle in 3D food printing, particularly in terms of 'food neophobia' (Lee et al., 2021; Tesikova et al., 2022). The consumer acceptance of 3D printed foods was detailed in the previous chapter. Regarding food safety, it is crucial to guarantee that all materials used are of food-grade quality, and that the printing process strictly follows well-defined hygiene standards to manufacture safe and quality products. An inherent challenge is the compatibility of ingredients; not all food items are available in a printable form, and attaining the appropriate uniformity and performance during printing is often a task. Likewise, preserving the texture and flavor of the printed food may be challenging due to the difficulties in controlling the quality of the end product using existing approaches. Printer constraints also present a challenge, as many current models suffer difficulties in speed and size, impeding accurate and efficient printing of larger or more complex objects; these factors are considered technical constraints associated with additive food manufacturing.

Moreover, it is essential to handle and store the ingredients correctly to prevent contamination and guarantee uniformity, introducing additional complications. The technical challenge covers all the material selection and characterization, particularly rheology complexities, post-printing quality and stability (Cheng et al., 2022). Besides, these common challenges also vary depending on their typical applications; for example, space food 3D printing faces more challenges than land-based 3D printing. This can be attributed to the microgravity conditions, the impact of cosmic radiation, and other such conditions (Santhoshkumar et al., 2023). Overall, many researchers have extensively reviewed the challenges associated with 3D printing (Agunbiade et al., 2022; Zhu et al., 2023). Nevertheless, it is interesting to note that several current studies focus on achieving large-scale industrial production capabilities, and the recent past has witnessed promising breakthroughs.

Advancements in 4D printing hold great promise for its application across various industries, particularly in food manufacturing. To ensure a successful future for this technology, it is crucial to develop innovative and customizable materials that can respond to different environmental stimuli for effective shape transformation. Additionally, there is a need for novel 4D printing software tailored to diverse methodologies. This software must consider the specific smart materials, printing techniques, geometrical and structural requirements, and the inherent shape-altering mechanisms involved in this printing. Also, 4D printing techniques should evolve to accommodate a wider range of materials rather than being limited to a single type. The potential for this printing is significant within the field of biomedical engineering, but considerable effort is required to develop faster and more cost-effective bioprinting methods.

10.5.1 Scalability

The potential for extensive production applications in additive food manufacturing is becoming progressively favorable as technology and industry requirements advance. This technique is particularly attractive for personalized nutrition, enabling the customization of products for distinct dietary requirements, age demographics, or health concerns, hence presenting significant prospects in the health and wellness industries. Manufacturers can control ingredient proportions and deposition precisely, ensuring the use of only necessary amounts, hence eliminating excess and spoiling. The method is compatible with alternative and sustainable ingredients such as insect proteins, algae, and lab-cultivated meat, reflecting the industry's transition towards environmentally conscious operations (Reuben et al., 2025). In large-scale applications, 3D printing facilitates food manufacturing by providing flexibility and responsiveness to demand, simplifying product lines' adaptation, introducing new items, and rapidly testing market prototypes, as setup periods and reconfiguration are generally lower than conventional production lines. Moreover, the automation and uniformity provided by 3D food printing may address labor shortages and enhance consistency in food production. In high-volume manufacturing environments, the meticulous regulation of ingredients, structure, and texture afforded by additive manufacturing can improve product quality and consistency, guaranteeing that each item adheres to stringent criteria.

On January 18, 2024, Austrian startup Revo Foods unveiled a novel 3D food printing technology intended for industrial use, demonstrated via a video release. This advanced technology incorporates a distinctive multinozzle system that facilitates uninterrupted food production, marking the inaugural feasible approach for mass manufacturing items through 3D food printing (Revo Foods, 2024). Similarly, many companies are interested in adapting additive manufacturing for food manufacturing, for instance, US-based companies 3D Systems, Modern Meadow, Materials Research Cooperation, BeeHex, Netherlands-based companies By flow, TNO, UK-based companies Nu Food, and Choc Edge. These developments mark a significant milestone in additive manufacturing for food, paving the way for large-scale production capabilities previously unattainable in 3D food printing. Nonetheless, for widespread adoption, improvements in speed, multimaterial and multidimensional printing capabilities, and cost-effectiveness are essential to rival conventional production. Currently, it is evident in most food printing laboratory studies that printing speeds and overall outputs are extremely low. Although this can directly link with the kind of research funding available, as these problems get resolved, the sector will witness the emergence of additive manufacturing as a feasible approach for the large-scale production of customized, sustainable, high-quality food products.

10.5.2 Piracy and Counterfeiting

Piracy in printing software presents a complex danger to the nascent business; interlinking intellectual property concerns, economic difficulties, and food printing safety and quality issues occur. For instance, with the increasing popularity of 3D food printers, corporations are investing in proprietary software to manage design, ingredient compatibility, temperature, and extrusion rates, thereby ensuring quality and safety. Nonetheless, piracy compromises this by providing illicit access to confidential data and proprietary algorithms, eroding organizations' competitive edge and discouraging innovation (Nachal et al., 2019). In the context of additive food manufacturing, legitimate software incorporates essential food safety processes, which pirated versions may omit, posing significant health concerns and noncompliance with regulatory standards set by international or national organizations (Zhu et al., 2023). Moreover, piracy leads to substantial revenue depletion, particularly for smaller firms that depend on software license payments for expansion. This financial pressure can hinder industry advancement and diminish consumer confidence if items are considered dangerous or inferior. Further, pirated software frequently experiences compatibility problems, which may lead to hardware failures or subpar print quality due to inadequate synchronization with particular printer models. Cybersecurity issues exacerbate the issue, as pirated versions sometimes lack essential upgrades, making users susceptible to cyberattacks and data breaches, particularly when the devices are interconnected. To mitigate these dangers, organizations in the additive manufacturing sector might implement licensing standards, encrypted access, and stringent authentication measures to safeguard their software.

10.5.3 Others

The challenges associated with 3D and 4D food printing are not limited to technical and scalability issues; they also encompass consumer acceptance and the adoption of new technology. The widespread adoption of 3D and 4D food printing may be impeded by the necessity of specialized training and skilled professionals to operate and maintain the equipment from a technological perspective. Furthermore, the technology's scalability is restricted by the high production costs, including printers, materials, and energy utilization, which render it less accessible to small businesses and startups. The lack of standardized regulations for additive manufacturing in the food industry further complicates global adoption, creating uncertainty around food safety, quality standards, and compliance (Santhoshkumar et al., 2023). In consumer acceptability, 3D and 4D food printing encounter obstacles in overcoming perceptions of artificiality and concerns regarding the naturalness of printed foods. Instilling trust and interest in consumers are imperative to educate them on the advantages of resource efficiency, sustainability, and customization. While navigating the complexities of integrating 3D and 4D printing into their existing supply chains, established food companies must adjust their marketing, branding, and pricing strategies to align with consumer preferences for personalized, ecofriendly, and innovative products (Rinshana et al., 2025). In a market where distinctive designs and formulas are critical differentiators, intellectual property theft and concerns regarding product authenticity also present risks. In order to accomplish widespread commercialization and consumer acceptance, additional research is required to develop clear regulatory frameworks, streamlined supply, distribution networks, and efficient business models. To successfully address these challenges, it will be necessary to collaborate across industries to reduce costs, improve consumer education, and cultivate confidence in the safety, quality, and sustainability of 3D and 4D printed foods. This will ultimately facilitate their integration into mainstream food production and consumption.

10.6 Conclusion

Food printing technologies are an emerging area of research focused on developing personalized, nutritious, and appealing food products. This chapter highlights the scientific contributions in this field, including key publications, prominent institutions, influential journals, patents, and leading researchers based on their collaborative networks and citation impact based on bibliometric and patent analyses from standard internationally recognized databases. It is evident that research funding is an underscored critical factor in advancing this research area, suggesting the significance of prioritizing funding for institutions and researchers to support high-quality studies that can drive transformative change on the continent. This chapter concludes with a presentation on current challenges, highlighting avenues for research and futuristic development. Nonetheless, it is evident that the benefits and the potential of the technology outweigh the current limitations, and the technology has promising prospects for the future.

References

Agunbiade, A. O., Song, L., Agunbiade, O. J., Ofoedu, C. E., Chacha, J. S., Duguma, H. T., Hossaini, S. M., Rasaq, W. A., Shorstkii, I., Osuji, C. M., Owuamanam, C. I., Okpala, C. O. R., Korzeniowska, M., & Guine, R. P. F. (2022). Potentials of 3D extrusion-based printing in resolving food processing challenges: A perspective review. *Journal of Food Process Engineering*, *45*(4). doi:10.1111/jfpe.13996

Anandharamakrishnan, C., Moses, J. A., & Anukiruthika, T. (2022). *3D printing of foods*. Wiley.

Anukiruthika, T., Moses, J. A., & Anandharamakrishnan, C. (2020). 3D printing of egg yolk and white with rice flour blends. *Journal of Food Engineering*, *265*. doi:10.1016/j.jfoodeng.2019.109691

Aria, M., & Cuccurullo, C. (2017). Bibliometrix: An R-tool for comprehensive science mapping analysis. *Journal of Informetrics*, *11*(4), 959–975. doi:10.1016/j.joi.2017.08.007

Baiano, A. (2022). 3D printed foods: A comprehensive review on technologies, nutritional value, safety, consumer attitude, regulatory framework, and economic and sustainability issues. *Food Reviews International*, *38*(5), 986–1016. doi:10.1080/87559129.2020.1762091

Bi, S., Pi, J., Chen, H., Zhou, Y., Liu, R., Chen, Y., Che, Q., Li, W., Gu, J., & Zhang, Y. (2024). Research landscape and trending topics on 3D food printing – a bibliometric review. *Rapid Prototyping Journal*, *30*(8), 1693–1706. doi:10.1108/RPJ-02-2024-0098

Cheng, Y., Fu, Y., Ma, L., Yap, P. L., Losic, D., Wang, H., & Zhang, Y. (2022). Rheology of edible food inks from 2D/3D/4D printing, and its role in future 5D/6D printing. *Food Hydrocolloids*, *132*, 107855. doi:10.1016/j.foodhyd.2022.107855

da Silva, T. H. H., & Sehnem, S. (2025). The utilization of Industry 4.0 technologies to enhance the circular economy through the engagement of stakeholders in Brazilian food technology companies. *Business Strategy and the Environment*, *34*(1), 129–161. doi:10.1002/bse.3954

de Farias, P. M., Matheus, J. R. V., Maniglia, B. C., Le-Bail, P., Le-Bail, A., Schmid, M., & Fai, A. E. C. (2024). Bibliometric mapping analysis of Pickering emulsion applied in 3D food printing. *International Journal of Food Science & Technology*, *59*(4), 2186–2196. doi:10.1111/ijfs.17040

Derossi, A., Caporizzi, R., Paolillo, M., Oral, M. O., & Severini, C. (2021). Drawing the scientific landscape of 3D Food Printing. Maps and interpretation of the global information in the first 13 years of detailed experiments, from 2007 to 2020. *Innovative Food Science & Emerging Technologies*, *70*, 102689. doi:10.1016/j.ifset.2021.102689

Derossi, A., Spence, C., Corradini, M. G., Jekle, M., Fahmy, A. R., Caporizzi, R., Devahastin, S., Moses, J. A., Le-Bail, A., Zhou, W., Zhang, M., Bhandari, B., & Severini, C. (2024). Personalized, digitally designed 3D printed food towards the reshaping of food manufacturing and consumption. *NPJ Science of Food*, *8*(1), 54. doi:10.1038/s41538-024-00296-5

Elshaer, I. A., Azazz, A. M. S., Zayed, M. A., Ameen, F. A., Fayyad, S., Fouad, A. M., Khalil, N. I., & Fathy, E. A. (2024). Innovating gastronomy through information technology: A bibliometric analysis of 3D food printing for present and future research. *Information*, *15*(8), 471. doi:10.3390/info15080471

Fasogbon, B. M., & Adebo, O. A. (2022). A bibliometric analysis of 3D food printing research: A global and African perspective. *Future Foods*, *6*, 100175. doi:10.1016/j.fufo.2022.100175

Fujiwara, K., Suzuki, Y., Toba, K., Ogawa, J., Furukawa, H., Hashizume, M., Noji, T., Teratani, K., & Ito, N. (2025). Multifunctional 3D food printer with quad-screw nozzle for four-color printing and dual ink mixing. *Journal of Food Engineering*, *388*, 112384. doi:10.1016/j.jfoodeng.2024.112384

Ghazal, A. F., Zhang, M., Mujumdar, A. S., & Ghamry, M.(2023). Progress in 4D/5D/6D printing of foods: applications and R&;D opportunities. *Critical Reviews in Food Science and Nutrition*, *63*(25), 7399–7422. doi:10.1080/10408398.2022.2045896

Hai Alami, A., Ghani Olabi, A., Khuri, S., Aljaghoub, H., Alasad, S., Ramadan, M., & Ali Abdelkareem, M. (2024). 3D printing in the food industry: Recent progress and role in achieving sustainable development goals. *Ain Shams Engineering Journal*, *15*(2), 102386. doi:10.1016/j.asej.2023.102386

Hassoun, A., Jagtap, S., Trollman, H., Garcia-Garcia, G., Duong, L. N. K., Saxena, P., Bouzembrak, Y., Treiblmaier, H., Para-López, C., Carmona-Torres, C., Dev, K., Mhlanga, D., & Aït-Kaddour, A. (2024). From Food Industry 4.0 to Food Industry 5.0: Identifying technological enablers and potential future applications in the food sector. *Comprehensive Reviews in Food Science and Food Safety*, *23*(6). doi:10.1111/1541-4337.70040

Hussain, S., Malakar, S., & Arora, V. K. (2022). Extrusion-based 3D food printing: Technological approaches, material characteristics, printing stability, and post-processing. *Food Engineering Reviews*, *14*(1), 100–119. doi:10.1007/s12393-021-09293-w

Keerthy, M. A., Dhanaselvam, K. R., Santhoshkumar, P., Ravikrishnan, V., & Moses, J. A. (2025). Improving the printability of watermelon rind using pectin: 3D printing optimization for the development of gummies. *Food Biomacromolecules*, *2*(2), 143–156. doi:10.1002/fob2.70008

Kolade, O., Adegbile, A., & Sarpong, D. (2022). Can university-industry-government collaborations drive a 3D printing revolution in Africa? A triple helix model of technological leapfrogging in additive manufacturing. *Technology in Society*, *69*, 101960. doi:10.1016/j.techsoc.2022.101960

Lee, K. H., Hwang, K. H., Kim, M., & Cho, M. (2021). 3D printed food attributes and their roles within the value-attitude-behavior model: Moderating effects of food neophobia and food technology neophobia. *Journal of Hospitality and Tourism Management*, *48*, 46–54. doi:10.1016/j.jhtm.2021.05.013

Market Research database. (2022). *3D food printing market research report*. https://www.marketsandmarkets.com/Market-Reports/3d-food-printing-market-267692011.html

Nachal, N., Moses, J. A., Karthik, P., & Anandharamakrishnan, C. (2019). Applications of 3D printing in food processing. *Food Engineering Reviews*, *11*(3), 123–141. doi:10.1007/s12393-019-09199-8

Neamah, H. A., & Tandio, J. (2024). Towards the development of foods 3D printer: Trends and technologies for foods printing. *Heliyon*, *10*(13), e33882. doi:10.1016/j.heliyon.2024.e33882

Rastogi, P., & Kandasubramanian, B. (2019). Breakthrough in the printing tactics for stimuli-responsive materials: 4D printing. *Chemical Engineering Journal*, *366*, 264–304. doi:10.1016/j.cej.2019.02.085

Reuben, S. J., Chandra, S., Praveena, B., Santhoshkumar, P., & Moses, J. A. (2025). 3D printing of alternative proteins: Approaches, challenges and advances. *Sustainable Materials and Technologies*, *43*, e01285. doi:10.1016/j.susmat.2025.e01285

Revo Foods. (2024). *3D food printing on an industrial scale enables a new food segment*. Revo Foods. https://revo-foods_1.prowly.com/286249-3d-food-printing-on-an-industrial-scale-enables-a-new-food-segment

Rinshana, P. F., Murugesan, B., Kim, Y. H., Alaguthevar, R., & Rhim, J.-W. (2025). Advances in 3D food printing technology: Innovation and applications in the food industry. *Food Science and Biotechnology*, *34*(2), 403–421. doi:10.1007/s10068-024-01779-7

Santhoshkumar, P., & Moses, J. (2025). Impact of konjac glucomannan on hot extrusion 3D printability of cake gel. *International Journal of Biological Macromolecules*, *305*, 141162. doi:10.1016/j.ijbiomac.2025.141162

Santhoshkumar, P., Negi, A., & Moses, J. A. (2023). 3D printing for space food applications: Advancements, challenges, and prospects. *Life Sciences in Space Research*, *40*. doi:10.1016/j.lssr.2023.08.002

Santhoshkumar, P., Raja, V., Priyadarshini, S. R., & Moses, J. A. (2024). Evaluating the 3D printability of pearl millet flour with banana pulp blends. *Journal of the Science of Food and Agriculture*, *104*(9). doi:10.1002/jsfa.13389

Santhoshkumar, P., Swathika, B. S., Negi, A., & Moses, J. A. (2025). Improving gel performance and hot extrusion three-dimensional printability of blueberry gel with guar gum. *Food Biophysics*, *20*(2), 1–15. doi:10.1007/s11483-025-09964-9

Scatto, M., Dolci, V., Venturini, K., Pini, B., Petroni, A., & Bigliardi, B. (2024). Additive manufacturing in space food: A bibliometric review. *Procedia Computer Science*, *253*, 2498–2507. doi:10.1016/j.procs.2025.01.310

Shahbazi, M., Jäger, H., Ettelaie, R., Chen, J., Mohammadi, A., & Kashi, P. A. (2024). 4D printing of poly-Pickering high internal phase emulsions to assembly a thermo-responsive shape-memory hierarchical macroporous scaffold. *Additive Manufacturing*, *81*, 103977. doi:10.1016/j.addma.2024.103977

Shao, J., Zheng, Z., Hu, J., Sriboonvorakul, N., & Lin, S. (2024). 3D-printed foods for dysphagia: a bibliometric review. *Foods*, *14*(12), 2058. doi:10.3390/foods14122058

Singh, D. (2023). 3D printing, insects and food: A bibliometric analysis. In D. Singh, R. Kumar, S. Singh, & S. Ramniwas (Eds.), *3D printing of sustainable insect materials* (pp. 123–134). Springer. doi:10.1007/978-3-031-25994-4_8

Sohel, A., Sahu, S., Mitchell, G. R., & Patel, M. K. (2025). 3D food printing: A comprehensive review and critical analysis on technologies, food materials, applications, challenges, and future prospects. *Food Engineering Reviews*. doi:10.1007/s12393-025-09400-1

Subhasri, D., Dutta, S., Leena, M. M., Moses, J. A., & Anandharamakrishnan, C. (2022). Gastronomy: An extended platform for customized nutrition. *Future Foods*, *5*, 100147. doi:10.1016/j.fufo.2022.100147

Teng, X., Zhang, M., & Mujumdar, A. S. (2021). 4D printing: Recent advances and proposals in the food sector. *Trends in Food Science & Technology*, *110*, 349–363. doi:10.1016/j.tifs.2021.01.076

Tesikova, K., Jurkova, L., Dordevic, S., Buchtova, H., Tremlova, B., & Dordevic, D. (2022). Acceptability analysis of 3D-printed food in the area of the Czech Republic based on survey. *Food*, *11*(20), 3154. doi:10.3390/foods11203154

Thorakkattu, P., Awasti, N., Sajith Babu, K., Khanashyam, A. C., Deliephan, A., Shah, K., Singh, P., Pandiselvam, R., & Nirmal, N. P. (2025). 3D printing: Trends and approaches toward achieving long-term sustainability in the food industry. *Critical Reviews in Biotechnology*, *45*(1), 48–68. doi:10.1080/07388551.2024.2344577

Ukkunda, N. S., Santhoshkumar, P., & Moses, J. A. (2025). Impact of different natural sweeteners on 3D printability and post-printing quality of marzipan. *Future Foods*, *11*, 100557. doi:10.1016/j.fufo.2025.100557

Varvara, R.-A., Szabo, K., & Vodnar, D. C. (2021). 3D food printing: Principles of obtaining digitally-designed nourishment. *Nutrients*, *13*(10), 3617. doi:10.3390/nu13103617

Wang, M., Bao, Y., Li, D., Bian, Y., Si, X., Gao, N., Cheng, Z., Gui, H., Dong, W., Jiang, H., & Li, B. (2024). Scientometrics and visualized analysis of 3D food printing: A boost to future food customized development. *Food Bioscience*, *58*, 103844. doi:10.1016/j.fbio.2024.103844

Zhu, W., Iskandar, M. M., Baeghbali, V., & Kubow, S. (2023). Three-dimensional printing of foods: A critical review of the present state in healthcare applications, and potential risks and benefits. *Foods*, *12*(17), 3287. doi:10.3390/foods12173287

11

Digitalization and the Future of Food Printing Technology

CHAPTER MENU

11.1 Introduction

A prosumer is a combination of the terms "producer" and "consumer," denoting persons who actively develop, customize, or alter the items or services they utilize (Lang et al., 2021). In contrast to conventional consumers who passively acquire completed things, prosumers engage actively, frequently providing ideas, feedback, or direct contributions to the design and functionality of products. In a digital environment, prosumers are frequently observed on platforms that enable customization, including 3D food printing, which allows them to consume and cocreate items (Halassi et al., 2019). The appeal of future food printers will be

4D Printing of Foods: Fundamentals, Applications and Prospects, First Edition.
C. Anandharamakrishnan, Jeyan A. Moses, and P. Santhoshkumar.

significantly influenced by the characteristics of prosumer platforms, according to qualitative evidence from interviews.

Desirability is perceived as a blend of prosumer experience and usability. A digital platform that is optimum for 3D food printing must offer prosumers a sufficient selection of customization options while also providing a user interface that is both intuitive and straightforward. This is crucial, as consumers may experience cognitive overflow due to an abundance of functions. Jayaprakash et al. (2019) have also described the distinct mechanisms by which prosumers are motivated to engage in the 3D printing of food. Also, the platform must be scalable to enable the seamless integration of new functionality by service providers in the future. It is imperative to monitor the consumption of caloric and nutrient content, as well as the printing processes and accessible culinary ingredients, in real time. Users must be able to effortlessly construct 3D designs with customizable predesigned templates and seamlessly submit independently crafted food designs. Incorporating user medical data and individual preferences into the printing process would be an additional advantage. The platform must also enable users to disseminate their recipes and 3D food designs with fellow prosumers.

Technological developments facilitating information acquisition, processing, and machine regulation are anticipated to enhance efficiency and sustainability in enterprises, a shift referred to in the food sector, and are part of Industry 4.0 and advanced concepts. This industrial evolution highlights the viability of customized food production to satisfy consumer needs, real-time data management, information visualization for superior decision-making, production process monitoring for increased productivity, and better resource utilization. It also includes cyber-physical systems, wherein computers, networking, big data analytics, and control technologies converge (Raja et al., 2022; Hassoun et al., 2024). Figure 11.1 illustrates the principal Industry 4.0 technologies pertinent to the food sector.

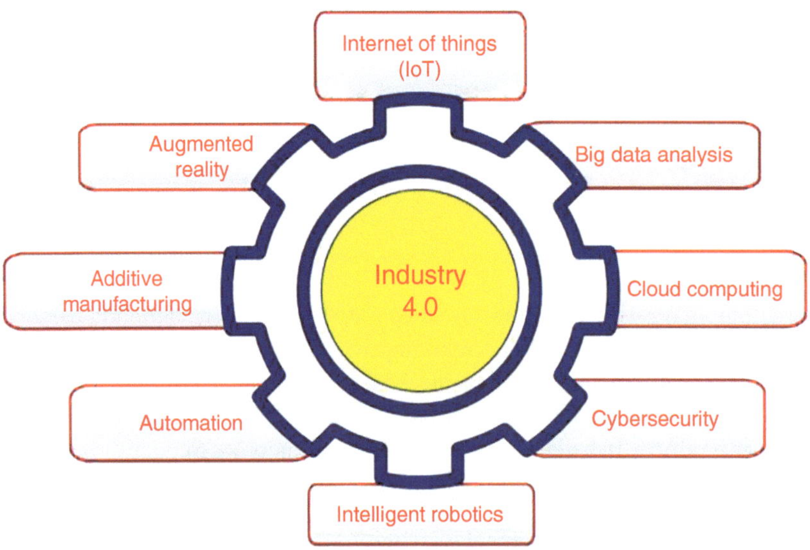

Figure 11.1 Overview of Industry 4.0 segments in the food industry. *Source*: Raja et al. (2022) / with permission of Elsevier.

The food Industry 4.0 integrates various advanced technologies, including digitalization such as artificial intelligence (AI), big data analytics, the Internet of Things (IoT), and blockchain, alongside other technological advancements such as smart sensors, robotics, digital twins, and cyber-physical systems (Hassoun et al., 2023a). Furthermore, novel culinary trends, such as 3D-printed foods, have taken root in Industry 4.0 (Djekić et al., 2023). These technologies are profoundly altering the food sector, yielding enormous environmental, economic, and human behavioral effects. Although each technology is significant, the most transformational and durable solutions often emerge from integrating various technologies (Jambrak et al., 2021). Digitalization in food processing encompasses more than mere analog-to-digital conversion; it signifies an interconnected network of interfaces, data management, and business operations, revolutionizing the food industry (Nugroho et al., 2023; Hassoun et al., 2024). The Food Industry 4.0 era has introduced novel challenges, opportunities, and trends, transforming tactics and future trajectories for food production and consumption and establishing the foundation for the transition to Industry 5.0 (Singh et al., 2025).

AI is extensively employed in healthcare, banking, transportation, manufacturing, and entertainment, augmenting efficiency, decreasing expenses, and boosting service quality (Rashid & Kausik, 2024). Fundamental AI technologies, machine learning (ML), deep learning (DL), natural language processing (NLP), and computer vision (CV) are widely used in several domains. Big data, encompassing high-velocity, high-volume, and high-dimensional structured and unstructured data from sensors and various sources, necessitate specialized storage, processing, and analysis technologies (Ding et al., 2023). In food systems, big data analytics have the capabilities to forecast market demand, enhance supply chain efficiency, elevate food safety and quality, and provide competitive benefits (Thirukumaran et al., 2024). Innovative technologies such as 3D-food printing and virtual reality (VR) create new opportunities for culinary advancement. 3D and 4D food printing have recently emerged in food industry applications (Nida et al., 2022). Such digitalization techniques integrated with additive manufacturing (AM) of foods enhance the production, quality, and safety and are detailed in this chapter.

11.2 Futuristic Applications – 4D Onward Technologies

The emergence of 4D and beyond their dimension in AM is set to transform food production, consumption, and interaction. 4D-printed foods utilize edible smart materials that can respond to internal or external stimuli by changing their color, shape, texture, and nutritional composition, facilitating highly personalized meals designed to meet individual preferences and requirements. Such aspects have been elaborated in previous chapters. Apart from foods, 4D printing has been employed in food packaging, utilizing intelligent materials that proactively maintain food quality by adjusting their characteristics to control temperature, humidity, and microbial proliferation, as well as indicate the quality of food through the intelligent function (also called intelligent packaging). Moreover, bioprinting has enabled the creation of alternative meat products, while sophisticated smart kitchen appliances are being developed to autonomously make meals according to user preferences and nutritional objectives, integrating real-time inventory management and expiration tracking. Using holographic interfaces and multisensory dining experiences further revolutionizes our culinary perception, providing interactive and immersive gourmet encounters. These inventions collectively seek to create a sustainable, personalized, and highly efficient global food ecology, with their dynamic applications detailed in the subsequent sections of this chapter.

11.2.1 Customized and Tailor-Made Foods

One of the most significant applications of AM in the culinary sector is food customization. Customized nutrition is a highly transformational element that establishes this technology as a rapidly advancing approach to food production, and this applies to 4D food printing. 4D food printing generates consumables and its capable of changing shape in response to environmental factors such as pH, moisture, and heat (Cen & Meng, 2024). For instance, flat-printed pasta may alter into unique shapes when boiled, which reduces the need for transportation and storage and provides convenience (Burke, 2024). Additionally, cutting-edge developments in personalized nutrition integrate AI and biotechnology to evaluate an individual's genetic profile, health information, and dietary habits. This enables foods to be carefully made to meet certain needs, similar to foods that boost immunity or control long-term diseases.

The smart feature of 4D-printed foods can join tailored mixes of nutrients, vitamins, and minerals to help specific health aims. 4D printing can raise the buyer's liking for other protein-based materials and grown meat. The making of "smart and responsive foods" that can watch their nutrition info and make on-the-spot changes to how they work and freshness is also helped by these technologies. The development foretells an upcoming revolution in culinary arts together with food technology, which will transform eating traditions into complete sensory encounters. Meals will gradually shift toward personalized and environmentally friendly encounters that serve specific customer requirements. The development path helps scientists create specialized nutritional food items, including boosters that strengthen immune systems in patients who need medical assistance. The intelligent functions of 4D-printed food enable personal healthcare goals through specially customized nutrient combinations of macronutrients, vitamins, and minerals. 4D printing helps consumers better accept alternative protein sources, including cultured meat products. Through their innovation researchers developed "intelligent and interactive foods," which possess autonomous nutritional content tracking and dynamic functionality based on user needs. These food technology breakthroughs constitute a transformative time that turns dining into accessibility for personalized health-oriented experiences.

11.2.2 Delivery Systems

The development of nutraceutical delivery systems through innovative technology and 4D printing creates flexible solutions with exact formulations to enhance health and wellness. Developing encapsulates, capsules, coatings, and matrices through this method allows scientists to enhance bioavailability. Next-generation technologies and 4D approaches encapsulate bioactive chemicals into intelligent carriers to improve chemical efficacy and stability (Zhou et al., 2020; Shen and Shen 2025). This advance enables more accurate nutraceutical delivery by triggering specific parts of the digestive tract, such as the stomach or intestines, using pH level changes. Active substances maintained through time-released mechanisms deliver a stable presence across daily segments (Melocchi et al., 2019). Efficient, tailored nutraceutical solutions can be developed through this approach. These technological devices prove more effective than traditional methods of nutraceutical administration. 4D printing enables the protected delivery of sensitive compounds by using complex 3D structures to achieve controlled phased drug releases (Yarali et al., 2024). Traditional nutrient delivery systems fail to match the superiority of these advanced systems because they use static designs that simultaneously release nutrients, leading to component interference and reduced efficiency. Several studies have examined 4D printing for pharmacological uses, but limited research exists about its application for delivering nutraceuticals. However, in many senses, logic would remain similar for both applications. Research needs additional attention because this gap presents challenges and opportunities for new scientific pursuits.

11.2.3 Vending Machines and Personalized Items

Advanced materials show potential to revolutionize vending machines alongside personalized food systems, creating interactive dispensing solutions for tailored needs through flexible customization options. Vending machines could use smart materials to integrate stimuli-sensitive system components. The implemented system enables these devices to change deliveries based on individual preferences along with health information obtained from smartphones and wearable technologies. Users could see holographic or augmented reality display features through the machines that enabled both direct customization during order placement and virtual product preview capabilities. The combination of advanced biometrics, together with AI and 4D printing technology in customized food systems, enables the creation of personalized nutritional experiences (Neamah & Tandio, 2024). Food products could contain active components that would release certain parts of the flavor profile, texture, and nutrients based on how consumers engage with their meals, thereby providing multiple sensory experiences. The incorporated systems can use individual biometric and DNA information to generate meals precisely customized to health needs as well as dietary boundaries and lifestyle preferences. A dependable interconnected platform requiring precise integration between AI, IoT, advanced materials, robotics, and blockchain needs to deliver full transparency to achieve effective technology implementation across these domains. Despite technical difficulties, the multifaceted potential for advancing vending machine technology and product customization into advanced nutrition-focused and interconnected systems remains significant.

11.2.4 Kitchen and Restaurant Printing

4D-printed foods can undergo a sensory transformation as part of preparation and presentation methods, which creates an innovative dining experience. 4D-printed spaghetti maintains flexible flat shapes that create twisted spirals during boiling, while heated pastries expand into complex designs. Uniform food changes allow clients to modify entry points through interactive controls for real-time flavor adjustment, shape transformations, and sensory enhancement (Subhasri et al., 2022). Restaurants leverage technology to produce instant made-to-order cuisine that caters to personalized dietary requirements, cultural food preferences, and esthetic tastes (Zoran et al., 2021). In addition to esthetics, 4D printing can enhance operational efficiency by optimizing food preparation and reducing the labor-intensive components of culinary arts. As technology progresses, possibilities expand beyond 4D, potentially leading to 5D and 6D printing with enhanced precision, AI-powered smart kitchens that analyze real-time consumer feedback to modify recipes and self-assembling foods utilizing nanotechnology and molecular gastronomy (Zatsu et al., 2024). These inventions can potentially transform eating and the fundamental notion of food, converting meals into participatory and sustainable experiences that merge science, art, flavor, fusion kitchen (particularly Nouvelle cuisine), and molecular gastronomy.

11.2.5 Digital Gastronomy

Integrating digital gastronomy with molecular gastronomy and modern 4D printing technologies signifies a revolutionary advancement in food creation, experience, and consumption. Molecular gastronomy integrates physics and chemistry to convert components into novel recipes, offering the scientific basis for these progressive innovations. By integrating molecular techniques and 4D food printing, chefs may create edible materials that undergo dynamic shape, texture, and flavor changes over time (Derossi et al., 2024). Incorporating augmented reality (AR)/virtual reality (VR) approaches and molecular principles amplifies the multisensory characteristics of food. Diners could encounter a molecular gastronomy creation, such as a nitrogen-infused foam dessert, while virtually enveloped in a snowy alpine environment that enhances the dish's frosty theme. Simultaneously, programmable meals developed through molecular modification and nanotechnology enable exact personalization (Roy et al., 2022). A single dish might undergo molecular transformations using molecular building blocks to provide varying textures, such as soft, crispy, or chewy, and to change in flavor from sweet to savory, customized to the user's preferences (Esmaeily et al., 2024).

Biometric and AI-driven technologies introduce additional sophistication. Real-time nutritional insights can be integrated with molecular gastronomy through wearable health monitors. A diner may relish the flavor of a chocolate mousse inspired by molecular gastronomy while consuming a calorie-free facsimile equipped with these advanced sensory technologies. Today's modern culinary revolution has established robotics together with autonomous systems as essential components (Derossi et al., 2023). Due to improved molecular techniques, robotic limbs can produce structured items from multilayer foams through edible films and liquid nitrogen spheres at accuracy levels beyond what humans can achieve. Virtual kitchens enable users to design their meals with molecular techniques for encapsulation and oxygenation before they send preparation orders for automated printing at home (Subhasri et al., 2022). Molecular gastronomy paired with 4D technology and digital engagement systems advances the dining experience into a personalized experience that combines sustainability with individual preference while remaining immersive.

11.2.6 Retail Packaging

4D-printed solutions transform retail packaging through sustainable and functional adaptations that react to environmental stimuli such as heat, moisture, and pressure. Through the heat-responsive action, the flat packet can shift into a bowl after hot water exposure and thus eliminate the requirement for additional containers and reduce packaging materials. 4D printing facilitates the utilization of sustainable materials through its ability to work with environmentally friendly starch-based products combined with mushroom composites, seaweed-derived films, and plant-based polymers (Danish et al., 2021). Through heat-responsive designs alongside light-sensitive animations and tactile textures, 4D printing increases packaging engagement with consumers and delivers personalized experiences promoting environmentally sustainable practices. Through 4D-printed designs, manufacturers can convert reusable packaging into multiple functional formats (Andrew & Dhakal, 2022). Furthermore, this technology can personalize packaging solutions according to each consumer's preferences. Overall, 4D printing is a key enabling technology for future retail packaging development because its ability to provide dynamic content engages users with personalized transformative designs that preserve the environment.

11.3 Information and Communication Technology (ICT)-Assisted Concepts

Advanced process control techniques such as computational modeling, simulation, ICT tools, and AI can be used in AM for food to optimize material characteristics, quality, and safety and even improve printing accuracy. More generally, on-the-fly AI-generated information enables immediate changes, which boosts personalization and alters the food design and processing as they become more efficient. These technological tools are set to dominate future food manufacturing processes.

11.3.1 For Food Preparation

In terms of 3D food printing, the main considerations are material selection, process improvement, waste reduction, equipment longevity, and research on the physicochemical changes that occur to food during processing (Anandharamakrishnan et al., 2022; Guo et al., 2019). By understanding the important physical processes in printing, modeling, and simulation, many "what if" scenarios can be explored, and the need for extensive testing and characterization can be reduced (Raja et al., 2023b). It is a detailed investigation into the thermomechanical properties, structural behaviors, and flow characteristics that will eventually determine the printability of a material. Besides this, many existing advanced instrumental techniques were adopted to predict printability in food materials, including rheology (steady shar, dynamic, three-interval thixotropic, creep recovery), crystallinity, and texture, especially in food. Assessing printability in food material can be time consuming since the material features and parameters influence print accuracy (Liu et al., 2020; Waseem et al., 2024). The construction of predictive models, particularly for food extrudability, frequently utilizes optimization techniques such as response surface methodology (RSM), which analyzes the impact of different printing factors on food printability (Lee & Hashimoto, 2024). For instance, an RSM-based simulation was adopted for textural-based 3D-printed food prediction, as presented in Figure 11.2. Nonetheless, these statistical models are generally tailored to specific materials, constraining their wider applicability across printing systems.

Various 3D printing techniques and material categories necessitate distinct modeling or simulation methodologies. For example, hot extrusion-based food printing utilizes thermomechanical models to analyze material behavior under thermal and mechanical stress.

The manufacturing of complex structures calls for combined printing methodologies that integrate extrusion processes with injectors. Assembling computational models with numerical simulations represents a technique for better prediction of print results (Raja et al., 2023b). 3D food printing requires testing multiple materials and design parameters and finishing processes to determine proper solutions because of its complex nature. Researchers have utilized computer simulations to monitor shear distributions, pressure patterns, and flow streams in real time (Ning et al., 2024). Real-time measurements of die expansion and remaining strains help estimate the quality standards of manufactured items throughout the printing process. Further model accuracy improvement could result

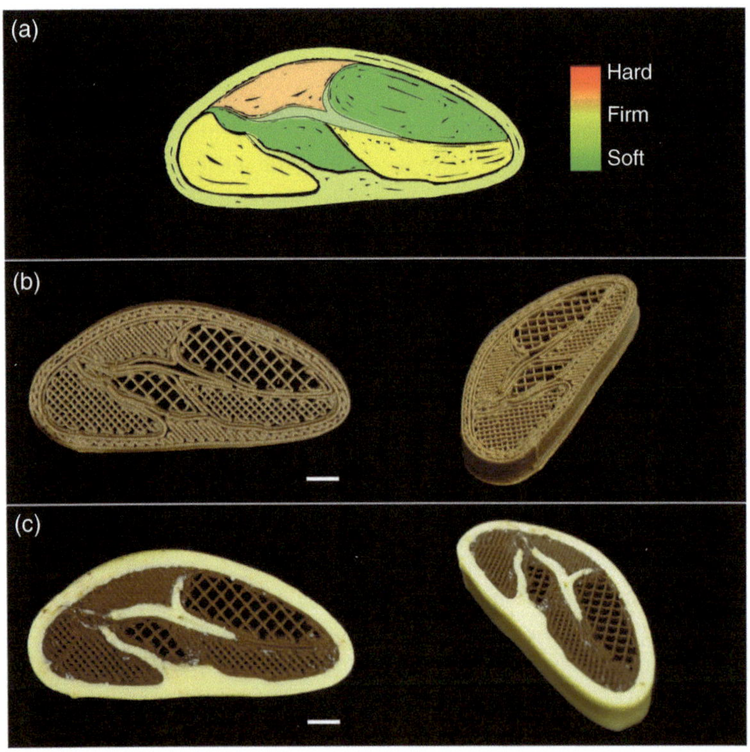

Figure 11.2 Direct ink writing-based 3D-printed cricket steak with different infill levels. (a) RSM-predicted model, (b) single material with different infill levels; (c) multi-material steak with milk as a fat substitute (scale bar: 5 mm). *Source*: Lee & Hashimoto (2024) / with permission of Elsevier / CC BY-NC-ND 4.0.

from hybrid modeling approaches that eliminate multiple initial conditions (Oyinloye & Yoon, 2021). 4D food printing technology represents an emerging field that uses 3D printing methods to create visually pleasing sustainable food products (Sajjad et al., 2023). The systems require keying appropriate response behaviors through modeling and simulation methods. To learn to respond appropriately to inputs, these systems will be keyed through modeling and simulation. The "perceptual-illusion control mechanism" basically posits that 3D food printers can use it to manipulate fullness. Food structures made with CAD could keep someone chewing longer in a manner that utilizes more of the jaw, creating greater fullness faster. This method provides a means of cutting back on food intake, which is especially helpful when people lead sedentary lives, which increases the likelihood of developing chronic diseases (Derossi et al., 2021).

11.3.1.1 AI

In the food industry, AI technologies such as expert systems, fuzzy logic, adaptive neuro-fuzzy inference systems (ANFIS), and intelligent sensors facilitate food classification, product development, marketing, quality enhancement, and supply chain management, thereby contributing to improved food safety (Mavani et al., 2022). AI now facilitates intelligent production and management in food quality assessment, production regulation, and safety surveillance (Zatsu et al., 2024). It emulates human cognition, allowing computers to engage in autonomous learning, reasoning, planning, and decision-making. AI employs mathematical models and statistical methods to forecast food processing and storage conditions, ensuring food quality (Taneja et al., 2023). The food industry benefits greatly from integrating AI with AM in food, improving production efficiency, raising product quality, and decreasing waste (Addanki et al., 2022). Printing parameters, nutritional quality, printing accuracy, product features, printing paste compositions, and faults in the finished printed objects can all be improved with insights given by AI. Minimizing the demand for prolonged testing, AI may find relationships between printing parameters and the end product's characteristics (Hassan et al., 2024). This may help to decrease food waste. With the advent of non-destructive testing, personalized meals may be created to suit specific dietary requirements, aligning with current trends toward eco-friendly, healthier food options. In addition, the fundamental connection between AM and AI is detailed by Yu et al. (2024).

The food industry must prioritize nutrient-rich, individually tailored products that are low impact due to the increasing complexity of consumer preferences. AI is a computing technology engineered to emulate human abilities in environmental perception, information processing, decision-making, and action execution to attain objectives. The diverge of applications in combining 3D food printing and AI is presented in Figure 11.3. Possible future developments in the 3D printing industry include combining AI with 3D printing of food and discovering novel protein sources (Reuben et al., 2025). The capacity to personalize protein content solutions is one way in which 3D food printing might take advantage of the improved formulations made possible by AI and ANNs. Genetic algorithms and artificial neural networks (ANNs) are two examples of AI's parameter optimization capabilities that have the potential to surpass conventional RSM. AI could also look at alternative energy sources that more closely match consumer preferences and demands (Abdelmoteleb et al., 2021). Interestingly, certain restaurants have begun using AI-powered

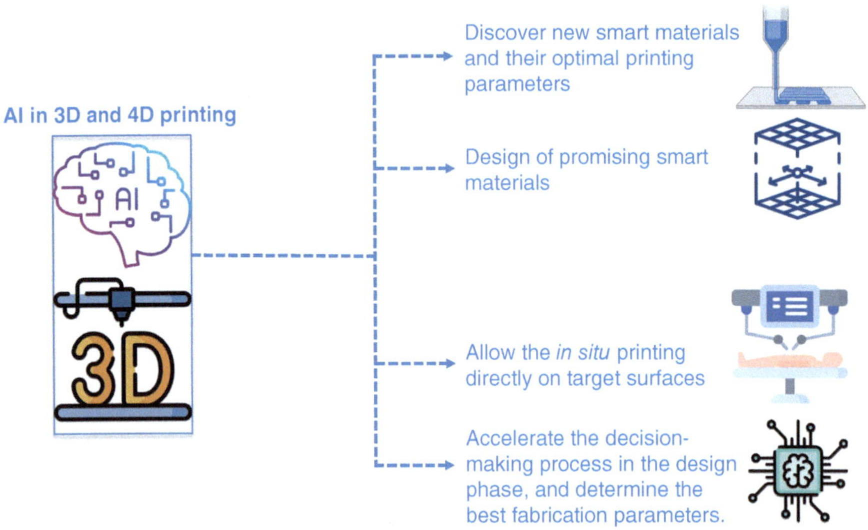

Figure 11.3 AI applications in additive manufacturing. *Source*: Pugliese & Regondi (2022) / MDPI / CC BY 4.0.

robotic chefs. For example, Sony created a robot to enhance food preparation and delivery using recipes emphasizing health and sustainability. Kentucky Fried Chicken (KFC) is investigating 3D-printed chicken cell-based goods to customize nuggets and minimize environmental effects (Bedoya et al., 2022).

Incorporating AI and 3D food printing in gastronomy can produce dishes that align with client preferences, mitigate allergen hazards, and enhance recipe and menu modification for the sustainable, prompt delivery of high-quality, individualized cuisine. A recent study on AI in food can create customized meal selections; ChatGPT and deep generative model were discovered to create an AI-based nutrition recommendation (Figure 11.4) (Papastratis et al., 2024). In brief, ML is advantageous for creating novel tailored foods that cater to nutritional requirements and sensory experiences. It can analyze consumer behaviors and

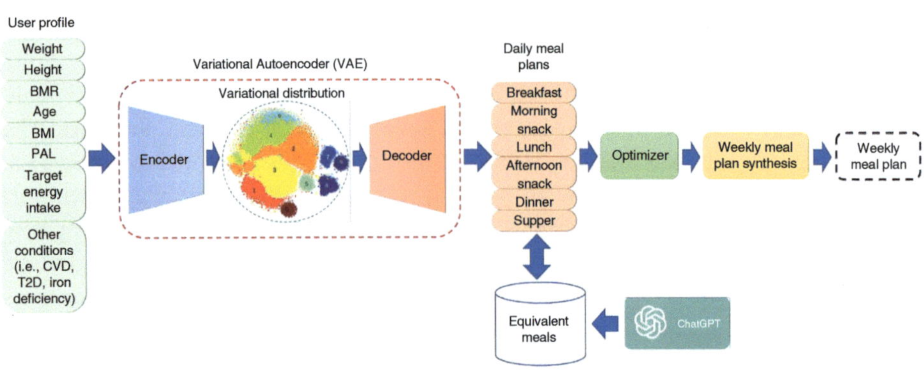

Figure 11.4 Overview of AI-based diet recommendation methods with ChatGPT. *Source*: Papastratis et al. (2024) / Springer Nature / CC BY 4.0.

be included in the design process, enabling the anticipation of foodmaterial properties and potentially creating new categories of food materials that fulfill design criteria, including sensory properties of the materials (Oral et al., 2021). Bedoya et al. (2022) exclusively documented ML approaches in food printing applications. Nevertheless, due to the nonlinear deformation properties of numerous food materials, which are challenging to model and predict using conventional methods, ML offers an alternative strategy called reinforcing learning algorithm for enhancing food material design, aiding in the comprehension of the underlying mechanisms in 4D printing materials (Mohammadi et al., 2024; Wang et al., 2024).

AI enhances the intelligent design of 4D food products by optimizing variables such as particle distribution, structure, texture, and stimuli responses; in the future, AI may replace or complement modeling and simulation tools used in 4D printing. Regarding material selection, AI assesses substances' chemical, physical, and nutritional properties to recommend optimal combinations that act like intelligent/smart materials. This guarantees that food retains its desired texture, flavor, and appearance during printing. It can predict the interactions of different materials, enabling the customization of food products for specific diets, such as vegetarian or allergen-free options. By emphasizing sustainability, AI advocates for environmentally friendly materials, rendering 4D food printing more ecologically responsible and versatile for future requirements (Yu et al., 2024). This advanced method makes 4D-printed food more adaptable to various consumer needs (Cao, 2022), including children, the elderly, and athletes, while reducing production errors and improving overall food quality (Yu et al., 2023). The integration of AI into 4D food printing has the potential to revolutionize the field. However, there are still issues to be solved, notably in real-time feedback and material customization; AI offers a pathway toward more personalized, sustainable, and efficient food production. Big data analysis can prove beneficial for optimizing the entire 4D food printing process. AI may use big data to refine print parameters, improve real-time control, and reduce costs by selecting the most efficient materials.

11.3.1.2 Robotic Printing

Robotics has been primarily implemented in packaging and food serving in the food business, and key factors such as economic efficiency, hygiene, robot-food interaction, and operational simplicity drive their integration. This offers several advantages in food processing, including 24/7 high-quality performance, synchronization, and process standardization. However, food scientists are needed to improve cost-effectiveness further and create worker-friendly environments (Derossi et al., 2023). AM, particularly 3D printing, constructs three-dimensional objects from digital files by layering materials. The four primary categories outlined by Bhatt et al. (2020) are multidirectional, conformal, supportless, and large-scale advanced robotic AM. This adaptable technology is employed across numerous industries to boost design flexibility, reduce lead times and costs, reduce waste, promote sustainability, and enable product customization. These find applications in aerospace, defense, automotive, medical, dental, consumer products, architecture, and food (Urhal et al., 2019). Developing complex kinematic joints, increasing the speed of building complicated structures, and enabling additive production of components such as electrical devices and energy storage solutions are all areas of current research focusing on improving functionality in AM (Yan et al., 2023). Recently, using mobile robots to transport components and deposit materials in a 3D printing setting can potentially improve the efficiency, economy, and waste reduction of large-scale construction projects (Zhang et al., 2018). Nevertheless, these mobile robots are not user friendly, and more research is required in food-based 3D printers.

Using robotics in the AM of food allows for the precise application of materials, making it possible to make unique cake decorations or hollow chocolates with intricate patterns, which are difficult to achieve with conventional procedures (Derossi et al., 2023). In 2018, the first 3D printing restaurant unveiled "ByFlow," a robotic arm-powered equipment that opens the door to digital gastronomy and facilitates simple recipe customization through its incredibly accurate food production. A recent report by Bezaleli Mizrahi et al. (2023) reported that a multiresolution-paste 3D printer significantly reduces printing times. This printer, featuring a universal robotic arm and various nozzle sizes, produced high-resolution structures within a controlled period. The researchers recommend the development of collision-free robots equipped with custom end-effectors for tasks such as printing, positioning, scanning, heating, and assembling meals from premade components, thus enhancing digital gastronomy workflows. Figure 11.5 (I) provides an overview of this mobile robotic-based 3D food printer system for restaurants.

Moreover, robotic 3D printing systems hold the potential for chefs to craft complex food shapes, add esthetic toppings, and create unique consumer experiences. These systems range from small-scale decorative units to larger machines capable of preparing entire dishes. In the future, high-performance systems could complete printing and postprocessing within minutes, parallel to conventional restaurant service times (Demei et al., 2022). With continuing technology improvement, customers may be able to tailor meals according to specific nutritional needs, such as protein or fat content, coupled with variations in shape, texture, and visual appeal. Bezaleli Mizrahi et al. (2023) presented an interesting proposed work for a digital gastronomy vision restaurant; digital tools involved customizing dishes to the consumers' flavors and nutritional preferences (Figure 11.5 (II)).

11.3.2 For Advanced Quality Assessment

Advanced quality evaluation is a significant component in 3D food printing; postprocessing quality evaluation can use CV and intelligent sensing technologies. This monitors texture, shape, and consistency in real time, guarantees accuracy, identifies flaws, and upholds superior quality standards during printing. Smart sensors and IoT devices are becoming increasingly vital in multiple areas of the food industry, encompassing transportation, storage, processing, and packaging phases (Hassoun et al., 2023b). These sensors facilitate the seamless incorporation of food traceability systems across the whole supply chain, enabling data gathered at each phase to yield insights into the condition and quality of food products (Ataei Kachouei et al., 2023). Landaluce et al. (2020) reported that these systems are mostly used for environmental monitoring, process and machinery regulation, facility automation, and food traceability. Similarly, IoT and sensor technologies integration allow for the automated measurement and recording of critical food processing properties, leading to a more accurate database with less opportunity for human error.

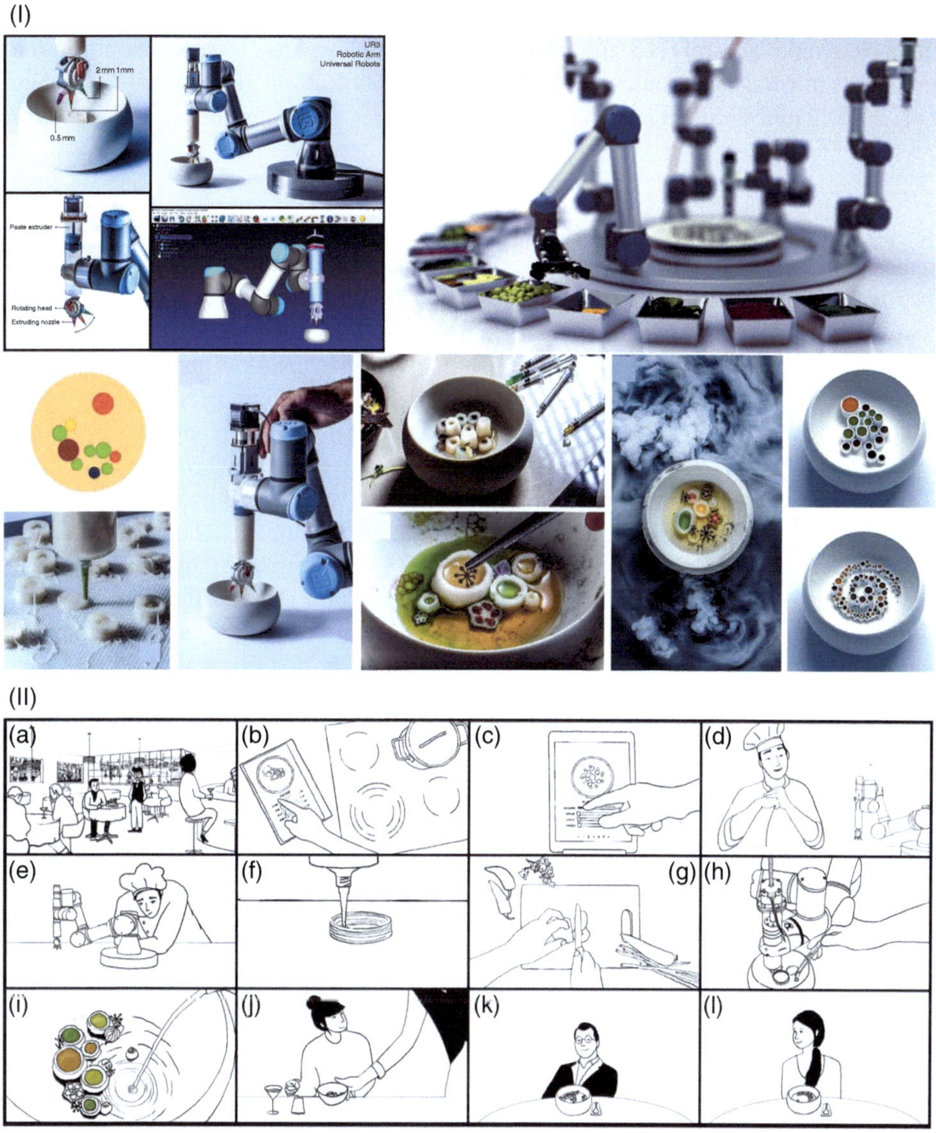

Figure 11.5 (I) Futuristic view of robotic food printing integrating gastronomy, (II) digital gastronomy vision restaurant: (a) Mobile robotics 3D printers in restaurants, (b) human machine interface tool, (c) customization of nutrients, (d) printer setting, (e) printer parameters setting, (f) printing food materials, (g) vegetable cutting, (h) adding ingredients and recipes along with the printed samples, (i) soups or recipes addition, and (j–l) serving to the consumers. *Source*: Bezaleli Mizrahi et al. (2023) / with permission of Elsevier.

3D-printed biosensors are used in food monitoring (Song et al., 2023), analysis (dos Santos et al., 2022), safety, and intelligent packaging functions (Tracey et al., 2022; Santhoshkumar et al., 2024b). In addition to that, biosensors are being engineered to integrate with ICT tools, relaying electrochemical data to enhance quality control in food processing. In recent times, 3D printing has also been used as a fabrication approach in edible electronics and robotics applications, and researchers have reported food printing approaches in edible electronics/nutritive electronics fabrications (Keller et al., 2016; Wu et al., 2020; Santhoshkumar et al., 2024b).

11.3.2.1 Computer Vision

The CV systems have several uses, including face identification, autonomous vehicles, medical imaging, and industrial quality control, because they mimic human vision and can identify patterns, objects, and sceneries (Csurka, 2017). This area integrates AI, image processing, and ML to aid computers in making educated decisions based on visual inputs. The goal is to create software to see like a human so machines can understand digital pictures, extract useful information, and make good decisions. Computers can now assess, evaluate, and comprehend visual data comparable to human perception, all due to sophisticated algorithms that use conventional and DL techniques (Javaid et al., 2024). Essentially, CV permits computers to grasp images at the pixel level using advanced software algorithms that retrieve, process, and interpret visual data accurately. The primary aim of the technology is to take photos or videos and analyze them to extract 3D information about the scene, encompassing activities such as image segmentation, classification, and detection (Ghazal et al., 2024). In the food industry, CV is extensively employed for automated inspection and quality control, including applications from food manufacturing line evaluations to quality assessments, including classifying barley flour by spatial pyramid segmentation, whereas the ANFIS weighing model for sweet oranges and ripeness in sweet peppers, image processing for quality control in mushrooms (Ding et al., 2023). Similarly, convolutional neural networks (CNNs) are proficient at feature extraction from images, and their utilization in food classification illustrates the analysis of gathered food photographs to distinguish various food categories. Image processing software manages data from images obtained from cameras or optical sensors. This procedure encompasses multiple methods, such as Fourier and harmonic analysis, wavelet decomposition, and curve fitting. Combined with ICT technologies such as ML, image-processing systems have demonstrated efficacy as decision-making aids in food supply chains (Raja et al., 2022). In 3D food printing, the shape fidelity of printed and postprocessed food was evaluated using image processing approaches in a recent decades (Raja et al., 2023a; Santhoshkumar et al., 2024a).

11.3.2.2 Sensors

Wireless sensing in the food sector needs robust radio technologies, efficient microprocessors, several input/output choices, reliable energy sources, a versatile design, and a robust software development platform. Colorimetric, metal oxide semiconductors, radio frequency identification, and biosensors are among the most prevalent sensor types in the food sector. These sensors can detect biochemical interactions during processing and storage (Raja et al., 2022). In addition, intelligent packaging, color changes considering the spoilage indication, and RFID monitoring the quality and sent data to respective sources regarding the products considering the smart sensing technologies in the food industry (Kalpana et al., 2019). In smart food packaging, the combination of IoT and sensors plays a crucial role in sharing, delivering and communicating data to consumers and merchants (Hassoun et al., 2023b). Similarly, AI and smart sensors combined real-time approaches to track and evaluate food safety and quality (Yu et al., 2022). These smart sensor systems include ambient intelligence, fuzzy logic, genetic algorithms, case-based reasoning, neural networks, knowledge-based systems, and automated knowledge acquisition. As an example of a traditional sensor, this method uses E-noses, E-tongues, and learning algorithms. Integrating these sensors with ANN improves the ability to analyze data, spot patterns, and decide (Tan & Xu, 2020). The 3D printing fabrication approach develops sensors and IoT-incorporated devices, including micro/nano level ranges (Hussan et al., 2024). Higher precision was achieved through these AM technologies, making these technologies used for microfluidic channel fabrication. Additionally, through these, the biocatalyst incorporation in microfluidic channels is widely used for healthcare applications (Mehta & Rath, 2021). However, the recent trends in microfluidic fabrication through 3D printing approaches were recently documented by Gonzalez et al. (2022). Sensor networks demonstrate efficacy in identifying toxins in food, while intelligent packaging solutions that integrate sensors are also becoming increasingly prevalent. Subsequently, 3D printing was investigated in the diverging of domains in tissue engineering scaffold fabrication (Felton et al., 2021), drug delivery (Ballacchino et al., 2021), nanomedicine fabrication (Kara et al., 2021), biosensors (Muñoz & Pumera, 2020; Remaggi et al., 2022), edible electronics (Santhoshkumar et al., 2024b), and circuit fabrication (Hamilton et al., 2018).

11.3.3 For Robust Supply Chains

Resilient supply chains in 3D food printing promote flexibility and sustainability by enabling on-demand production and lowering reliance on traditional logistics. The convergence of blockchain technology is transforming 3D food printing almost endlessly, enhancing trust, transparency, and traceability in production. This technology employs a decentralized ledger to permanently document the entire food production process, from ingredient sourcing through packaging. As a result, businesses monitor transit, verify quality, and regulate suppliers. Additionally, smart contracts on the blockchain allow for the automation of transactions for integrity in compliance. Customers will also be more inclined to trust and be loyal to such businesses, provided that third-party verification on sustainability, ingredient quality, and responsible sourcing is available (Thirukumaran et al., 2024). They are revolutionizing the food industry by making it more open, robust, and environmentally friendly with the help of modern blockchain technology and 3D food printing. 3D printing gives unprecedented advantages for food processing due to recent developments in bioprinting, flexible materials, and better second-level interfaces with blockchain and AI. Combined, these innovations hold the potential to profoundly improve food production, customization, and consumption with their efficiency, sustainability, and individualism (Küfeoğlu, 2022; Padhiary et al., 2024). Pant et al. (2024) reported that combining 3D printing systems with blockchain technology and AI capabilities significantly enhances food manufacturing process. AI-driven research examines extensive datasets through sophisticated analysis methods to boost 3D printing operations while predicting the food market forecasting and making personalized food selections. The research by Haleem et al. (2022) shows that 3D printing systems directed by AI exploit personal health details to manufacture specific food products that address particular nutritional requirements in custom meal plans. The methodology used in the study supports the validity of the proposed concept.

11.3.3.1 Big Data

Big data, often referred to as large-scale data, is defined by its massive volume, rapid generation speed, and varied characteristics. These factors make traditional methods of data acquisition, storage, management, and analysis challenging within typical time constraints (Sivarajah et al., 2017). The potential of big data to enhance food safety remains restricted because of inadequate data collection and poor management practices (Liu et al., 2018). A typical big data analysis workflow includes four main stages: The big data workflow includes stages that focus on data collection followed by steps for data storage and transfer and data analysis featuring techniques such as text mining and neural networks as well as Bayesian methods before concluding with data visualization elements that utilize R and Ciros Gephi and Cytoscape tools (Li et al., 2023). This approach helps identify correlations among datasets, revealing which data points have the most significant influence on food safety (Ding et al., 2023). Real-time monitoring and traceability systems in food production and supply chains produce substantial data. Managing this data is intricate and necessitates more than basic mathematical analysis; it frequently depends on sophisticated computer capabilities (Rejeb et al., 2022).

Big data and cloud computing are essential in ICT, allowing businesses to convert extensive data streams into actionable insights for operational and strategic decision-making (Biermann et al., 2021). This analytical methodology has revealed new opportunities in chemical processing, pharmaceuticals, and food sectors, where significant real-time data acquisition is critical. As economic and technical advancements transform the food sector, consumer expectations for food quality have intensified, emphasizing flavor and nutrition more (Nugroho et al., 2023; Raja et al., 2022). Conventional food science technologies inadequately address the varied culinary preferences of different locations and individuals. Supply chain management and food safety are two areas where new technologies, such as big data and AI, profoundly impact the food industry. In many food-processing enterprises, random sensors have emerged as the most straightforward sources of big data analysis. Jin et al. (2020) and Misra et al. (2022) cite significant improvements in food safety by integrating big data, AI, and IoT. Kazama et al. (2018), neural network modeling has been developed to work using local ingredients and big data, hoping to push the consumption of international cuisines. Huang et al. (2023) and Nunes et al. (2023) reported AI and big data as significantly improving the position of local producers by providing what will usually make it to the market. Small-scale manufacturers, on their part, could stand to gain through blockchain-enabled traceability systems that give their customers insight into the product (Thirukumaran et al., 2024). Automation systems for 3D printing foods will be impacted by the combination of AM and food data science, providing insights into how people use specific culinary items.

11.3.3.2 Cloud Computing

Cloud computing offers a streamlined computational framework for the analysis of large datasets. The primary distinction between cloud computing and big data analytics is that cloud computing is the infrastructure for data storage and access. In contrast, big data analytics refers to the methodology of analyzing that data (Berisha et al., 2022). Simultaneous big data processing necessitates various innovative cloud-based technologies, as managing substantial volumes of data concurrently presents significant challenges. New cloud-based technologies must be adopted to manage big data for concurrent processing. Cloud computing is increasingly vital in food safety and management systems for storing information related to each product, making it accessible to retailers and customers (Rejeb et al., 2020). Cloud computing technology integrates diverse components of a particular industry with minimal resources. It comprises three service models: Software as a Service, Platform as a Service, and Infrastructure as a Service, which are provided to enterprises using a service-oriented architecture. Cloud deployment is often categorized into three types: private, public, and hybrid cloud. Food industry applications utilize cloud computing technology to digitize operations and assess carbon footprints across supply chain networks (Raja et al., 2022). Combining IoT and cloud computing allows improved inter-network collaboration as these platforms provide new data collection approaches besides storage, distribution, and computational power (Mededjel et al., 2017).

Comparable food traceability systems first receive data from sensors that undergo storage procedures in cloud and fog storage solutions. Platform instances such as Microsoft Azure provide cloud platforms to store electrical signals collected by food processing sensors and traceability actuators (Sergi et al., 2021). Cloud computing technologies dress up IoT systems by providing essential computational power and data storage facilities for efficient filtering, processing, and data retrieval and storage functionalities. Modernizing food sciences depends crucially on individualized nutrient customization achieved through the application of AM. Cloud systems operate as food supply monitoring tools by tracking environmental conditions in physical products throughout the supply chain. The processed data supports regulatory adherence together with decision-making based on analysis.

11.3.4 For Consumer Interfacing

From the consumer application perspective, AM offers the potential to visualize and modify food design models in real time using AR, VR, and mixed-reality (MR) technologies. These support the user experience by enabling texture, shape, and aroma customization immediately before printing.

AR disrupts 3D food printing by adopting new methodologies that take design creation and user interaction to a different level. Users can now see and personalize their food designs by projecting virtual models onto the printing surface before the actual printing process starts. This feature, due to the possibility of precise control over surface roughness, size, and shape, helps produce intricate and unique culinary designs. In turn, AR can offer real-time advice on setting up the printer, loading ingredients, carrying out maintenance, or even getting nutritional information. This enables users to navigate the process with greater precision and fewer errors (Chai et al., 2022). Renner et al. (2015) observed that the rapid advancement of 3D food printing is significantly influenced by VR, which provides a stimulating design environment that fosters collaboration and innovation. Users can efficiently create and modify food designs in this interactive environment, leading to visually appealing and complex final products (Çöl et al., 2023). Furthermore, VR facilitates remote collaboration, allowing teams to generate ideas from various locations and enabling chefs to engage in the design phase, thus aligning products more closely with customer preferences (Holuša et al., 2023). This technology improves culinary education and aids

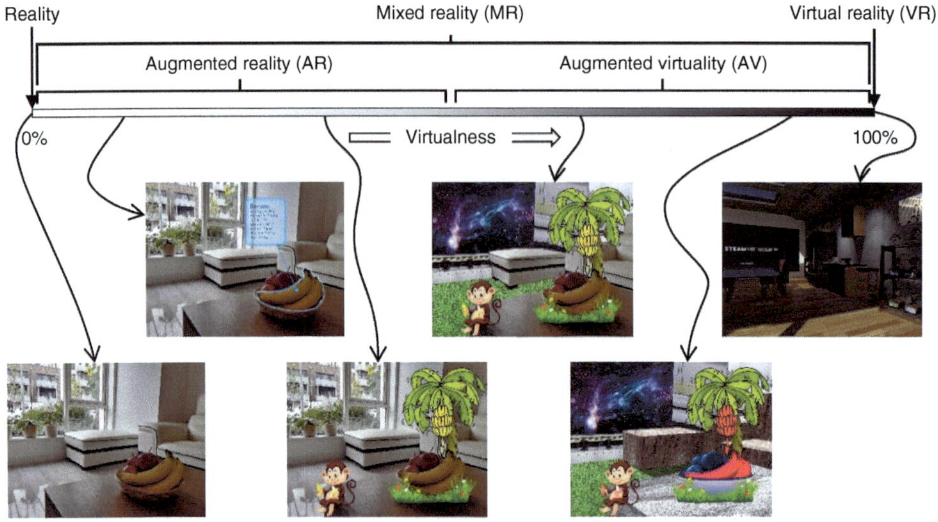

Figure 11.6 Overview of AR, MR, and VR. *Source*: Chai et al. (2022) / Elsevier / CC BY 4.0.

consumers in making informed design choices through virtual cooking lessons and simulated food tastings. VR aids in informed culinary decision-making by integrating nutritional information into food models (Oliver & Hollis, 2021). AR allows users to experience 25% virtual content alongside 75% real content, whereas VR provides 75% virtual content and 25% real content, according to Çöl et al. (2023).

MR facilitates the seamless interaction of virtual objects with the physical environment by integrating virtual and physical realms (Figure 11.6). MR systems facilitate physical interaction and manipulation of virtual elements, whereas AR focuses on observing virtual overlays or objects. AR and MR are fundamentally different because of their interactive features, and they are often used interchangeably due to the absence of a universally accepted definition for MR. Despite this ambiguity, research and development in MR have advanced considerably over the past decade, evidenced by an increase in implementations across various sectors, including the food industry (Chai et al., 2022). Korsgaard et al. (2020) conducted a study utilizing MR to develop an immersive dining experience. This experience enabled users to digitally share meals with remote companions, thereby improving social interaction and the overall dining experience.

11.4 Sustainability in Food Printing Technology

The AM of foods prioritizes minimizing environmental consequences, improving resource efficiency, and advancing global sustainability objectives. 3D printing facilitates exact regulation of portion sizes and ingredient utilization, thereby reducing food waste and transforming wastes into nutrient-dense edible inks in accordance with circular economy concepts. These printing technologies enable localized, on-demand food production, reducing transit requirements, and storage needs while promoting the utilization of sustainable components, including plant-based proteins, algae, and insect-derived powders.

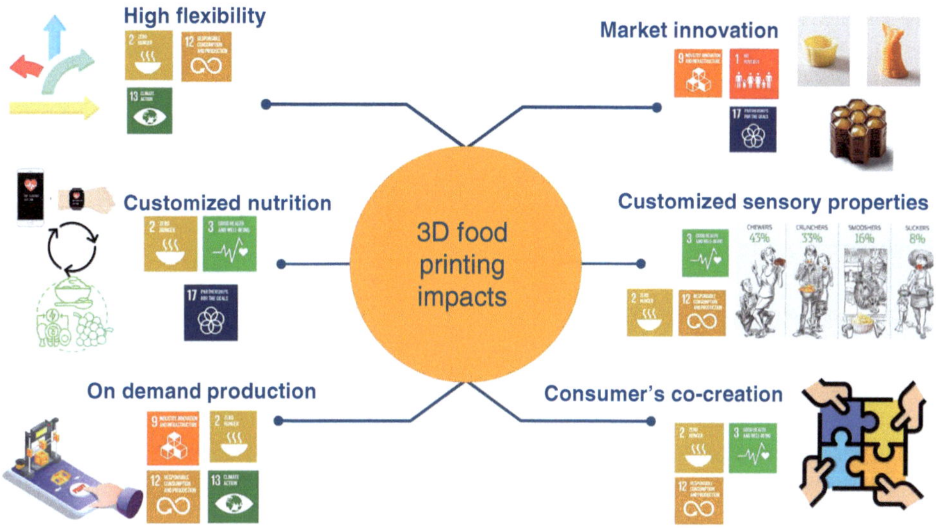

Figure 11.7 Additive manufacturing of foods contribution to SDGs. *Source*: Derossi et al. (2024) / with permission of Springer Nature.

Moreover, 4D printing imparts dynamic characteristics, establishing AM as essential for developing sustainable and innovative food systems. Achieving sustainable development goals (SDGs) such as eradicating hunger (SDG2) and enhancing health and well-being for all (SDG3) by 2030 necessitates significant advancements in research and innovation (Figure 11.7). Additionally, the FAO's foresight about new food systems has examined its benefits, such as the creation of distinctive food shapes and structures that augment sensory appeal (Derossi et al., 2024). Food printing can enhance the acceptability and enjoyment of otherwise unpalatable ingredients or byproducts by integrating nutrients or chemicals, utilizing visual presentation and sensory design to affect meal preferences. These considerations combined highlight the potential of 3D and 4D food printing in tackling sustainability and global food issues.

11.5 Conclusion

A dramatic change in the field known as Industry 4.0 is occurring because traditional manufacturing processes are not keeping up with the high demands of a dynamic environment. Innovations in digital manufacturing technology have brought about changes in connectivity, intelligence, and adaptable automation. New manufacturing technologies make revolutionary solutions that bridge the digital and physical worlds possible. As a pioneer in this shift, AM is reshaping manufacturing processes while vastly improving the customer experience. The ability of 3D food printing to incorporate sustainable practices into food production, introduce unique features, and boost personalization makes it stand out among the groundbreaking technologies poised to transform our interaction with food. Keeping mass-produced food goods sustainable, with an emphasis on practicality and ecological responsibility, is vital as the revolution moves from 4.0 to 5.0. Exciting new possibilities for investigating alternate food sources and customizing products to suit specific dietary needs emerge when food printing techniques and AI are integrated. Research and innovation must be continuous to tackle substantial scientific, technological, economic, and social issues to achieve a food system that is both sustainable and health focused. The goal is to make food printing accessible to everybody.

References

Abdelmoteleb, M., Zhang, C., Furey, B., Kozubal, M., Griffiths, H., Champeaud, M., & Goodman, R. E. (2021). Evaluating potential risks of food allergy of novel food sources based on comparison of proteins predicted from genomes and compared to www.AllergenOnline.org. *Food and Chemical Toxicology*, *147*, 111888. doi:10.1016/j.fct.2020.111888

Addanki, M., Patra, P., & Kandra, P. (2022). Recent advances and applications of artificial intelligence and related technologies in the food industry. *Applied Food Research*, *2*(2), 100126. doi:10.1016/J.AFRES.2022.100126

Anandharamakrishnan, C., Moses, J. A., & Anukiruthika, T. (2022). *3D printing of foods*. John Wiley & Sons Ltd.

Andrew, J. J., & Dhakal, H. N. (2022). Sustainable biobased composites for advanced applications: Recent trends and future opportunities–A critical review. *Composites Part C: Open Access*, *7*, 100220. doi:10.1016/j.jcomc.2021.100220

Ataei Kachouei, M., Kaushik, A., & Ali, M. A. (2023). Internet of things-enabled food and plant sensors to empower sustainability. *Advanced Intelligent Systems*, *5*(12). doi:10.1002/aisy.202300321

Ballacchino, G., Weaver, E., Mathew, E., Dorati, R., Genta, I., Conti, B., & Lamprou, D. A. (2021). Manufacturing of 3D-printed microfluidic devices for the synthesis of drug-loaded liposomal formulations. *International Journal of Molecular Sciences*, *22*(15), 8064. doi:10.3390/ijms22158064

Bedoya, M. G., Montoya, D. R., Tabilo-Munizaga, G., Pérez-Won, M., & Lemus-Mondaca, R. (2022). Promising perspectives on novel protein food sources combining artificial intelligence and 3D food printing for food industry. *Trends in Food Science & Technology*, *128*, 38–52. doi:10.1016/j.tifs.2022.05.013

Berisha, B., Mëziu, E., & Shabani, I. (2022). Big data analytics in Cloud computing: an overview. *Journal of Cloud Computing*, *11*(1), 24. doi:10.1186/s13677-022-00301-w

Bezaleli Mizrahi, A., Lachnish, A. Z., & Zoran, A. (2023). Digital gastronomy testcase: A complete pipeline of robotic induced dish variations. *International Journal of Gastronomy and Food Science*, *31*, 100625. doi:10.1016/J.IJGFS.2022.100625

Bhatt, P. M., Malhan, R. K., Shembekar, A. V., Yoon, Y. J., & Gupta, S. K. (2020). Expanding capabilities of additive manufacturing through use of robotics technologies: A survey. *Additive Manufacturing*, *31*, 100933. doi:10.1016/J.ADDMA.2019.100933

Biermann, O., Koya, S. F., Corkish, C., Abdalla, S. M., & Galea, S. (2021). Food, big data, and decision-making: A scoping review–the 3-D commission. *Journal of Urban Health*, *98*(S1), 69–78. doi:10.1007/s11524-021-00562-x

Burke, R. M. (2024). New dimensions in dining: A review of the gastronomic evolution of 3D printed foods from 2013 to 2024 and beyond. *International Journal of Gastronomy and Food Science*, *38*, 101027. doi:10.1016/j.ijgfs.2024.101027

Cao, L. (2022). A new age of AI: Features and futures. *IEEE Intelligent Systems*, *37*(1), 25–37. doi:10.1109/MIS.2022.3150944

Cen, S., & Meng, Z. (2024). 4D printing: A novel application for structuring oils with fat-analog characteristics. *Trends in Food Science & Technology*, *153*, 104729. doi:10.1016/j.tifs.2024.104729

Chai, J. J. K., O'Sullivan, C., Gowen, A. A., Rooney, B., & Xu, J.-L. (2022). Augmented/mixed reality technologies for food: A review. *Trends in Food Science & Technology*, *124*, 182–194. doi:10.1016/j.tifs.2022.04.021

Çöl, B. G., İmre, M., & Yıkmış, S. (2023). Virtual reality and augmented reality technologies in gastronomy: A review. *EFood*, *4*(3). doi:10.1002/efd2.84

Csurka, G. (Ed.). (2017). *Domain adaptation in computer vision applications*. Springer International Publishing. https://doi.org/10.1007/978-3-319-58347-1

Danish, M., Vijay Anirudh, P., Karunakaran, C., Rajamohan, V., Mathew, A. T., Koziol, K., Thakur, V. K., Kannan, C., & Balan, A. S. S. (2021). 4D printed stereolithography printed plant-based sustainable polymers: Preliminary investigation and optimization. *Journal of Applied Polymer Science*, *138*(36). doi:10.1002/app.50903

Demei, K., Zhang, M., Phuhongsung, P., & Mujumdar, A. S. (2022). 3D food printing: Controlling characteristics and improving technological effect during food processing. *Food Research International*, *156*, 111120. doi:10.1016/J.FOODRES.2022.111120

Derossi, A., Bhandari, B., van Bommel, K., Noort, M., & Severini, C. (2021). Could 3D food printing help to improve the food supply chain resilience against disruptions such as caused by pandemic crises? *International Journal of Food Science & Technology*, *56*(9), 4338–4355. doi:10.1111/ijfs.15258

Derossi, A., Di Palma, E., Moses, J. A., Santhoshkumar, P., Caporizzi, R., & Severini, C. (2023). Avenues for non-conventional robotics technology applications in the food industry. *Food Research International*, *173*, 113265. doi:10.1016/j.foodres.2023.113265

Derossi, A., Spence, C., Corradini, M. G., Jekle, M., Fahmy, A. R., Caporizzi, R., Devahastin, S., Moses, J. A., Le-Bail, A., Zhou, W., Zhang, M., Bhandari, B., & Severini, C. (2024). Personalized, digitally designed 3D printed food towards the reshaping of food manufacturing and consumption. *NPJ Science of Food*, *8*(1), 54. doi:10.1038/s41538-024-00296-5

Ding, H., Tian, J., Yu, W., Wilson, D. I., Young, B. R., Cui, X., Xin, X., Wang, Z., & Li, W. (2023). The application of artificial intelligence and big data in the food industry. *Foods*, *12*(24), 4511. doi:10.3390/foods12244511

Djekić, I., Velebit, B., Pavlić, B., Putnik, P., Šojić Merkulov, D., Bebek Markovinović, A., & Bursać Kovačević, D. (2023). Food quality 4.0: Sustainable food manufacturing for the twenty-first century. *Food Engineering Reviews*, *15*(4), 577–608. doi:10.1007/s12393-023-09354-2

dos Santos, D. M., Cardoso, R. M., Migliorini, F. L., Facure, M. H. M., Mercante, L. A., Mattoso, L. H. C., & Correa, D. S. (2022). Advances in 3D printed sensors for food analysis. *TrAC Trends in Analytical Chemistry*, *154*, 116672. doi:10.1016/j.trac.2022.116672

Esmaeily, R., Razavi, M. A., & Razavi, S. H. (2024). A step forward in food science, technology and industry using artificial intelligence. *Trends in Food Science & Technology*, *143*, 104286. doi:10.1016/j.tifs.2023.104286

Felton, H., Hughes, R., & Diaz-Gaxiola, A. (2021). Negligible-cost microfluidic device fabrication using 3D-printed interconnecting channel scaffolds. *PLoS One*, *16*(2), e0245206. doi:10.1371/journal.pone.0245206

Ghazal, S., Munir, A., & Qureshi, W. S. (2024). Computer vision in smart agriculture and precision farming: Techniques and applications. *Artificial Intelligence in Agriculture, 13*, 64–83. doi:10.1016/j.aiia.2024.06.004

Gonzalez, G., Roppolo, I., Pirri, C. F., & Chiappone, A. (2022). Current and emerging trends in polymeric 3D printed microfluidic devices. *Additive Manufacturing, 55*, 102867. doi:10.1016/j.addma.2022.102867

Guo, C.-F., Zhang, M., & Bhandari, B. (2019). A comparative study between syringe-based and screw-based 3D food printers by computational simulation. *Computers and Electronics in Agriculture, 162*, 397–404. doi:10.1016/j.compag.2019.04.032

Halassi, S., Semeijn, J., & Kiratli, N. (2019). From consumer to prosumer: A supply chain revolution in 3D printing. *International Journal of Physical Distribution & Logistics Management, 49*(2), 200–216. doi:10.1108/IJPDLM-03-2018-0139

Haleem, A., Javaid, M., Pratap Singh, R., & Suman, R. (2022). Medical 4.0 technologies for healthcare: Features, capabilities, and applications. *Internet of Things and Cyber-Physical Systems, 2*, 12–30. doi:10.1016/j.iotcps.2022.04.001

Hamilton, C. A., Alici, G., & in het Panhuis, M. (2018). 3D printing vegemite and marmite: Redefining "breadboards.". *Journal of Food Engineering, 220*, 83–88. doi:10.1016/j.jfoodeng.2017.01.008

Hassan, M., Misra, M., Taylor, G. W., & Mohanty, A. K. (2024). A review of AI for optimization of 3D printing of sustainable polymers and composites. *Composites Part C: Open Access, 15*, 100513. doi:10.1016/j.jcomc.2024.100513

Hassoun, A., Aït-Kaddour, A., Abu-Mahfouz, A. M., Rathod, N. B., Bader, F., Barba, F. J., Biancolillo, A., Cropotova, J., Galanakis, C. M., Jambrak, A. R., Lorenzo, J. M., Måge, I., Ozogul, F., & Regenstein, J. (2023a). The fourth industrial revolution in the food industry–Part I: Industry 4.0 technologies. *Critical Reviews in Food Science and Nutrition, 63*(23), 6547–6563. doi:10.1080/10408398.2022.2034735

Hassoun, A., Jagtap, S., Trollman, H., Garcia-Garcia, G., Abdullah, N. A., Goksen, G., Bader, F., Ozogul, F., Barba, F. J., Cropotova, J., Munekata, P. E. S., & Lorenzo, J. M. (2023b). Food processing 4.0: Current and future developments spurred by the fourth industrial revolution. *Food Control, 145*, 109507. doi:10.1016/j.foodcont.2022.109507

Hassoun, A., Jagtap, S., Trollman, H., Garcia-Garcia, G., Duong, K., Saxena, L. N. P., Bouzembrak, Y., Treiblmaier, H., Para-López, C., Carmona-Torres, C., Dev, K., Mhlanga, D., & Aït-Kaddour, A. (2024). From Food Industry 4.0 to Food Industry 5.0: Identifying technological enablers and potential future applications in the food sector. *Comprehensive Reviews in Food Science and Food Safety, 23*(6), e370040. doi:10.1111/1541-4337.70040

Holuša, V., Vaněk, M., Beneš, F., Švub, J., & Staša, P. (2023). Virtual reality as a tool for sustainable training and education of employees in industrial enterprises. *Sustainability, 15*(17), 12886. doi:10.3390/su151712886

Huang, Y., Liu, Y., Yu, X., Li, K., Li, G., Zhu, B., & Dong, X. (2023). Rapid screening of sensory attributes of mackerel using big data mining techniques and rapid sensory evaluation methods. *Journal of Texture Studies, 54*(6), 872–884. doi:10.1111/jtxs.12776

Hussan, K. S. J., Subramaniam, M. P., Kenz, K. T. M., Sreeram, P., Parvathi, S., Sari, P. S., Pullanchiyodan, A., Mulhivill, D. M., & Raghavan, P. (2024). Fabrication and challenges of 3D printed sensors for biomedical applications-Comprehensive review. *Results in Engineering, 21*, 101867. doi:10.1016/j.rineng.2024.101867

Jambrak, A. R., Nutrizio, M., Djekić, I., Pleslić, S., & Chemat, F. (2021). Internet of nonthermal food processing technologies (Iontp): Food industry 4.0 and sustainability. *Applied Sciences*, *11*(2), 1–20. doi:10.3390/app11020686

Javaid, M., Haleem, A., Singh, R. P., & Ahmed, M. (2024). Computer vision to enhance healthcare domain: An overview of features, implementation, and opportunities. *Intelligent Pharmacy*, *2*(6). doi:10.1016/j.ipha.2024.05.007

Jayaprakash, S., Ituarte, I. F., & Partanen, J. (2019). Prosumer-driven 3D food printing: Role of digital platforms in future 3D food printing systems. In *Fundamentals of 3D food printing and applications* (pp. 331–354). Elsevier. https://doi.org/10.1016/B978-0-12-814564-7.00011-0

Jin, C., Bouzembrak, Y., Zhou, J., Liang, Q., van den Bulk, L. M., Gavai, A., Liu, N., van den Heuvel, L. J., Hoenderdaal, W., & Marvin, H. J. P. (2020). Big Data in food safety–A review. *Current Opinion in Food Science*, *36*, 24–32. doi:10.1016/j.cofs.2020.11.006

Kalpana, S., Priyadarshini, S. R., Maria Leena, M., Moses, J. A., & Anandharamakrishnan, C. (2019). Intelligent packaging: Trends and applications in food systems. *Trends in Food Science & Technology*, *93*, 145–157. doi:10.1016/j.tifs.2019.09.008

Kara, A., Vassiliadou, A., Ongoren, B., Keeble, W., Hing, R., Lalatsa, A., & Serrano, D. R. (2021). Engineering 3D printed microfluidic chips for the fabrication of nanomedicines. *Pharmaceutics*, *13*(12), 2134. doi:10.3390/pharmaceutics13122134

Kazama, M., Sugimoto, M., Hosokawa, C., Matsushima, K., Varshney, L. R., & Ishikawa, Y. (2018). A neural network system for transformation of regional cuisine style. *Frontiers in ICT*, *5*. doi:10.3389/fict.2018.00014

Keller, A., Stevens, L., Wallace, G. G., In Panhuis, M., & Het. (2016). 3D printed edible hydrogel electrodes. *MRS Advances*, *1*(8), 527–532. doi:10.1557/ADV.2015.10

Korsgaard, D., Bjorner, T., Bruun-Pedersen, J. R., Sorensen, P. K., & Perez-Cueto, F. J. A. (2020). *Eating together while being apart: a pilot study on the effects of mixed-reality conversations and virtual environments on older eaters' solitary meal experience and food intake*. 2020 IEEE Conference on Virtual Reality and 3D User Interfaces Abstracts and Workshops (VRW), 365–370. https://doi.org/10.1109/VRW50115.2020.00079

Küfeoğlu, S. (2022). *Emerging technologies: value creation for sustainable development*. Springer International Publishing. doi:10.1007/978-3-031-07127-0

Landaluce, H., Arjona, L., Perallos, A., Falcone, F., Angulo, I., & Muralter, F. (2020). A review of IoT sensing applications and challenges using RFID and wireless sensor networks. *Sensors*, *20*(9), 2495. doi:10.3390/s20092495

Lang, B., Dolan, R., Kemper, J., & Northey, G. (2021). Prosumers in times of crisis: Definition, archetypes and implications. *Journal of Service Management*, *32*(2), 176–189. doi:10.1108/JOSM-05-2020-0155

Lee, C. P., & Hashimoto, M. (2024). Prediction of textural properties of 3D-printed food using response surface methodology. *Heliyon*, *10*(7), e27658. doi:10.1016/j.heliyon.2024.e27658

Li, J., Herdem, M. S., Nathwani, J., & Wen, J. Z. (2023). Methods and applications for Artificial Intelligence, Big Data, Internet of Things, and Blockchain in smart energy management. *Energy and AI*, *11*, 100208. doi:10.1016/j.egyai.2022.100208

Liu, G., Li, G., Yang, R., & Guo, L. (2018). Improving food safety in supply chain based on Big Data. *E3S Web of Conferences*, *53*, 03084. doi:10.1051/e3sconf/20185303084

Liu, L., Yang, X., Bhandari, B., Meng, Y., & Prakash, S. (2020). Optimization of the formulation and properties of 3D-printed complex egg white protein objects. *Foods*, *9*(2), 164. doi:10.3390/foods9020164

Mavani, N. R., Ali, J. M., Othman, S., Hussain, M. A., Hashim, H., & Rahman, N. A. (2022). Application of artificial intelligence in food industry–a guideline. *Food Engineering Reviews*, *14*(1), 134–175. doi:10.1007/s12393-021-09290-z

Mededjel, M., Belalem, G., & Neki, A. (2017). Towards a traceability system based on cloud and fog computing. *Multiagent and Grid Systems*, *13*(1), 47–68. doi:10.3233/MGS-170261

Mehta, V., & Rath, S. N. (2021). 3D printed microfluidic devices: A review focused on four fundamental manufacturing approaches and implications on the field of healthcare. *Bio-Design and Manufacturing*, *4*(2), 311–343. doi:10.1007/s42242-020-00112-5

Melocchi, A., Uboldi, M., Inverardi, N., Briatico-Vangosa, F., Baldi, F., Pandini, S., Scalet, G., Auricchio, F., Cerea, M., Foppoli, A., Maroni, A., Zema, L., & Gazzaniga, A. (2019). Expandable drug delivery system for gastric retention based on shape memory polymers: Development via 4D printing and extrusion. *International Journal of Pharmaceutics*, *571*, 118700. doi:10.1016/j.ijpharm.2019.118700

Misra, N. N., Dixit, Y., Al-Mallahi, A., Bhullar, M. S., Upadhyay, R., & Martynenko, A. (2022). IoT, big data, and artificial intelligence in agriculture and food industry. *IEEE Internet of Things Journal*, *9*(9), 6305–6324. doi:10.1109/JIOT.2020.2998584

Mohammadi, M., Kouzani, A. Z., Bodaghi, M., Long, J., Khoo, S. Y., Xiang, Y., & Zolfagharian, A. (2024). Sustainable robotic joints 4D printing with variable stiffness using reinforcement learning. *Robotics and Computer-Integrated Manufacturing*, *85*, 102636. doi:10.1016/j.rcim.2023.102636

Muñoz, J., & Pumera, M. (2020). 3D-printed biosensors for electrochemical and optical applications. *TrAC Trends in Analytical Chemistry*, *128*, 115933. doi:10.1016/j.trac.2020.115933

Neamah, H. A., & Tandio, J. (2024). Towards the development of foods 3D printer: Trends and technologies for foods printing. *Heliyon*, *10*(13), e33882. doi:10.1016/j.heliyon.2024.e33882

Nida, S., Moses, J. A., & Anandharamakrishnan, C. (2022). Emerging applications of 5D and 6D printing in the food industry. *Journal of Agriculture and Food Research*, *10*, 100392. doi:10.1016/j.jafr.2022.100392

Ning, X., Devahastin, S., Wang, X., Wu, N., Liu, Z., Gong, Y., Zhou, L., Huo, L., Ding, W., Yi, J., Guo, C., & Hu, X. (2024). Understanding 3D food printing through computer simulation and extrusion force analysis. *Journal of Food Engineering*, *370*, 111972. doi:10.1016/j.jfoodeng.2024.111972

Nugroho, G., Tedjakusuma, F., Lo, D., Romulo, A., Pamungkas, D. H., & Kinardi, S. A. (2023). Review of the application of digital transformation in food industry. *Journal of Current Science and Technology*, *13*(3), 774–790. doi:10.59796/jcst.V13N3.2023.1285

Nunes, C. A., Ribeiro, M. N., de Carvalho, T. C., Ferreira, D. D., de Oliveira, L. L., & Pinheiro, A. C. (2023). Artificial intelligence in sensory and consumer studies of food products. *Current Opinion in Food Science*, *50*, 101002. doi:10.1016/j.cofs.2023.101002

Oliver, J. H., & Hollis, J. H. (2021). Virtual reality as a tool to study the influence of the eating environment on eating behavior: A feasibility study. *Foods*, *10*(1), 89. doi:10.3390/foods10010089

Oral, M. O., Derossi, A., Caporizzi, R., & Severini, C. (2021). Analyzing the most promising innovations in food printing. Programmable food texture and 4D foods. *Future Foods*, *4*, 100093. doi:10.1016/j.fufo.2021.100093

Oyinloye, T., & Yoon, W. (2021). Application of computational fluid dynamics (CFD) simulation for the effective design of food 3D printing (a review). *Processes*, *9*(11), 1867. doi:10.3390/pr9111867

Padhiary, M., Barbhuiya, J. A., Roy, D., & Roy, P. (2024). 3D printing applications in smart farming and food processing. *Smart Agricultural Technology*, *9*, 100553. doi:10.1016/j.atech.2024.100553

Pant, R., Singh, R., Gehlot, A., Akram, S. V., Gupta, L. R., & Thakur, A. K. (2024). A systematic review of additive manufacturing solutions using machine learning, internet of things, big data, digital twins and blockchain technologies: A technological perspective towards sustainability. *Archives of Computational Methods in Engineering*, *31*(8). doi:10.1007/s11831-024-10116-4

Papastratis, I., Konstantinidis, D., Daras, P., & Dimitropoulos, K. (2024). AI nutrition recommendation using a deep generative model and ChatGPT. *Scientific Reports*, *14*(1), 14620. doi:10.1038/s41598-024-65438-x

Pugliese, R., & Regondi, S. (2022). Artificial intelligence-empowered 3D and 4D printing technologies toward smarter biomedical materials and approaches. *Polymers*, *14*(14), 2794. doi:10.3390/polym14142794

Raja, V., Krishnamoorthy, S., Moses, J. A., & Anandharamakrishnan, C. (2022). ICT applications for the food industry. In *Future foods* (pp. 613–626). Elsevier. doi:10.1016/B978-0-323-91001-9.00001-3

Raja, V., Moses, J. A., & Anandharamakrishnan, C. (2023a). Effect of 3D printing conditions and post-printing fermentation on pearl millet fortified idli. *Journal of the Science of Food and Agriculture*, *103*(5), 2401–2412. doi:10.1002/jsfa.12410

Raja, V., Nimbkar, S., Moses, J. A., Ramachandran Nair, S. V., & Anandharamakrishnan, C. (2023b). Modeling and simulation of 3D food printing systems–scope, advances, and challenges. *Foods*, *12*(18), 3412. doi:10.3390/foods12183412

Rashid, A. B., & Kausik, M. A. K. (2024). AI revolutionizing industries worldwide: A comprehensive overview of its diverse applications. *Hybrid Advances*, *7*, 100277. doi:10.1016/j.hybadv.2024.100277

Rejeb, A., Keogh, J. G., & Rejeb, K. (2022). Big data in the food supply chain: A literature review. *Journal of Data, Information and Management*, *4*(1), 33–47. doi:10.1007/s42488-021-00064-0

Rejeb, A., Keogh, J. G., Zailani, S., Treiblmaier, H., & Rejeb, K. (2020). Blockchain technology in the food industry: A review of potentials, challenges and future research directions. *Logistics*, *4*(4), 27. doi:10.3390/logistics4040027

Remaggi, G., Zaccarelli, A., & Elviri, L. (2022). 3D printing technologies in biosensors production: Recent developments. *Chemosensors*, *10*(2), 65. doi:10.3390/chemosensors10020065

Renner, A., Holub, J., Sridhar, S., Evans, G., & Winer, E. (2015, August 2). A virtual reality application for additive manufacturing process training. Volume 1A: 35th Computers and Information in Engineering Conference. https://doi.org/10.1115/DETC2015-47807

Reuben, S. J., Chandra, S., Praveena, B., Santhoshkumar, P., & Moses, J. A. (2025). 3D printing of alternative proteins: Approaches, challenges and advances. *Sustainable. Materials and Technologies*, e01285. doi:10.1016/j.susmat.2025.e01285

Roy, R., Marakkar, S., Vayalil, M. P., Shahanaz, A., Anil, A. P., Kunnathpeedikayil, S., ... Yadav, K. K. (2022). Drug-food interactions in the era of molecular big data, machine intelligence, and personalized health. *Recent Advances in Food Nutrition & Agriculture*, *13*(1), 27–50. doi:10.2174/2212798412666220620104809

Sajjad, R., Chauhdary, S. T., Anwar, M. T., Zahid, A., Khosa, A. A., Imran, M., & Sajjad, M. H. (2023). A review of 4D printing – Technologies, shape shifting, smart polymer based materials, and biomedical applications. *Advanced Industrial and Engineering Polymer Research*, *7*(1), 20–36. doi:10.1016/j.aiepr.2023.08.002

Santhoshkumar, P., Raja, V., Priyadarshini, S. R., & Moses, J. A. (2024a). Impact of baking and steaming of 3D printed Pearl millet–Banana pulp blends on physicochemical and sensory attributes. *International Journal of Gastronomy and Food Science*, *37*, 100962. doi:10.1016/j.ijgfs.2024.100962

Santhoshkumar, P., Ramu, D., Mahalakshmi, L., & Moses, J. A. (2024b). 3D printed edible electronics: Components, fabrication approaches and applications. *Biosensors and Bioelectronics*, *117059*. doi:10.1016/j.bios.2024.117059

Sergi, I., Montanaro, T., Benvenuto, F. L., & Patrono, L. (2021). A smart and secure logistics system based on IoT and cloud technologies. *Sensors*, *21*(6), 2231. doi:10.3390/s21062231

Shen, C., & Shen, A. (2025). 4D printing: innovative solutions and technological advances in orthopedic repair and reconstruction, personalized treatment and drug delivery. *BioMedical Engineering OnLine*, *24*(1), 5. doi:10.1186/s12938-025-01334-3

Singh, K. A., Patra, F., Ghosh, T., Mahnot, N. K., Dutta, H., & Duary, R. K. (2025). Advancing food systems with industry 5.0: A systematic review of smart technologies, sustainability, and resource optimization. *Sustainable. Futures*, *9*, 100694. doi:10.1016/j.sftr.2025.100694

Sivarajah, U., Kamal, M. M., Irani, Z., & Weerakkody, V. (2017). Critical analysis of Big Data challenges and analytical methods. *Journal of Business Research*, *70*, 263–286. doi:10.1016/j.jbusres.2016.08.001

Song, D., Chen, X., Wang, M., Wu, Z., & Xiao, X. (2023). 3D-printed flexible sensors for food monitoring. *Chemical Engineering Journal*, *474*, 146011. doi:10.1016/j.cej.2023.146011

Subhasri, D., Dutta, S., Leena, M. M., Moses, J. A., & Anandharamakrishnan, C. (2022). Gastronomy: An extended platform for customized nutrition. *Future Foods*, *5*, 100147. doi:10.1016/j.fufo.2022.100147

Tan, J., & Xu, J. (2020). Applications of electronic nose (e-nose) and electronic tongue (e-tongue) in food quality-related properties determination: A review. *Artificial Intelligence in Agriculture*, *4*, 104–115. doi:10.1016/j.aiia.2020.06.003

Taneja, A., Nair, G., Joshi, M., Sharma, S., Sharma, S., Jambrak, A. R., Roselló-Soto, E., Barba, F. J., Castagnini, J. M., Leksawasdi, N., & Phimolsiripol, Y. (2023). Artificial intelligence: Implications for the agri-food sector. *Agronomy*, *13*(5), 1397. doi:10.3390/agronomy13051397

Thirukumaran, R., Priya, V. K. A., Raja, V., Nimbkar, S., Moses, J. A., & Anandharamakrishnan, C. (2024). Blockchain technology and advancements in the agri-food industry. *Journal of Biosystems Engineering*. doi:10.1007/s42853-024-00221-4

Tracey, C. T., Predeina, A. L., Krivoshapkina, E. F., & Kumacheva, E. (2022). A 3D printing approach to intelligent food packaging. *Trends in Food Science & Technology*, *127*, 87–98. doi:10.1016/j.tifs.2022.05.003

Urhal, P., Weightman, A., Diver, C., & Bartolo, P. (2019). Robot assisted additive manufacturing: A review. *Robotics and Computer-Integrated Manufacturing*, *59*, 335–345. doi:10.1016/j.rcim.2019.05.005

Wang, N., Li, R., Wang, X., & Yang, X. (2024). 4D food printing: Key factors and optimization strategies. *Trends in Food Science & Technology*, *145*, 104380. doi:10.1016/j.tifs.2024.104380

Waseem, M., Tahir, A. U., & Majeed, Y. (2024). Printing the future of food: The physics perspective on 3D food printing. *Food Physics*, *1*, 100003. doi:10.1016/j.foodp.2023.100003

Wu, Y., Ye, D., Shan, Y., He, S., Su, Z., Liang, J., Zheng, J., Yang, Z., Yang, H., Xu, W., & Jiang, H. (2020). Edible and nutritive electronics: Materials, fabrications, components, and applications. *Advanced Materials Technologies*, *5*(10), 2000100. doi:10.1002/admt.202000100

Yan, Y., Jiang, Y., Ng, E. L. L., Zhang, Y., Owh, C., Wang, F., Song, Q., Feng, T., Zhang, B., Li, P., Loh, X. J., Chan, S. Y., & Chan, B. Q. Y. (2023). Progress and opportunities in additive manufacturing of electrically conductive polymer composites. *Materials Today Advances*, *17*, 100333. doi:10.1016/J.MTADV.2022.100333

Yarali, E., Mirzaali, M. J., Ghalayaniesfahani, A., Accardo, A., Diaz-Payno, P. J., & Zadpoor, A. A. (2024). 4D printing for biomedical applications. *Advanced Materials*, *36*(31), 2402301. doi:10.1002/adma.202402301

Yu, Q., Zhang, M., Bhandari, B., & Li, J. (2023). Future perspective of additive manufacturing of food for children. *Trends in Food Science & Technology*, *136*, 120–134. doi:10.1016/j.tifs.2023.04.009

Yu, Q., Zhang, M., Mujumdar, A. S., & Li, J. (2024). AI-based additive manufacturing for future food: Potential applications, challenges and possible solutions. *Innovative Food Science & Emerging Technologies*, *92*, 103599. doi:10.1016/j.ifset.2024.103599

Yu, Z., Jung, D., Park, S., Hu, Y., Huang, K., Rasco, B. A., Wang, S., Ronholm, J., Lu, X., & Chen, J. (2022). Smart traceability for food safety. *Critical Reviews in Food Science and Nutrition*, *62*(4), 905–916. doi:10.1080/10408398.2020.1830262

Zatsu, V., Shine, A. E., Tharakan, J. M., Peter, D., Ranganathan, T. V., Alotaibi, S. S., Mugabi, R., Muhsinah, A. B., Waseem, M., & Nayik, G. A. (2024). Revolutionizing the food industry: The transformative power of artificial intelligence–a review. *Food Chemistry: X*, *24*, 101867. doi:10.1016/j.fochx.2024.101867

Zhang, X., Li, M., Lim, J. H., Weng, Y., Tay, Y. W. D., Pham, H., & Pham, Q. C. (2018). Large-scale 3D printing by a team of mobile robots. *Automation in Construction*, *95*, 98–106. doi:10.1016/J.AUTCON.2018.08.004

Zhou, W., Qiao, Z., Nazarzadeh Zare, E., Huang, J., Zheng, X., Sun, X., Shao, M., Wang, H., Wang, X., Chen, D., Zheng, J., Fang, S., Li, Y. M., Zhang, X., Yang, L., Makvandi, P., & Wu, A. (2020). 4D-printed dynamic materials in biomedical applications: Chemistry, challenges, and their future perspectives in the clinical sector. *Journal of Medicinal Chemistry*, *63*(15), 8003–8024. doi:10.1021/acs.jmedchem.9b02115

Zoran, A., Gonzalez, E. A., Mizrahi, A. B., & Lachnish, A. Z. (2021). Cooking with computers: The vision of digital gastronomy. In *Gastronomy and food science* (pp. 35–53). Elsevier. 10.1016/B978-0-12-820057-5.00003-0

Index

4D Printing of Foods: Fundamentals, Applications and Prospects, First Edition.
C. Anandharamakrishnan, Jeyan A. Moses, and P. Santhoshkumar.

d

e

f

m

n

o

p